Preservative-Free and Self-Preserving Cosmetics and Drugs

COSMETIC SCIENCE AND TECHNOLOGY

Series Editor

ERIC JUNGERMANN

Jungermann Associates, Inc.
Phoenix, Arizona

1. Cosmetic and Drug Preservation: Principles and Practice, *edited by Jon J. Kabara*
2. The Cosmetic Industry: Scientific and Regulatory Foundations, *edited by Norman F. Estrin*
3. Cosmetic Product Testing: A Modern Psychophysical Approach, *Howard R. Moskowitz*
4. Cosmetic Analysis: Selective Methods and Techniques, *edited by P. Boré*
5. Cosmetic Safety: A Primer for Cosmetic Scientists, *edited by James H. Whittam*
6. Oral Hygiene Products and Practice, *Morton Pader*
7. Antiperspirants and Deodorants, *edited by Karl Laden and Carl B. Felger*
8. Clinical Safety and Efficacy Testing of Cosmetics, *edited by William C. Waggoner*
9. Methods for Cutaneous Investigation, *edited by Robert L. Rietschel and Thomas S. Spencer*
10. Sunscreens: Development, Evaluation, and Regulatory Aspects, *edited by Nicholas J. Lowe and Nadim A. Shaath*
11. Glycerine: A Key Cosmetic Ingredient, *edited by Eric Jungermann and Norman O. V. Sonntag*
12. Handbook of Cosmetic Microbiology, *Donald S. Orth*
13. Rheological Properties of Cosmetics and Toiletries, *edited by Dennis Laba*
14. Consumer Testing and Evaluation of Personal Care Products, *Howard R. Moskowitz*
15. Sunscreens: Development, Evaluation, and Regulatory Aspects. Second Edition, Revised and Expanded, *edited by Nicholas J. Lowe, Nadim A. Shaath, and Madhu A. Pathak*
16. Preservative-Free and Self-Preserving Cosmetics and Drugs: Principles and Practice, *edited by Jon J. Kabara and Donald S. Orth*

ADDITIONAL VOLUMES IN PREPARATION

Hair and Hair Care, *edited by Dale H. Johnson*

Preservative-Free and Self-Preserving Cosmetics and Drugs

Principles and Practice

edited by

Jon J. Kabara

Technology Exchange, Inc.
Galena, Illinois, and
Michigan State University
East Lansing, Michigan

Donald S. Orth

Neutrogena Corporation
Los Angeles, California

Marcel Dekker, Inc. New York•Basel•Hong Kong

Library of Congress Cataloging-in-Publication Data

Preservative-free and self-preserving cosmetics and drugs:
principles and practice / edited by Jon J. Kabara, Donald S. Orth.
p. cm.— (Cosmetic science and technology series; v. 16)
Includes index.
ISBN 0-8247-9366-8 (hardcover: alk. paper)
1. Cosmetics. 2. Chemical preservatives. I. Kabara, Jon J.
II. Orth, Donald S. III. Series.
TP983.P8917 1996
668'.55—dc21

96-47675
CIP

The publisher offers discounts on this book when ordered in bulk quantities. For more information, write to Special Sales/Professional Marketing at the address below.

This book is printed on acid-free paper.

Marcel Dekker, Inc.
270 Madison Avenue, New York, New York 10016

Current printing (last digit):
10 9 8 7 6 5 4 3 2 1

PRINTED IN THE UNITED STATES OF AMERICA

About the Series

The Cosmetic Science and Technology series was conceived to permit discussion of a broad range of current knowledge and theories of cosmetic science and technology. The series is made up of books written by one or more authors or edited volumes with a number of contributors. Authorities from industry, academia, and the government are participating in writing these books.

The aim of this series is to cover the many facets of cosmetic science and technology. Topics are drawn from a wide spectrum of disciplines ranging from chemistry, physics, biochemistry, and analytical and consumer evaluations to safety, efficacy, toxicity, and regulatory questions. Organic, inorganic, physical, and polymer chemistry, emulsion technology, microbiology, dermatology, and toxicology all play a role in cosmetic science.

There is little commonality in the scientific methods, processes, or formulations required for the wide variety of cosmetics and toiletries manufactured. Products range from hair care, oral care, and skin care preparations to lipsticks, nail polishes and extenders, deodorants, body powders and aerosols to over-the counter products, such as antiperspirants, dandruff treatments, antimicrobial soaps, and acne and sunscreen products.

Cosmetics and toiletries represent a highly diversified field with many subsections of science and "art." Indeed, even in these days of high technology, "art" and intuition continue to play an important part in the development and evaluation of formulations and the selection of raw materials. There is a move toward more sophisticated scientific methodologies in the fields of preservative efficacy testing, claim substantiation, safety testing, product evaluation, and chemical analyses.

Emphasis in the Cosmetic Science and Technology series is placed on reporting the current status of cosmetic technology and science in addition to historical reviews. The series includes books on safety, efficacy testing, oral hygiene, scientific and regulatory foundations, cosmetic product testing, analytical methods and techniques, antiperspirants and deodorants, glycerine, rheological properties, cosmetic microbiology, sunscreens, methods of cutaneous investigation, and consumer testing and evaluation of personal care products. Several of the books have found a wider audience and have been translated into Japanese. Contributions range from highly sophisticated and scientific treatises to primers, practical applications, and pragmatic presentations. Authors are encouraged to present their own concepts as well as established theories. Contributors have been asked not to shy away from fields that are still in a state of transition, nor to hesitate to present detailed discussions of their own work. Altogether, we intend to develop in this series a collection of critical surveys and ideas covering diverse phases of the cosmetic industry.

Preservative-Free and Self-Preserving Cosmetics and Drugs is the sixteenth book in the series. It investigates the concept of self-preserving formulations and promotes the idea of either eliminating or greatly reducing the use of chemical preservatives. In so doing, this volume introduces a new paradigm in the formulation approach to cosmetic and drug products. The book is divided into eleven chapters ranging from basic discussions of the principles of self-preserving products, survival strategies of microorganisms, and "water activity" to a review of different classes of chemicals that can provide antimicrobial activity. Included are certain surfactants, fatty acids and their esters, phospholipids, "aroma" chemicals, chelating agents, and antioxidants. This is the first book published that brings together the principles of preservation and different classes of formulation components.

I want to thank the editors, Dr. Jon J. Kabara and Dr. Donald S. Orth, for collaborating in the writing and editing of this book. It is the second time that both have contributed to the Cosmetic Science and Technology series. Dr. Kabara edited *Cosmetic and Drug Preservation*, and Dr. Orth wrote the *Handbook of Cosmetic Microbiology*. Special recognition is also due to Sandra Beberman and the editorial staff at Marcel Dekker, Inc. In addition, I would like to thank my wife, Eva, without whose constant support and editorial help I would never have undertaken this project.

Eric Jungermann, Ph.D.
Series Editor

Preface

The term *preservative-free* means "without preservative chemicals." Usually, multiple-use aqueous consumer products in their current containers cannot be made preservative-free merely by removing preservatives from the formula. It is possible to have preservative-free aqueous products only if they are sterilized and packaged and stored appropriately.

The term *self-preserving* is more appropriate than preservative-free for most aqueous products containing a chemical system that kills microorganisms and/or prevents their growth. The concept of "preservative-free" is controversial; however, the emphasis of this book is to focus on the principles of preservation that may be used to make products that are microbiologically safe at the time of manufacture and during use by the consumer. No other publication presently available addresses this subject to the depth presented in this book.

Preservative-free and self-preserving products may be developed by understanding and applying the principles of preservation as they relate to each product. Some of the principles discussed are water activity, pH, special lipids, surfactants, antioxidants (phenols), aroma chemicals, chelators, and packaging. Products that are adequately preserved and microbiologically safe may be developed by selection of chemicals to create conditions that have static or cidal effects on microorganisms.

Although we could continue to use standard preservative systems for domestic products, globalization requires that we think beyond traditional formula development. Self-preserving products offer many advantages for formulation because they reduce and/or eliminate chemical preservatives that are a source of skin irritation and contact sensitization; they meet the demands of a growing segment of consumers who want natural products; they may encourage the use of contamination-resistant packaging; they allow the use of global formulas without the regulatory issues surrounding the use of chemical preservatives, and they may reduce formula costs by avoiding addition of preservatives.

Bacteria, yeasts, and molds cope with stressful conditions by using a wide range of survival strategies designed to prevent or repair injury so that they can survive and grow. Several mechanisms are discussed to illustrate how microorganisms deal with unfavorable conditions that are used to limit their growth in products. Consumer-friendly, environment-friendly products made without preservatives are possible, but this may require modification of current formulations and packaging. Opportunities exist for innovative workers who want to look beyond current technologies and apply the principles of preservation in developing novel, preservative-free and self-preserving cosmetics and drugs. We hope this book will be a platform for future discoveries.

We wish to acknowledge the efforts of our contributing authors and the cooperation of the staff at Marcel Dekker, Inc. We are pleased to recognize Betty Kabara and Barbara Jo Orth for their patience, understanding, encouragement, and continued support because this greatly helped our writing and editing.

Suggestions and comments about the book are welcomed.

Jon J. Kabara, Ph.D.
Donald S. Orth, Ph.D.

Contents

Contributors

A. Larry Branen, Ph.D. Department of Food Science and Toxicology, University of Idaho, Moscow, Idaho

Daniel K. Brannan, Ph.D. Department of Biology, Abilene Christian University, Abilene, Texas

Oreste Cozzoli, Dr. Stazione Sperimentale per le Industrie degli Oli e dei Grassi (SSOG), Milan, Italy

P. Michael Davidson, Ph.D. Department of Food Science and Toxicology, University of Idaho, Moscow, Idaho

Davin C. Enigl, M.S., M.E.A.S. Hazard Analysis Critical Control Points, HACCP Validations—sm, Sacramento, California

Dennis L. Fost, Ph.D. Marketing and Commercial Development, Mona Industries, Inc., Paterson, New Jersey

Jon J. Kabara, Ph.D. Technology Exchange, Inc., Galena, Illinois, and Professor Emeritus, Michigan State University, East Lansing, Michigan

Donald S. Orth, Ph.D. Research and Development Department, Neutrogena Corporation, Los Angeles, California

Kent M. Sorrells, M.B.A., Ph.D. Quality Assurance, National Steak and Poultry, Owasso, Oklahoma

John I. Yablonski, M.S. Bio-Control Consultants, Inc., Westfield, New Jersey

1

Principles for Product Preservation

Jon J. Kabara
Technology Exchange, Inc.
Galena, Illinois
Michigan State University
East Lansing, Michigan

Donald S. Orth
Neutrogena Corporation
Los Angeles, California

Preservatives are used in aqueous cosmetics and drugs to prevent microbial contamination. While this has solved the problem of contamination in most instances, it has created other problems. No preservative is free of allergenic potential, and many consumers want "natural" products that do not contain preservatives. Consequently, views on cosmetic preservation and "the greening of cosmetics" are on a collision course.

In recent years, there has been increasing interest in developing and marketing natural cosmetics, including those that are "preservative-free" (i.e., products that do not contain ingredients commonly recognized as antimicrobial preservatives), to meet the consumer demand for natural products. A glance at product labeling in local retail outlets reveal scores of cosmetics and over-the-counter (OTC) drugs with labeling that indicates these products are "natural," "fragrance-free," "hypoallergenic," "noncomedogenic," "PABA-free," "chemical free," etc.

"Preservative-free" has been added to the labeling of a few products. What is the scientific basis for the creation of this new class of products? Is there a need for them?" Are they safe? Are they satisfactorily preserved? Do they have suitable "cosmetic elegance"?

The term *preservative-free* means without preservative chemicals. It is possible to have preservative-free aqueous products if they are sterilized and packaged/stored appropriately. The term *self-preserving* is more appropriate than preservative-free for most aqueous cosmetic and OTC drug products in multiple-use containers because formulas without any form of preservation will spoil if they are not packaged or stored in such a manner that microbial contamination and growth are prevented. Most aqueous products have a "preservative system" created by the physicochemical composition of the formula (with or without the addition of chemical preservatives) that kills microorganisms and/or prevents their growth. Kabara (1) discussed formulation of self-preserving products using a systems approach to cosmetic preservation.

Although cosmetics and OTC drugs are not necessarily intended to be sterile, satisfactorily preserved aqueous formulas in multiple-use containers have a preservative system that makes them self-sterilizing (2). A self-preserving formula is one that either prevents microbial growth (i.e., is bacteriostatic) or kills microorganisms (i.e., is bactericidal/microbicidal). A self-preserving formula may or may not contain antimicrobial preservatives.

Relatively anhydrous products (e.g., talcum powder, cornstarch, mineral oil, and stick deodorants) do not require preservatives because they do not have sufficient water content to support microbial growth. However, manufacturers may incorporate preservatives into these products to prevent microbial growth that may occur due to inadvertent addition of water by consumers during use (or abuse).

Aqueous products, on the other hand, require preservatives unless their composition, packaging, or storage conditions prevent microbial contamination and growth. The physicochemical make-up of the product, including low water activity (a_w), high or low pH, and multifunctional materials that have antimicrobial properties (i.e., alcohols, surfactants, quaternary ammonium compounds), determines whether a formula is "self-preserving." Aqueous products in multiple-use containers generally require the addition of preservatives if they do not prevent microbial growth.

The type of packaging and consumer use habits determine the preservative requirements of a formula. The required D-value (RDV) concept was introduced to relate the preservative system of the formula with product packaging and consumer use/abuse (2,3). Although an unpreserved formula in a multiple-use container would not be acceptable because it may become contaminated during use, the formula would be acceptable if it were manufactured, sterilized, and supplied in a unit-dose container. The same formula in a multiple-use container may be suitable for restricted use if it were handled as we now handle perishable foods, such as milk, which is refrigerated when not being used. Alternately, unpreserved products could be frozen and thawed immediately before use.

Currently available antimicrobial preservatives may cause some adverse reactions including skin irritation, eye irritation, and contact sensitization (2). The North American Contact Dermatitis Group identified preservatives as a group of ingredients most commonly associated with skin irritation due to the use of cosmetics (4,5). Recognizing that preservatives may be a source of irritation provides a driving force for attempting to find milder, less irritating preservative systems.

Although it would be desirable to eliminate preservatives that cause adverse reactions (e.g., skin and eye irritation, contact sensitization) in some individuals, this is not possible with most existing formulations in their current form/packaging. Preservatives are used in these products because self-preserving products may be unattainable without accepting specific formula and/or packaging constraints. This chapter will introduce topics covered in detail in later chapters. The material presented includes 1) a discussion of the principles of preservation providing background information on the control of microorganisms, 2) a discussion of the preservative system concept, which shows how the principles of preservation are used to design self-preserving products, 3) an overview of survival strategies of microorganisms to illustrate mechanisms used by microorganisms to survive in hostile environments, and 4) some advantages and disadvantages of self-preserving products.

1.1 PRINCIPLES OF PRESERVATION

Bacteria, yeasts, and molds are ubiquitous in nature. Cosmetic raw materials and finished products provide a wide spectrum of inorganic and organic compounds, which are available for microbial growth when sufficient water is present. Microbial growth will occur in aqueous systems held at ambient temperatures found in manufacturing plants, in trade channels, and in consumers' homes if control measures are not instituted (2). The principles of preservation are used globally to control microbial growth in foods, drugs, and cosmetics. These principles are:

1. Asepsis (keeping microorganisms out of the process stream)
   - adherence to current good manufacturing practices (GMPs)
   - microbiological validation of processes
   - microbiological validation of cleaning and sanitization programs
2. Removal of microorganisms
   - trimming and washing
   - filtration and clarification
   - sedimentation and centrifugation
3. Retarding growth or killing microorganisms
   - high temperatures (heat sterilization, pasteurization)
   - low temperatures (refrigeration, freezing)

drying conditions (low a_w)
high/low pH
removal of substrates
anaerobic atmosphere (CO_2, N_2 gas packing)
preservatives/biocides
irradiation
mechanical disruption (grinding, high pressure)

The quality of finished products and the microbiological problems that occur during manufacturing may be influenced by the kinds and numbers of microorganisms present. Quality defects including altered color, flavor, odor, viscosity, and texture become evident when sufficient microbial growth has occurred. Preventing unnecessary microbial contamination of the process stream is an important consideration in making products in accordance with GMPs. Frequently, the extent of microbial contamination of in-process materials and finished products depends on the *microbial load* of raw materials and conditions that permit growth of microorganisms (2). This is particularly true for "natural" raw materials.

The principles of preservation are applied to manufacturing and storage of products to minimize the microbial load and to retard growth or kill microorganisms that may gain access to the process stream and finished products. Microorganisms have physical and chemical requirements for growth including suitable temperature, pH, water (osmotic pressure or a_w), nutrients, minerals, oxygen [or the proper oxidation/reduction (O/R) potential], organic growth factors, and freedom from harmful radiation and chemicals (preservatives, antibiotics). It is evident that microbial growth may be prevented by interfering with these requirements.

The principles of preservation utilized most successfully in self-preserving products are low or high pH, low a_w, multifunctional materials that have antimicrobial activity, and combinations of these. Generally, microorganisms of interest in raw materials or in cosmetic products grow best around neutrality (pH 7.0). The growth rate of microorganisms generally decreases as the pH departs from neutrality (or from the optimum pH for growth for each different organism). Although many yeasts and molds are able to tolerate acid pH conditions (i.e., pH < 4.0), many microorganisms are metabolically injured or "stressed" by extreme pH conditions—where the pH is <4 or >10 (6). It should be noted that the excess acidity or alkalinity in low- or high-pH products may make these products harsh or irritating to the skin.

1.2 INJURY OF MICROORGANISMS

When conditions are suitable, a population of microorganisms utilizes available substrates and grows. When conditions are unfavorable, the population stops growing. Individual cells within the population may die or become metabolically injured or "stressed" as a result of exposure to adverse physical and chemical conditions. Cellular injury may be caused by altered membrane permeability, altered enzyme function, degradation of ribosomes, and increased sensitivity to selective agents (6–8).

Sublethally injured microorganisms may be more susceptible than noninjured microorganisms to secondary stresses, such as elevated temperatures, low a_w, and the presence of inhibitory chemicals; consequently, they are unable to grow under certain environmental conditions. The significance of this is that stressed cells may escape detection during routine microbiological testing, when in fact they may be capable of recovery and growth after an extended lag period or after removal of the adverse conditions. Injury can be reversible, and injured microorganisms may revive and grow (6).

Injured microorganisms can be detected in several ways. One of the most convenient means is to compare the aerobic plate count (APC) on a noninhibitory growth medium with the APC on a selective medium (6,9). After a period of exposure to stress, some cells in the population lose their characteristic tolerance to the selective agent—with concomitant decrease in the observed APC on the selective medium. Nevertheless, the APC on nonselective media remains the same, unless cell death has occurred.

A hypothetical example of the effect of injury on the APC during the lag phase of growth is illustrated in Figure 1.1. Here, a suspension of 10^5 bacteria/ml has been injured by exposure to low pH. All viable cells retain their ability to grow on the nonselective plating medium (open circles), but only a portion of the cells (2.5×10^2 bacteria/ml) is able to form colonies on the selective medium (closed circles). The difference in APCs on the selective and nonselective media reflects the extent of injury of cells in the population. In this case, more than 99% of the cells have been injured.

After removal of the source of stress (i.e., by neutralizing the acid pH in this example), the injured cells must recover before they can begin dividing. This recovery period requires resuscitation and/or repair and causes an *extended lag phase*—a period in which the population appears to be dormant because it is not growing (6). Cellular repair may take a period of hours in a favorable environment and longer when conditions are not optimum. After recovery, the cells are able to grow on the selective medium and the APCs become the same on both the selective and nonselective media. After all cells have undergone the repair process, the population goes into a log growth phase (Fig. 1.1).

Microorganisms in self-preserving cosmetic and OTC drug products may be exposed to several different kinds of physical and chemical stresses including low pH, low a_w, chelating agents, surface active agents, free fatty acids, and nutrient depletion (starvation). Stressed microorganisms that are viable but nonculturable represent a hazard because they may recover and grow in the product at some later time (6).

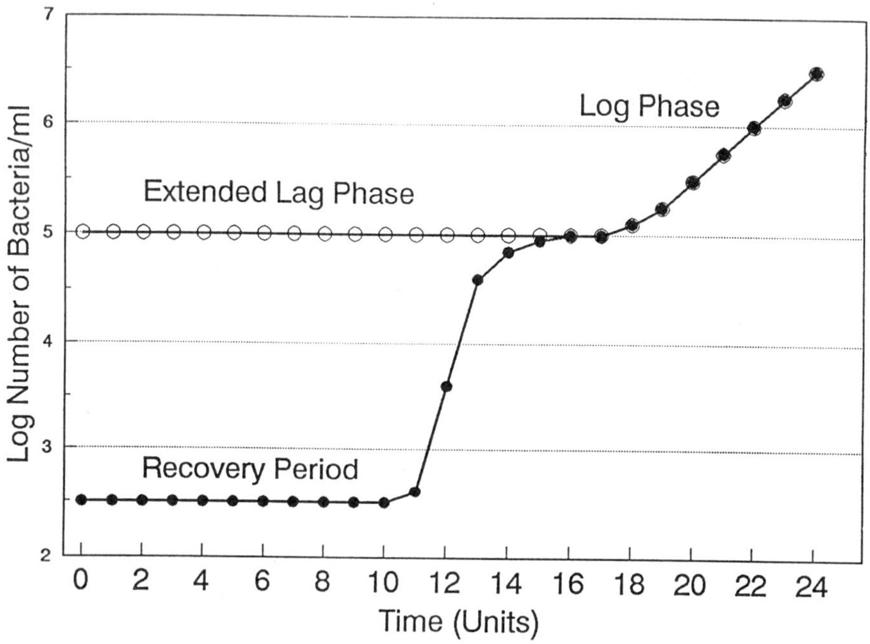

FIGURE 1.1 Hypothetical example of the effect of sublethal injury on the ability of bacteria to grow. This figure represents the first portion of the idealized bacterial growth curve. Here, a portion of the population was injured, as determined by plating on a nonselective medium (open circles) and on a selective medium (closed circles). The APCs on the nonselective medium remain at the same level until injured members of the population recover. The APCs on the selective medium are lower than those on the nonselective medium until the recovery period is complete. This recovery period results in an extended lag phase. The population goes into the log phase of growth at the completion of the lag phase.

1.3 PRESERVATION OF PRODUCTS WITH LOW OR HIGH pH

The pH of a product may be lowered using mineral acids or organic acids. Freese et al. (10) reported that lipophilic acids (e.g., lauric acid, myristic acid) are antimicrobial because they inhibit membrane transport of oxidizable substrates. Kabara (11) examined the structure-function relationships of anionic surfactants as antimicrobial agents and reported the minimum inhibitory concentration (MIC) of free fatty acids for gram-positive bacteria. He found the most active chain length for saturated fatty acids to be C_{12} (lauric acid), the most active monounsaturated fatty acid C_{16} (palmitoleic acid), and the most active polyunsaturated fatty acid $C_{18:2}$ (linoleic acid) when using pneumonococci, group A streptococci, and β-hemolytic streptococci (non–group A) as test organisms. Kabara reported that free fatty acids do not affect gram-negative bacteria except for short-chain fatty acids with fewer than eight carbon atoms. This work suggests that free fatty acids will contribute to the antimicrobial action of a formula, especially at pH values acid to the pKa of carboxylic acids (i.e., pH < 4.5).

1.4 PRESERVATION OF PRODUCTS WITH LOW WATER ACTIVITY

Water is a requirement for growth of all known living organisms, from bacteria to humans. As would be expected, different microorganisms have different minimum water requirements. The water requirement of microorganisms may be expressed in terms of available water, or a_w. The a_w is the vapor pressure of the solution (solutes in water) divided by the vapor pressure of the solvent (water). The a_w of pure water would be 1.00, and a 1.0 M solution of an ideal solute would have an a_w of 0.9832. If a solute such as glucose is dissolved in water, it will lower the water vapor pressure. One mole of an ideal solute dissolved in one kilogram of water will depress the vapor water by 1.77%, and the solution will have an aqueous vapor pressure that is 98.23% of the vapor pressure of pure water (without solutes) at the same temperature. The atmosphere above this solution will have a relative humidity (R.H.) of 98.23%, and the solution will have an a_w of 0.9823. At equilibrium, the a_w is a numerical expression equal to the R.H./100% (2,12).

Several generalizations may be made about the moisture requirements of microorganisms as follows:

Each microorganism has a characteristic optimal a_w and range of a_w for growth for a given set of environmental conditions.

Factors affecting the a_w requirements of microorganisms include the kind of solutes present, the nutritional value of the substrates available, temperature, O_2 supply, pH, and the presence of inhibitors.

An unfavorable a_w will decrease the growth rate and the maximum yield of cells.

The lag time before initiation of growth or germination of spores is increased as the a_w becomes more unfavorable.

Bacteria generally have higher a_w requirements than yeasts, and yeasts generally have higher a_w requirements than molds.

Different classes of microorganisms have different tolerance to low a_w, and this determines whether a particular bacterium, yeast, or mold is able to grow in a specific product. The reported lower limits of a_w for several microorganisms of interest to the cosmetic industry are presented in Table 1.1. Although the reported a_w requirements for specific microorganisms depend on the experimental conditions and the physicochemical conditions of the environment, studies indicate that the a_w requirements for *Pseudomonas* species are quite high (minimum $a_w = 0.97$). Addition of solute to decrease the a_w to <0.96 may be a convenient means of

TABLE 1.1 Lowest Reported a_w Values Permitting Growth of Microorganisms

Microorganism	Minimum a_w value
Many bacteria	0.91
Pseudomonas spp.	0.97
Escherichia coli	0.95
Bacillus subtilis	0.945
Staphylococcus aureus	0.86
Halophilic bacteria	0.75
Many yeasts	0.88
Osmophilic yeasts	0.60
Many molds	0.80
Xerophilic fungi	0.65

Source: Adapted from Ref. 2.

controlling the growth of pseudomonads in many aqueous systems (including cosmetic products). *Escherichia coli* and *Enterobacter aerogenes* have minimum a_w requirements around 0.95. This suggests that the growth of most coliforms and pseudomonads may be prevented by decreasing the a_w to <0.94. Review of the data in Table 1.1 reveals that the moisture requirements of *Staphylococcus aureus* are much lower ($a_w = 0.86$) than are those of the above gram-negative organisms. As noted above, the moisture requirements for any given microorganism depend on other factors, including nutrient availability, the type of solute used to decrease the a_w, the pH, and other physicochemical factors.

Certain raw materials decrease the a_w of a formulation. Humectants such as glycerol, propylene glycol, and sorbitol and other water-soluble materials decrease the a_w and increase the stress imposed on microorganisms. Inorganic salts, acids, and alkalis used to adjust the viscosity and pH decrease the a_w. Curiously, some raw materials contain a considerable amount of salt. For example, protein hydrolysates may contain >15% salt so that use of 1–5% protein hydrolysate will add up to 0.75% NaCl. Hydrocolloids (xanthan gum, guar gum, etc.) used for increasing viscosity and alcohols (SD alcohol 40, isopropyl alcohol) used for enhancing the penetration of specific ingredients, decreasing viscosity, or solubilizing ingredients may also decrease the a_w. Greater than 15% alcohol is effective in inhibiting bacteria, yeasts, and molds, especially when the pH is acidic (pH < 4.5). Higher concentrations of alcohol destabilize cell membranes, denature proteins, and kill microorganisms. Lower concentrations of alcohol contribute to the preservative system of the product.

The above discussion reveals that the a_w of a product has broad implications because it determines whether microbial growth is possible. Although determination of moisture content is useful for understanding the composition of the product, the moisture content may tell us very little about the chemical or microbial dynamics of the system—unless the moisture available for microbial growth is determined. For example, a 25% NaCl solution has an a_w of about 0.75, whereas a 25% sucrose solution has an a_w of about 0.98. The differences in a_w are due to differences in molar concentrations obtained with 25% NaCl and sucrose. Although both solutions have the same moisture content, their ability to support microbial growth is different. Many bacteria, yeasts, and molds are able to grow in the presence of 25% sucrose, but only the halophilic bacteria, osmophilic yeasts, and xerophilic fungi are able to grow in the presence of 25% NaCl (2).

Low a_w by itself may not be sufficient to kill microorganisms. For example, an a_w of 0.90 will not permit *E. coli* to grow, but viable cells may persist for weeks or months in a *dormant state*. The survival of the *E. coli* population depends on other stresses imposed by the formula (acidic pH, lack of nutrients, membrane-destabilizing surfactants or chelating agents, etc.) and storage conditions including temperature. Microorganisms are preserved by lyophilization (freeze-drying), and it is reasonable to think that products with very low a_w values, such as powders, may allow microorganisms to survive for extended periods. Thus, cells may enter a dormant state because removal of virtually all water causes metabolism to cease.

1.5 PRESERVATION OF PRODUCTS WITH CHELATING AGENTS

Removal of substrates and essential nutrients helps prevent the growth of microorganisms. Chelating agents sequester divalent and trivalent metal ions and frequently are used to bind metal ions that are prooxidants and cause deterioration in fragrance, color, and appearance of cosmetic formulations. Sequestration of metal ions interferes with microbial growth because divalent metal ions are required as enzyme cofactors and are required to maintain the integrity of the gram-negative bacterial cell envelope. Chelating agents commonly used in cosmetic products include tetrasodium EDTA, trisodium EDTA, tetrasodium etidronate, and citric acid. Several reports indicate that chelating agents function as preservative system potentiators and/or synergists (2,13–15). It is not possible to control microbial growth using levels of chelating agents normally found in cosmetic products (i.e., $\leq$1%) because some microorganisms accumulate reserve materials, including calcium, which may be complex with poly β-hydroxybutyric acid (PHB) or dipicolonic acid (6,13).

1.6 PRESERVATION OF PRODUCTS WITH SURFACTANTS

Many components in cosmetic products have the potential to interfere with microbial growth if they are present in high enough concentrations or when their action is combined with that of other formula ingredients. Anionic surfactants such as ammonium lauryl sulfate (ALS) and sodium lauryl ether sulfate are commonly used in shampoos, bath gels, and cleansers because they help solubilize lipids (e.g., sebum on the scalp and hair, greasy residues). When present in low concentrations, various surfactants may be used as substrates for energy and growth, which is why aqueous anionic surfactants require preservation. At higher concentrations, anionic surfactants destabilize cell membranes and the lipopolysaccharide (LPS) cell envelope of gram-negative bacteria. Different species of bacteria differ in their tolerance to anionic surfactants. Gram-positive bacteria such as the staphylococci generally cannot grow in 10% surfactant, some species of spore-forming bacilli and *E. coli* have moderate tolerance to surfactants, and some strains of *Pseudomonas aeruginosa* are able to grow in 28% ALS. Low concentrations of surfactant (1–5%) may add to the stress imposed on microorganisms when other factors in their environment are unfavorable. The antimicrobial effects of surfactants, low pH, low a_w, chelating agents, etc. are additive or synergistic in some cases.

Phosphate esters, glyceryl stearate, and glyceryl dilaurate generally have no significant antimicrobial activity at levels used for emulsifying creams and lotions. Kabara reported that esterification of a fatty acid to a anhydric alcohol yields an ester with no antimicrobial activity, whereas esterification to a polyhydric alcohol produces an antimicrobial compound (16). Glyceryl monolaurate has antibacterial activity against gram-positive bacteria and fungi at use levels of 0.1–2.0% (1,14). Antibacterial activity against gram-negative bacteria is achieved by combining glyceryl monolaurate with a chelating agent such as disodium EDTA.

Quaternary ammonium compounds (quats) are used for hair conditioning, skin feel, emulsification, and their antimicrobial activity. Different quats possess different antimicrobial activities. Benzalkonium chloride (BAC) is used as an antimicrobial agent and is found in mouthwashes and contact lens cleaning solutions. This quat is effective against most microorganisms at a concentration of 0.1%; however, gram-negative bacteria—especially the pseudomonads—are more resistant than are the gram-positive bacteria (17). Linoleamidopropyl PG-dimonium chloride phosphate is a primary cationic emulsifier that is reported to leave a smooth after-feel on the skin, reduce tackiness, and have an MIC of ≤300 ppm for gram-positive bacteria, *C. albicans*, and molds. Cocamidopropyl PG-dimonium chloride phosphate is reported to contribute to foaming and detergency of a formula, to condition the skin, and to have MICs of ≤300 ppm for gram-positive bacteria, ≤1250 ppm for gram-negative bacteria, and ≤5000 ppm for yeasts and molds (18). Ethyl *N*-cocoyl-L-arginate PCA salt is a cationic surfactant that is reported to have a MIC of ≤500 ppm for bacteria and yeast (19).

1.7 PRESERVATION WITH MULTIFUNCTIONAL MATERIALS

Each ingredient in a formula may have more than one effect on the physical and/or chemical properties of the formula. As noted by Kabara (1), multifunctional ingredients may be used to help preserve cosmetic and drug products because they affect the ability of microorganisms to grow. Although antioxidants are used to stabilize fragrances and colors, the phenolic antioxidants are reported to potentiate the antimicrobial action of other formula components (2,14,20,21). Fragrances and essential oils may contribute to the preservative action of the formula because they contain alcohols, phenols, esters, organic acids, aldehydes, and terpenes, which may have antimicrobial activity. Short-chain alcohols (e.g., ethyl alcohol and isopropyl alcohol) are used to solubilize ingredients, to give fragrances a "lift," and to contribute to the antimicrobial action of the formula.

1.8 PRESERVATIVE SYSTEM

The preservative system of a product (2,14) includes both compounds with known antimicrobial activity and formula components that may contribute directly or indirectly to antimicrobial activity. The physicochemical make-up of a formula, which includes factors such as product pH, a_w, nutrient availability, surfactant concentration, sequestering agents, alcohols, interfering materials, etc., determines whether a formula is self-preserving and/or whether specific preservatives are needed for adequate preservation.

Some familiar products use low pH as part of their preservative system. The low pH and low a_w obtained using 25% aluminum chlorohydrate in a roll-on antiperspirant helps create conditions unfavorable to microbial growth—especially gram-negative organisms like *P. aeruginosa* and *E. coli*. The acid pH of cationic hair conditioners (around pH 4) and cationic conditioning agents (quats) contribute to the antimicrobial action of these products.

Liquid soaps with an alkaline pH (pH 9.5–10.5) present an unfavorable environment for the growth of some microorganisms due to the chaotrophic (i.e., membrane destabilizing) effects of ionized fatty acids and the free alkalinity due to NaOH. Liquid soaps may contain substantial quantities of dissolved solutes (e.g., glycerol, salts), which decrease the a_w. Chelating agents and phenolic antioxidants, such as butylated hydroxyanisole (BHA), become part of the product preservative system.

1.9 SURVIVAL STRATEGIES OF MICROORGANISMS

Microorganisms respond to the physical and chemical conditions in their environment. The global responses of microorganisms are regulated and coordinated with all metabolic responses to provide bacteria, yeasts, and molds with efficient means of dissimilating substrates for use as sources of energy or as nutrients for growth. The processes by which microorganisms respond to conditions in the environment are termed survival strategies (6). The survival strategies discussed in Chapter 2 include acid tolerance, osmoregulation, and the production of siderophores.

1.10 ADVANTAGES AND DISADVANTAGES OF PRESERVATIVE-FREE AND SELF-PRESERVING PRODUCTS

One of the primary objectives in creating preservative-free and self-preserving products is to minimize the irritation potential of a formula. Whether a reduction in irritation could be observed by consumers depends on the severity of the undesirable reactions caused by the preservatives in a specific formula. A second benefit of preservative-free and self-preserving products is in marketing, because this class of products would fill a real and/or perceived need on the part of some consumers.

It must be recognized that use of low pH and low a_w imposes physicochemical constraints on self-preserving formulas, and many products in their current form may not be preserved adequately without the use of preservatives. Decreasing the pH and increasing inorganic salts to nonionic emulsions that use acrylic acid co-/homopolymers, magnesium aluminum silicate, or gums as thickening agents may cause substantial decreases in viscosity or alter the solubility of other formula components. Modification of the product with alternative thickening agents may affect skin feel and/or the kinetics of percutaneous absorption. Decreasing the a_w by addition of nonionic materials (e.g., sugars, glycerol, propylene glycol, butylene glycol, sorbitol) may create problems on product application (e.g., sticky, tacky, dragy skin feel).

As noted above, self-preserving products may be microbiostatic—not microbicidal. This means that contaminants would be prevented from growing, but that they may remain in the product for extended periods. The acceptance criteria for preservative efficacy testing of self-preserving products would not necessarily be the same as for conventionally preserved products. Testing would need to establish that:

Both routine test organisms and contaminants reasonably expected to be introduced during manufacturing and/or use by the consumer are unable to grow.

Microorganisms do not change the physicochemical composition of the product (raise the pH by use of the arginine deiminase system, decrease the a_W as a result of uptake or utilization of salts and/or water-soluble solutes).

Mixed flora do not have associative growth effects so that one organism could alter the system, thereby allowing another to grow.

Microorganisms do not grow following abuse studies that subject the product to conditions reasonably expected from normal use and abuse, including dilution with water.

Packaging prevents contamination during consumer use/abuse.

The safety of self-preserving products will have to be evaluated on an individual product basis. Although contamination with nonobjectionable microorganisms may not present a consumer health hazard (providing they were not capable of growing or rendering the product susceptible to contamination and growth by other microorganisms), opportunistic pathogens including *P. aeruginosa* or *C. albicans* may present an unacceptable risk to consumers. The hazard would depend on the kinds and numbers of contaminating microorganisms, on the site/frequency of use, and on an individual's susceptibility to the microorganism and its metabolic by-products (enzymes, toxins, etc.). For example, low levels of contamination by these microorganisms may present minimal risks for products applied to normal skin; however, the risks increase when products are applied to damaged skin (e.g., skin with cuts and abrasions, chapped skin with cracks and fissures) or to sensitive areas (e.g., around the eyes, on baby skin). In some instances, creative new approaches to product packaging may be needed to protect the formula from microbial contamination. As in determining RDVs, the type of packaging determines the preservation requirements of a product. Careful studies are required to assess the safety of aqueous products in multiple-use containers.

1.11 SUMMARY

Preservative-free and self-preserving products are an emerging class of cosmetics intended to provide the same performance benefits as existing formulations *and* reduced irritation potential. Many contemporary haircare, skincare, and eyecare formulas may not simply be converted to preservative-free products by elimination of preservative chemicals or to self-preserving products by formula changes that produce low pH and/or low a_w because these products may have unacceptable "cosmetic elegance." Preservative-free and self-preserving products may be developed by understanding and applying the principles of preservation as they relate to each product. This may require modification of current formulations, more protective packaging, and testing to ensure that microbial survival strategies will not allow bacteria, yeasts, or molds to grow. The development of these new products will require creativity and cooperation among formulators, microbiologists, engineers, and package designers.

Technology has developed to the point where we understand microbial physiology and are able to use the principles of preservation to create products that are adequately preserved. Consumer-friendly, environment-friendly products made without preservatives are possible. The trend to use of multipurpose formula components to create self-preserving products is the wave of the future. The following chapters provide insights into the creation of preservative-free and self-preserving products.

REFERENCES

1. J. J. Kabara. Food-grade chemicals in a systems approach to cosmetic preservation. In: J. J. Kabara (Ed.). *Cosmetic and Drug Preservation: Principles and Practice.* Marcel Dekker, New York, 1984, pp. 339–356.
2. D. S. Orth. *Handbook of Cosmetic Microbiology.* Marcel Dekker, New York, 1993, pp. 1–118.
3. D. S. Orth. The required D-value: Evaluating product preservation in relation to packaging and consumer use/abuse. *Cosmet. Toiletr. 107*(12):39–43 (1992).
4. H. J. Eiermann et al. Prospective study of cosmetic reactions: 1977–1980. *J. Am. Acad. Dermatol. 6*(5):909–917 (1982).
5. E. Schöpf and A. Baumgartner. Contact allergies to preservatives in cosmetic preparations. *J. Appl. Cosmetol. 8*:39–49 (1990).
6. D. S. Orth. *Handbook of Cosmetic Microbiology.* Marcel Dekker, New York, 1993, pp. 119–219.
7. B. Ray and M. L. Speck. Repair of injury induced by freezing *Escherichia coli* as influenced by recovery medium. *Appl. Microbiol. 24*:258–263 (1972).
8. A. Hurst. Bacterial injury: a review. *Can. J. Microbiol. 23*:936–944 (1977).
9. F. F. Busta. Introduction to injury and repair of microbial cells. *Adv. Appl. Microbiol. 23*:195–201 (1978).
10. E. Freese, C. W. Sheu, and E. Galliers. Functions of lipophilic acids as antimicrobial food additives. *Nature 241*:321–325 (1973).
11. J. J. Kabara. Medium-chain fatty acids and esters. In: A. L. Branen and P. M. Davidson, eds. *Antimicrobials in Foods.* Marcel Dekker, New York, 1984, pp. 305–322.
12. W. C. Frazier and D. C. Westhoff. *Food Microbiology.* 3rd ed. McGraw-Hill Book Company, New York, 1978, pp. 3–16.
13. D. S. Orth, C. M. Lutes Anderson, D. K. Smith, and S. R. Milstein. Synergism of preservative system components: use of the survival curve slope method to demonstrate anti-*Pseudomonas* synergy of methyl paraben and acrylic acid homopolymer/copolymers *in vitro. J. Soc. Cosmet. Chem. 40*:347–365 (1989).
14. J. J. Kabara. Multifunctional food-grade preservatives in cosmetics. *Drug Cosmet. Ind. 125*(4):60,62,64,66,68,70,75,76,140,142,144,145 (1979).
15. J. R. Hart. Chelating agents as preservative potentiators. In: J. J. Kabara, ed. *Cosmetic and Drug Preservation: Principles and Practice.* Marcel Dekker, New York, 1984, pp. 323–337.
16. J. J. Kabara. Structure-function relationships of surfactants as antimicrobial agents. *J. Soc. Cosmet. Chem. 29*:733–741 (1978).
17. M. T. Parker. The clinical significance of the presence of micro-organisms in pharmaceutical and cosmetic preparations. *J. Soc. Cosmet. Chem. 23*:415–426 (1972).
18. Technical Bulletins 1016b & 1019. Mona Industries, Paterson, NJ, 1991.
19. Technical Bulletin CAE-8301. Ajinomoto Co., Inc., Japan, 1991.
20. A. Lamikanra and T. A. Ogunbayo. A study of the antibacterial activity of butyl hydroxyanisole (BHA). *Cosmet. Toiletr. 100*(10):69–71,74 (1985).
21. J. J. Zeelie and T. J. McCarthy. The antimicrobial activity of phenolic antioxidants in aqueous solution. *Cosmet. Toiletr. 97*(11):61–64 (1984).

2

The Effect of Acid pH on Microorganisms and Survival Strategies that Permit Growth in Products

Donald S. Orth
Neutrogena Corporation
Los Angeles, California

In recent years, there has been increasing interest in developing and marketing natural cosmetics, including those that are "preservative-free" (i.e., products that do not contain ingredients commonly recognized as antimicrobial preservatives), to meet the consumer demand for natural products. Although cosmetics and over-the-counter (OTC) drugs are not necessarily intended to be sterile, satisfactorily preserved aqueous formulas in multiple-use containers must have a preservative system that makes them self-sterilizing (i.e., self-preserving) or they must be packaged/stored in such a manner that microbial contamination and growth is not possible.

A self-preserving formula is one that prevents microbial growth (i.e., is bacteriostatic or fungistatic) or kills microorganisms (i.e., is bactericidal or fungicidal). A self-preserving formula may or may not contain antimicrobial perservatives. It is believed that the term "self-preserving" more accurately describes the action of the formula than "preservative-free," except in those instances where the formula does not have a preservative system and would spoil if it were not packaged or stored in such a manner to prevent microbial contamination and growth.

Adequate preservation of a product involves more than the antimicrobial effectiveness of the formula. The type of packaging and consumer use habits affect preservative requirements. The required D-value (RDV) concept was introduced to relate the preservative system of the formula with product packaging and consumer use/abuse (1,2). Although an unpreserved formula in a multiple-use container would not be acceptable because it could become contaminated during use, the formula would be acceptable if it were manufactured, sterilized, and supplied in a unit-dose container. The same formula in a multiple-use container may be suitable for restricted use if it were handled as we now handle perishable foods (i.e., refrigeration and freezing).

Several chapters in this book discuss principles of preservation and how these are used to create self-preserving products. Adverse physical and chemical conditions in self-preserving products include dehydration [i.e., low water activity (a_w)], extreme pH conditions, oxidative stress (i.e., free radicals), fluctuations in oxidation/reduction potential (i.e., the availability of O_2 for aerobic growth or the lack of it for anaerobic growth), and starvation (i.e., lack of nutrients).

Bacteria, yeasts, and molds cope with stressful conditions using a wide range of mechanisms designed to prevent or repair injuries so that they can survive and grow. *Survival strategies* are the morphological features and physiological processes by which microorganisms respond to changes in the environment (5). This chapter presents information on the effect of acid pH on microorganisms and discusses six survival strategies—capsules, the bacterial cell wall and outer membrane of gram-negative bacteria, acid tolerance, adaptation, osmoregulation, and siderophores—and explores how these strategies increase the difficulty of controlling microbial growth and persistence in self-preserving products.

2.1 THE EFFECT OF ACID pH ON MICROORGANISMS

The preservative system of a product includes compounds with known antimicrobial activity and formula components that may contribute directly or indirectly to antimicrobial activity (3,4). The physicochemical make-up of a formula, which includes factors such as product pH, a_w, nutrient availability, surfactant concentrations, sequestering agents, alcohols, interfering materials, etc., determines whether a formula is self-preserving and/or whether specific preservatives are needed for adequate preservation.

Some familiar products use low pH as part of their preservative systems. Most people are familiar with foods that use acid pH to help prevent microbial growth. These include pickles, yogurt, vinegar, and cheese, which contain short-chain fatty acids (i.e., acetic, lactic, propionic acids), and beer and wine, which also contain organic acids (i.e., malic, citric, and tartaric acids) and ethanol.

Several cosmetic and OTC drug products have acidic pH (pH 3–5), which contributes to the preservative systems of these products. Salicylic acid is allowed for use in OTC drug products that treat acne and dandruff. Alpha-hydroxy acid (AHA) products that contain glycolic, lactic, and other AHAs have become popular for treating signs of skin aging. The low pH and low a_w obtained using 25% aluminum chlorohydrate in roll-on antiperspirants help create conditions unfavorable for microbial growth. Cationic surfactants, which are added to provide desirable skin-feel, may contribute to the preservative system of antiperspirants. Hair conditioners generally contain cationic conditioning agents and are acidic—typically around pH 4–5. The acidic pH often is obtained by adding citric acid. The acidic pH and the chelating action of citric acid enhance the preservative system of the formula. In addition, alcohol and propylene glycol may be used to help solubilize formula components and adjust the viscosity to the desired range. The combined effects of low pH, low a_w, cationics, chelators, and/or solubilizers contribute to the preservative systems of these products. (See Sec. 2.4 for a discussion of acid tolerance of microorganisms.)

Some products use alkaline pH as part of their preservative system. Liquid soaps with an alkaline pH (pH 9.0–10.5) present an environment unfavorable for the growth of some microorganisms due to the chaotrophic (i.e., membrane destabilizing) effects of the free alkalinity. Liquid soaps may contain substantial quantities of dissolved solutes (e.g., glycerol, inorganic salts, and salts of free fatty acids) which decrease the a_w. In addition, these products frequently contain chelating agents and phenolic antioxidants, such as butylated hydroxy toluene (BHT), which are part of the product preservative system (3,4). The combined effects of high pH, low a_w, high soap/surfactant level, and antioxidants contribute to the preservative systems of these products. Unused bar soaps are also self-preserving because of their alkaline pH and low a_w.

Relatively anhydrous products (e.g., talcum powder, cornstarch, mineral oil, and stick deodorants) do not require preservatives because they do not have sufficient water to support microbial growth. Even though the a_w of these products may be too low to permit growth, manufacturers may incorporate preservatives into these products to prevent microbial growth that may occur due to inadvertent addition of water by consumers during use (or abuse).

2.2 CAPSULES (EXOPOLYSACCHARIDE)

Many bacteria produce large amounts of extracellular polymer during growth under natural conditions. In nearly all bacteria, this polymer is composed of high molecular weight acidic exopolysaccharide (EPS) such as dextran (glucose polymer), levan (fructose polymer), hyaluronic acid (*N*-acetylglucosamine and glucuronic acid polymer), cellulose (glucose polymer), and complex polysaccharides. Complex EPS heteropolymers have an organized, usually repeating structure and may contain branched, highly substituted polysaccharides composed of various combinations of glucose, rhamnose, glucuronic acid, galactose, *N*-acetylglucosamine, and uronic acids with different glycosidic linkages (6,7).

When microscopic examination reveals that the EPS forms a condensed, well-defined layer closely surrounding the cell, it is classed as a *capsule*. The polymer is called a *glycocalyx* when it forms a loose meshwork of fibrils extending outward from the cell. The term *slime layer* may be used to describe masses of polymer that appear to be totally detached from the cells, but in which viable cells may be entrapped.

The EPSs of pseudomonads (especially alginate of *P. aeruginosa*) have been studied extensively because they are both virulence factors and survival strategies (5,6). This section reviews the structure and functions of *P. aeruginosa* EPSs, some environmental signals that induce EPS synthesis, and mechanisms by which EPS enhances the survival of *P. aeruginosa* in natural environments and in cosmetic products.

2.2.1 Structure and Function of *Pseudomonas* EPS

Mucoid strains of *P. aeruginosa* produce copious amounts of alginate on complex agar media after overnight growth. The EPS produced by pseudomonads generally is negatively charged due to the presence of uronic acids (β-D-mannuronate and α-L-guluronate). *P. aeruginosa* alginate is a linear copolymer of D-mannuronate and variable amounts of the C-5 epimer L-guluronate (Fig. 2.1). Additional negative charges may be introduced by substitution of the monosaccharides with pyruvyl, succinyl, acetyl, or formyl groups. The high molecular weight and negative charge of the polysaccharides frequently enables the formation of highly hydrated gel-like networks in which a comparatively small amount of EPS interacts with a large volume of water. The gel matrix is stabilized by chelation of divalent cations (Ca^{2+} ions), which cross-links adjacent polysaccharide chains (7,8).

Fluorescent pseudomonads (*P. aeruginosa*, *P. fluorescens*, *P. viridiflava*, and *P. putida*) can produce alginate EPS both in vitro and when growing in association with green plants. Fett et al. (9) reported that approximately 10% of the fluorescent pseudomonads from plants produced mucoid growth on agar media. They analyzed purified EPS preparations and found that the EPS of *P. aeruginosa* consists of

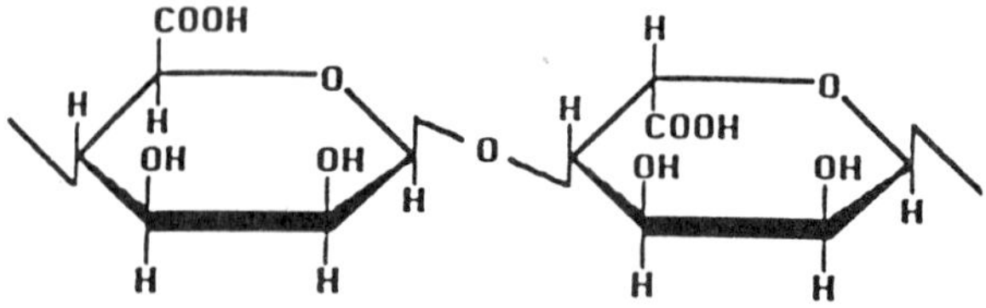

FIGURE 2.1 Structure of alginate showing β-1,4-linked D-mannuronic acid and L-guluronic acid.

alginate. The strains of *P. fluorescens* studied produced three different EPSs: (1) levan (polymer of fructose) and alginate, (2) marginalan (an acidic EPS composed of glucose and galactose in a 1:1 molar ratio with pyruvate and succinate substituents), and (3) levan and a novel EPS that contained the neutral sugars rhamnose, mannose, and glucose in a 1:1:1 molar ratio substituted with pyruvate and acetate. These alginates were acetylated and contained 7–12% acetate; the marginalans contained pyruvate (11–22%) and succinate (15–26%). The types of EPSs found in one strain of *P. aeruginosa*, 18 strains of *P. fluorescens*, and one strain of *P. viridiflava* appear in Table 2.1.

In view of the fact that the number of strains examined by Fett and coworkers was limited (20 strains), it is quite possible that different fluorescent pseudomonads produce EPSs with different patterns of levan, alginate, novel EPS, and marginalan. Alginate production is influenced by environmental factors (8,10). Fluorescent pseudomonads often produce alginate when grown on culture media containing glucose or gluconate (11). When sucrose was used as the carbon source, levan, levan and alginate, or alginate alone was produced, depending on the strain. Gacesa and Russell (7) noted that O-acetylated alginate is the major component of the slime of *P. aeruginosa* and that a particular strain has a constant mannuronate:guluronate ratio, regardless of the growth conditions. Aeration, a lack of iron, surfactants, and possibly low a_w may also affect EPS production (6,8,12,13). Curiously, the mucoid phenotype is relatively unstable in vitro and rapidly reverts to the nonmucoid phenotype on subculturing (6,12). This may be due, in part, to the regulation of alginate production by the presence of iron in the growth medium because production of alginate is under the control of iron (see discussion below).

2.2.2 Role of EPS in Adhesion and Biofilms

The glycocalyx is involved in the adhesion of microorganisms to surfaces (14–17). Clean surfaces placed in aquatic ecosystems are rapidly colonized by *P. aeruginosa*. Initial adhesion is mediated by alginate fibers (14). Adherent cells produce

TABLE 2.1 EPSs Produced by Fluorescent Pseudomonads

Pseudomonas spp.	Levan	Alginate	Novel EPS	Marginalan
P. aeruginosa	–	+	–	–
P. fluorescens	+	+	–	–
	–	–	–	+
	+	–	+	–
P. viridiflava	–	+	–	–

large amounts of alginate, and they divide to form microcolonies surrounded by EPS (e.g., a biofilm). The alginate gel functions as an ion-exchanger to trap nutrients within this biofilm. The biofilm also protects viable cells from the antibacterial action of colicins, surfactants, biocides, heavy metals, and antibiotics (11,14,15,18).

2.2.3 Role of EPS in Protecting Microorganisms

Smith and Simpson (19) reported that alginate is capable of scavenging free radicals and that mucoid *P. aeruginosa* showed greater resistance to killing by hypochlorite than did a nonmucoid revertant. Zevenhuizen and Ebbink (20) postulated that EPS of *Achromobacter* spp. formed complexes with heavy metals (through the pyruvate carboxyl group) as a means of detoxification. Pyruvate is known to function as a free radical scavenger and prevent active oxygen species from damaging cells contained within the EPS (5,20). The occurrence of nearly 22% pyruvate in the marginalan capsule of fluorescent pseudomonads and perhaps lower levels in the EPSs of other gram-negative bacteria would appear to be a significant means of protecting these bacteria from external reactive oxygen species.

Ohman and Goldberg (21) noted that *P. aeruginosa* alginate may protect cells from desiccation. It is likely that the gel matrix surrounding cells tends to resist removal of water, which would provide a means of survival—although not necessarily growth—under low a_w conditions. The negatively charged EPS may function as an ion exchanger and attract oppositely charged molecules. It is possible that it could make additional nutrients available to cells in this manner. Finally, it would seem that the EPS may function as a reservoir of nutrients to be used by cells under nutrient-limiting conditions.

2.2.4 Effect of EPS on Microbial Survival in Cosmetics

The numerous functions of EPS (and the EPSs of pseudomonads, in particular) reviewed in this section provide insight into the role of EPS as a survival strategy. EPS helps microorganisms resist the antimicrobial effects of hostile product environments including desiccation, nutrient depletion, and free radical damage. The EPS surrounding cells tends to resist removal of water, which helps provide microorganisms with a means of survival—although not necessarily growth—under low a_w conditions. Other unfavorable conditions, such as low pH, lack of nutrients, and the presence of surfactants, increase the stress on microorganisms and decrease the likelihood of their persistence and survival. Nevertheless, once microorganisms have formed capsules/biofilms, they may be more resistant to disinfectants/sanitizers in manufacturing plants and to preservative systems in finished products.

2.3 CELL WALL (ENVELOPE)

The cell wall (cell envelope) surrounds the cell, gives it shape, and protects the cell from adverse environmental conditions. The cell wall is essential for survival under most environmental conditions. Cell-wall–deficient bacteria (L forms) are gram-negative or gram-positive bacteria that have lost their cell walls. When special precautions are taken, L forms can be grown on culture media to prevent lysis of the fragile cells. The role of the cell wall in protecting cells from lysis in hypotonic solutions is well known to microbiologists, so the discussion in this section focuses on other mechanisms by which the bacterial cell envelope protects cells and enables them to survive when exposed to adverse physical and chemical conditions.

2.3.1 Structure of Bacterial Cell Walls

The layers of the cell envelope between the cytoplasmic membrane and the capsule are considered to be the cell wall. The gram-positive bacterial cell wall consists primarily of peptidoglycan and teichoic acids. The gram-negative bacterial cell envelope contains three outer layers consisting of a peptidoglycan layer, a phospholipid-protein (lipoprotein) layer, and the outermost lipopolysaccharide (LPS) layer (5). The peptidoglycan (also called murein or mucopeptide) gives the cell wall the high tensile strength required to resist internal osmotic pressures ranging from 5 to 20 atmospheres as a result of solute concentration via active transport (23). Also, diaminopimelic acid is a unique amino acid found in bacterial cell walls.

The gram-negative bacterial peptidoglycan layer is much thinner than that of gram-positive bacteria, and this accounts for differences in dye retention and staining character when performing the Gram stain. Components of the cell envelopes of gram-positive and gram-negative bacteria are illustrated in Figures 2.2 and 2.3.

Lipoprotein is the most abundant protein of gram-negative bacteria. It functions to stabilize the outer membrane and to anchor it to the peptidoglycan layer. The periplasmic space (PS) is between the cytoplasmic membrane/peptidoglycan layer and the outer membrane (22). The PS in most gram-negative bacteria is filled with a hydrated peptidoglycan gel that contains proteins (including hydrolytic enzymes) and oligosaccharides. The outer membrane has proteinaceous pores, or porins, which make it permeable to low molecular weight molecules; however, it prevents nonselective passage of large molecules such as proteins and antibiotics. This molecular sieving helps prevent the loss of periplasmic proteins due to nonspecific mechanisms (e.g., leakage). As could be expected, the perme-

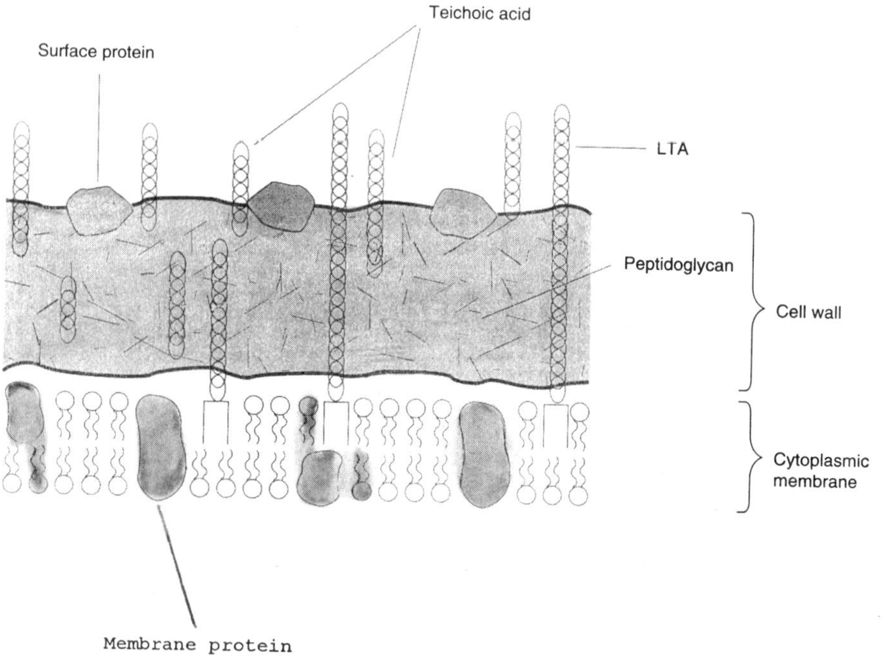

FIGURE 2.2 Schematic representation of cell wall of gram-positive bacteria.

ability of the outer membrane varies from one species to the next. The porins of enterobacteria (e.g., *Escherichia coli* and *Salmonella typhimurium*) have exclusion limits of about 600 daltons, in contrast to the porins of *P. aeruginosa*, which may have exclusion limits of 3000 daltons (23).

2.3.2 The Lipopolysaccharide of Gram-Negative Bacteria

The LPS of gram-negative bacteria is composed of Lipid A, polysaccharide, and a terminal series of repeat carbohydrate units. The negatively charged carbohydrates are noncovalently cross-bridged by divalent cations. This stabilizes the outer membrane and provides a barrier to hydrophobic molecules (5). Chelating agents and/or polycationic antibiotics, such as the polymyxins and aminoglycosides, remove cations and reduce the integrity of the LPS. This increases the permeability of the outer membrane to large hydrophobic molecules, antibiotics, and preservatives. The increase in antimicrobial activity produced by a chelating agent working with an antibiotic or a preservative is called "permeabilization synergy" (24).

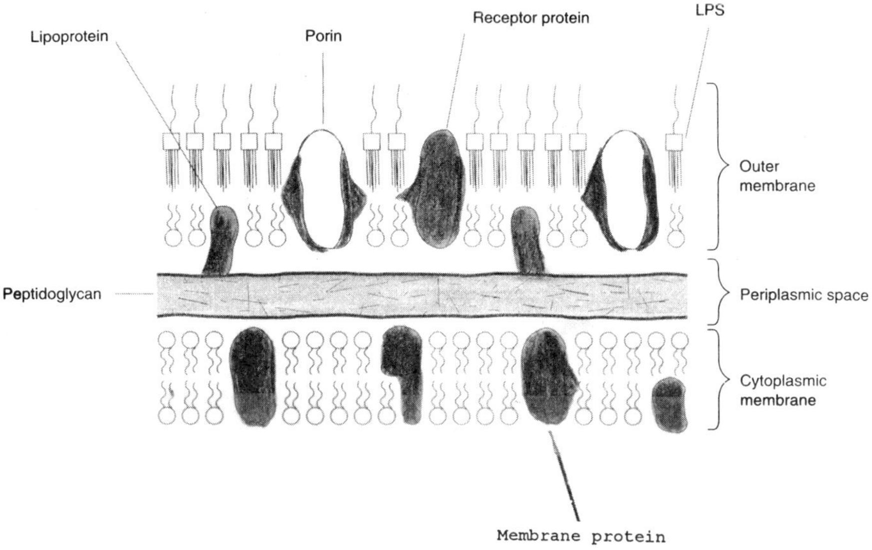

FIGURE 2.3 Schematic representation of cell wall of gram-negative bacteria.

2.3.3 Role of LPS in Protecting Microorganisms

The LPS provides a physical/chemical barrier through which reactive oxygen species generated outside of the cell must pass to interact with targets (i.e., cytoplasmic membrane, DNA, etc.). Dahl et al. (25) reported that some strains of bacteria that do not produce a large amount of LPS have greater sensitivity to exogenous singlet oxygen than do strains that have a larger portion of LPS. Gram-positive bacteria do not have a protective structure analogous to the gram-negative bacterial outer membrane, with its associated PS and LPS. It is possible that the outer membrane neutralizes the effect of singlet oxygen because it contains unsaturated fatty acids and proteins—compounds known to react with singlet oxygen (25).

The outer membrane and LPS of gram-negative bacteria are not considered to be vital targets for reactive oxygen species; consequently, reaction of free radicals with the outer membrane may protect cells and spare vital targets, such as the cytoplasmic membrane or DNA (5,25). Although the peptidoglycan cell wall of gram-positive bacteria affords some protection, that provided by the gram-negative bacterial cell envelope is greater in most cases.

Dahl and coworkers noted that reaction of singlet oxygen with the outer membrane could result in the production of other reactive oxygen species (e.g., peroxy radicals), so the toxicity would be the sum all reactive oxygen species reaching the cytoplasmic membrane and intracellular structures. In addition, carotenoid pigments in the cytoplasmic membrane may react with and help detoxify reactive oxygen species (25). Discussion of carotenoid pigments as a survival strategy was presented elsewhere (5).

2.3.4 Effect of Cell Walls on Microbial Survival in Cosmetics

The cell envelope is the interface between the cell and its environment. The bacterial cell wall provides shape to microorganisms and prevents destructive swelling and lysis in hypotonic environments. The cell envelope contains porins which regulate entry of molecules into the periplasmic space. The structural components and enzymes of the envelope may interact with toxic chemicals and reactive oxygen species, thereby reducing their concentration/toxicity and delaying their arrival to the cytoplasmic membrane and cytoplasm. In some instances, this may enable the bacteria to adapt—to change levels and/or activities of enzymes and induce stress proteins—so that they may be better prepared to resist changes in their environment.

The complex organization of the gram-negative bacterial cell envelope, with its associated peptidoglycan, PS, outer membrane, and LPS molecules, appears to offer a broader range of protective mechanisms than are available in the larger peptidoglycan cell envelope of gram-positive bacteria. These mechanisms may help explain why gram-negative bacteria have been so troublesome in manufacturing plants and cosmetic products and why they may be difficult to control in self-preserving products.

2.4 ACID TOLERANCE

All living microorganisms have characteristic ranges of temperature, pH, O/R potential, and a_w that permit growth. As noted above, microorganisms of interest to the cosmetic and pharmaceutical industries generally are aerobic and have pH optima for growth around neutrality (around pH 5–8). Microbial cellular function is dependent on the maintenance of appropriate intracellular pH (pH_i). Although the relationship between pH homeostasis and acid tolerance in bacteria is not completely understood, it is believed that maintenance of pH_i within a fairly narrow range is essential for acid tolerance. As the pH becomes more extreme, microorganisms must expend a greater amount of energy to maintain pH and ionic strength gradients across the cytoplasmic membrane (5).

2.4.1 Mechanisms for Maintaining Intracellular pH in Bacteria

Bacteria have mechanisms that operate to maintain the pH_i at a relatively constant pH—even though the pH outside of the cell (pH_o) fluctuates. These pH homeostasis mechanisms are discussed in the following paragraphs.

Cytoplasmic Membranes

Cell membranes are believed to play a role in acid tolerance of microorganisms. Membranes are selectively permeable—they allow movement of specific ions and compounds into and out of the cell—which enables viable microorganisms to establish chemical and electrical gradients across the cytoplasmic membrane. Sublethal injury is frequently accompanied by impaired membrane integrity, as reflected by leakage of 260- and 280-nm–absorbing materials into the culture medium (5). If membranes are a site of injury, repair involves reestablishment of normal membrane function, including the Na^+/K^+ pump (ATPase).

Membrane ATPase

The primary mechanism of pH maintenance in microorganisms utilizes the expulsion of protons from the cytoplasm by ATPases at the expense of ATP (26). Modulation of the cytochromes (the primary cellular proton pumps in prokaryotic organisms) and the sodium/proton and potassium/proton antiport systems (the membrane ATPases) enable the cell to maintain a ΔpH ($pH_i - pH_o$) of about 2 pH units, with the interior of the cell being more alkaline than the external environment (27). This constitutive pH-maintenance system handles acid stress above pH 4.0 in enterobacteriaceae such as *Salmonella* spp. (27). Bender et al. (28) reported that acid tolerance and proton permeability were dependent on the pH profiles of the membrane ATPases of various strains of oral streptococci. Besides expending energy (ATP) to pump protons out of the cells, the ATPase can operate in reverse and couple H^+ ion movement into the cell with ATP synthesis. Adaptation to acid stress involves modulation of membrane ATPase activity (27–31).

Acid Tolerance Response

Foster (27) reported that the salmonellae can survive pH_O 4.0 for at least 4 hours with virtually 100% viability; however, they rapidly lose their ability to survive when the pH decreases below 4. The *S. typhimurium* strain studied was able to adapt to acid stress when it was exposed to culture medium adjusted to pH 5.8 for one generation. These adapted cells had a 100- to 1000-fold higher number of survivors when the population was then exposed to pH 3.3 (compared to unadapted salmonellae). Foster termed this adaptation an *acid tolerance response* (ATR).

Foster (27) believed that the ATR system of *S. typhimurium* has two forms of pH homeostasis. The first is a constitutive system that handles pH stress above pH 4.0 and includes ATPases that are known to be inducible in some bacteria (27,29). The second involves pH-regulated gene expression that produces acid-shock proteins. Foster observed that some of the acid-shock proteins were heat-shock proteins, such as chaperonins specified by DnaK and GroEL genes, and that these were different from the ATR proteins. Chaperonins help maintain the native conformation of proteins and help renature proteins that have been denatured by acid (or heat). Such stabilizing effects may minimize DNA and internal protein denaturation, which may occur when the pH_i is <5.5.

Acid-Labile Enzymes

It has been known for many years that bacteria alter their metabolism in response to changes in environmental pH. Cellular physiology requires groups of enzymes working coordinately in sequential chains to carry out a given metabolic process, such as glycolysis—the anaerobic conversion of glucose to lactic acid. Allosteric regulatory enzymes have both an active (catalytic) site and an allosteric site for binding to the regulating metabolite (the effector or modulator of the enzyme). Covalently regulated enzymes are pacemaker enzymes that are modulated by interconversion of their active and inactive forms. Orth (5,31) proposed that pH-induced depolymerization of regulatory enzymes may help explain the response of microorganisms subjected to low pH environments. For example, cultures of *Streptococcus faecalis* in complex media with excess glucose stop growing when the pH of the culture drops to about pH 4.9. Neutralization of the acid permits the cells to resume glycolysis and growth (32). The major end product of glycolysis in streptococci is lactic acid, so acid-lability of regulatory enzymes involved with its production will prevent cells from making so much lactic acid that they literally pickle themselves. It is believed that this mechanism is, at least in part, responsible for cessation of metabolism of cells when they are exposed to extremely acidic environments, such as may be used in self-preserving products. It is possible that the effect of pHi on regulatory enzymes is, in part, responsible for some metabolic changes observed when microorganisms are in an acidic environment.

Extracellular Enzymes

Microorganisms have enzyme systems that help raise the pH of acid environments. Slonczewski (30) observed that both *E. coli* and *Micrococcus luteus* produce amino acid decarboxylases at low pH_i. These enzymes work to neutralize the pH_o. The acid-inducible locus of *exa* is an allele of *cadA*, which encodes for lysine decarboxylase (30). Decarboxylation of lysine forms cadaverine and CO_2, and cadaverine is a basic amine that raises the pH. Slonczewski (30) also noted that both *E. coli* and *M. luteus* produce amino acid deaminases at high pH_i. Deamination of amino acids results in the formation of organic acids and ammonia. For example, deamination of alanine results in the formation of lactic acid and ammonia. This results in a net increase in the pH of the culture medium (or product) and may provide sufficient neutralization of the pH to allow growth.

The Arginine Deiminase System

The arginine deiminase system has been studied in a variety of bacteria including lactic acid bacteria, mycoplasmas, halobacteria, *Pseudomonas* spp., and *Bacillus* spp. (31–34). This system provides energy (ATP) through the catabolism of arginine to ornithine, CO_2, and ammonia. Marquis et al. (32) reported that some strains of streptococci had unusual acid tolerance and that arginine could be catabolized at pH values well below the minima for growth or glycolysis (pH 3.5). Although extremely acidic environments can be lethal to microorganisms, catabolism of arginine and the production of ammonia raises the pH and may enable cells to survive. Marquis and coworkers demonstrated that *P. aeruginosa* could produce ammonia at initial pH values as low as pH 3.0. Ammonia raised the pH so that growth could resume. Although the streptococci studied were less acid tolerant than *P. aeruginosa*, they could catabolize arginine and produce ammonia at initial pH values around pH 4.0.

Arginine deiminase (EC 3.5.3.6), ornithine carbamoyltransferase (EC 2.1.3.3), and carbamate kinase (EC 2.7.2.2) are the principal enzymes of the arginine deiminase system. In addition, there may be an inducible transport system for ornithine and arginine across the cytoplasmic membrane (33). The primary reactions occurring in the arginine deiminase system are shown in Figure 2.4. The arginine deiminase system is believed to be inducible and under the control of catabolite repression (32). Bacteria use this system under conditions of limited oxygen and, to a lesser extent, when carbon and energy sources have been depleted (34). *P. aeruginosa* is capable of anaerobic respiration—anaerobic growth using a terminal electron acceptor other than molecular oxygen (e.g., nitrate). The arginine deiminase system enables this organism to grow anaerobically with L-arginine as the only energy source—even when there are no exogenous electron acceptors. The enzymes involved in the arginine deiminase reaction are not acid-labile at pH 3.0 in some bacteria (32).

The arginine deiminase system is a survival strategy because it plays an important role in the acid–base physiology of bacteria in which it is present. In addition, it may enable oxidative pseudomonads to grow anaerobically without respiration (32), because ATP provides energy for cellular reactions including maintenance of pH_i. The catabolism of arginine with the concomitant rise in pH could help bacteria modulate an unfavorably low pH_o and recover from acid stress sufficiently severe to stop glycolysis and growth. Survival strategies that enable microorganisms to increase the pH_i could permit survival and growth in acid products.

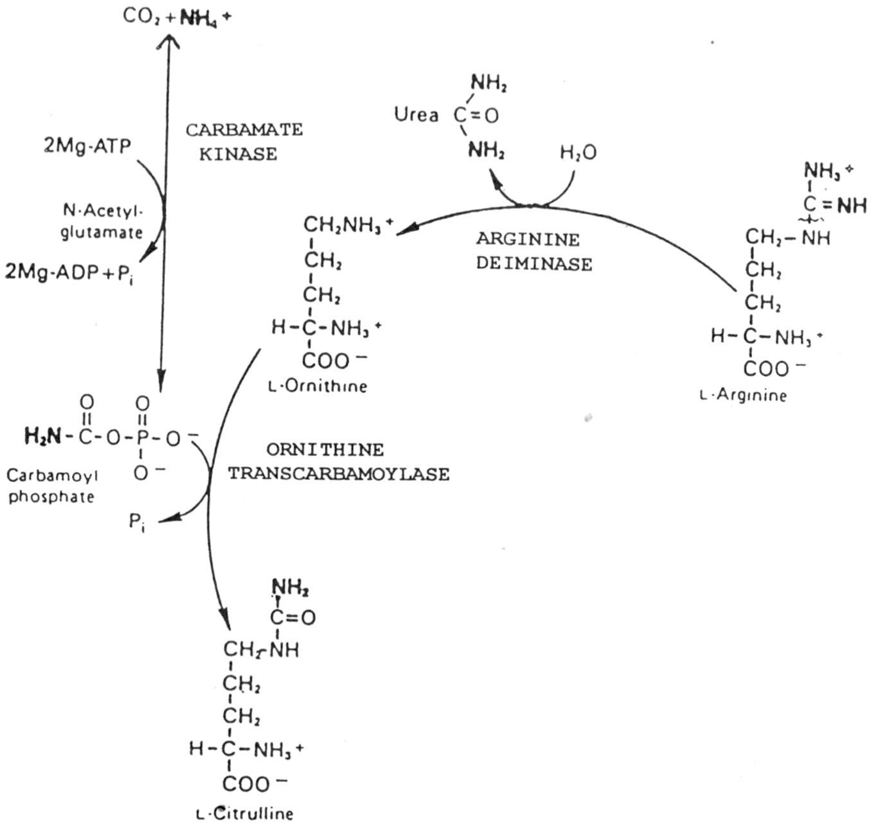

FIGURE 2.4 Enzyme reactions of the arginine deiminase system.

Metabolism of Organic Acids

Some microorganisms can modulate unfavorable pH_o by converting organic acids in the medium into neutral compounds. Blomqvist et al. (35) reported that some bacteria can utilize pyruvate via α-acetolactate to form the neutral compound 2,3-butanediol (2,3-BD). Acetic acid is also able to induce 2,3-BD (36,37). The production of 2,3-BD is enhanced when oxygen is limited and when the pH is lowered (38,39). Johansen et al. (38) found that when the pH is as low as pH 5.8, 55% of the pyruvate goes into 2,3-BD in *Enterobacter aerogenes.*

Blomqvist and coworkers reported optimal production of 2,3-BD under slightly acidic (pH 6), oxygen-limited conditions and in the presence of acetate. They concluded that 2,3-BD production prevented acetate overproduction and subsequent intracellular acidification.

The reduction of acetoin to 2,3-BD regenerates NAD^+ from NADH. It appears that production of neutral compounds, such as 2,3-BD, may play a role in preventing intracellular acidification by modulation of the metabolism from acid production to the formation of neutral compounds and/or in regulation of the intracellular $NADH/NAD^+$ ratio (35–37).

2.4.2 Effect of Acid Tolerance on Microbial Survival in Cosmetics

In many instances, acidic conditions help control microbial survival and growth in conventional and self-preserving cosmetic products. Antiperspirants, hair conditioners, and alpha-hydroxy acid products may have a pH of ≤3.0; however, this low pH does not guarantee that the product is microbicidal because microorganisms may be able to modify the pH of their environment. The examples discussed above reveal some mechanisms by which bacteria are able to modulate the pH of their environment.

2.5 ADAPTATION

When taken collectively, bacteria, yeasts, and molds have diverse metabolic capabilities. Under the proper conditions, microorganisms are capable of utilizing virtually any organic compound for growth. In order to survive, microorganisms must respond to changes in their environment so that they are able to compete with other living cells for the available nutrients. Adaptation is a survival strategy and consists of both genetic and metabolic modulation in response to physical and chemical changes in the environment.

The modulation of bacterial metabolism in response to substrates in the environment—including cosmetic products—reflects the ability of microorganisms to adapt (5). Adapted microorganisms are those that have been "trained" by exposure to diluted product and/or raw materials, such that these microorganisms are able to grow in that product (even though it is adequately preserved) and/or raw materials. A number of publications describe microbial adaptation and/or development of resistance to cosmetic products/preservatives (40–45).

Microorganisms adapt to changes in their environment by altering the expression of genes. Levy (42) noted that the capacity of microorganisms to degrade substrates can be adapted by modification of control and regulatory mechanisms for enzyme production and/or specificity and from unique enzymatic activities due to gene recruitment and/or genetic rearrangements.

The primary protein structure (i.e., amino acid sequence) of an enzyme is determined by the genetic information contained in the triplet codes in the DNA. This information is passed on to mRNA, which takes the coded information to the ribosome—the site of protein synthesis. Physical and chemical treatment of living microorganisms may cause mutations—sudden, heritable changes in the nucleotide sequence in the DNA. If this altered DNA is not removed by repair mechanisms, it may pass the altered information to the mRNA, which causes the resultant proteins to have an altered primary structure. Mutations may cause partial or total loss of enzymatic activity, depending on the site of the mutation in the DNA and the effect it has on the structure of the enzyme (whether it is at an "active site" or a "binding site" of the enzyme). Mutations rarely enhance catalytic activity of an altered enzyme (5).

2.5.1 Adaptation by Enzyme Induction

Microorganisms can produce specific enzymes in response to specific inducers. An example frequently given in beginning microbiology classes is the induction of β-galactosidase. *E. coli* grown on media containing glucose will not ferment lactose at any appreciable rate due to the low basal level of β-galactosidase. If lactose or some other β-galactosides are added to the growth medium, β-galactosidase is induced (i.e., the level increases significantly above the basal level) and the culture of *E. coli* is then able to ferment lactose. Examples of enzyme induction include the induction of stress proteins and ATPase with low pH_i, induction of 2,3-BD by acetic acid, and induction of lysine decarboxylase by acidic pH (presented above).

2.5.2 Adaptation by Enzyme Repression and Derepression

Repression occurs when the presence of a metabolite in the medium reduces new synthesis of that metabolite. For example, addition of the amino acid histidine to a culture of *S. typhimurium* represses synthesis of all enzymes in the histidine biosynthetic pathway. Derepression occurs when enzyme biosynthesis occurs following removal or depletion of an essential biosynthetic intermediate from the growth medium.

The production of pyocyanin—an iron-binding molecule produced by fluorescent pseudomonads when the concentration of iron is low—is derepressed when the iron concentration is low (see below). Curiously, many enzymes are induced along with other enzymes that may have related functions. For example, pyocyanin production and the production of the transport protein(s) responsible for internalizing pyocyanin-iron complexes are derepressed together. In the same manner, low iron levels derepress genes responsible for the production of elastase (a proteolytic enzyme capable of degrading elastin) and alginate. Coordinate induction occurs when several enzymes are turned on by a single inducer. Similarly, coordinate derepression occurs when expression of several enzymes is reduced by a single repressor, as with pyocyanin.

2.5.3 Adaptation by Regulation of Enzyme Activities

The survival of microorganisms depends on efficient mechanisms for utilizing substrates for energy and growth. The global response of microorganisms to environmental stimuli involves complex metabolic regulation of gene expression, which is mediated by the regulation of enzyme-catalyzed reactions in living cells. The regulation of enzymes may be influenced by changing the absolute quantity of enzyme, by altering the pool size of reactants other than the enzyme, and by altering the catalytic activity of the enzyme (5).

The total quantity of any given enzyme in a cell is determined by its rate of synthesis and the rate of its degradation. The quantity of an enzyme may be increased by increasing its rate of synthesis (i.e., by induction or derepression) or by decreasing its rate of degradation. On the other hand, the quantity of an enzyme may be reduced by decreasing its rate of synthesis (by repression) or by increasing the rate of its degradation.

The metabolic reactions occurring within microbial cells are governed by the enzymes present, the substrates available, and the intra- and extracellular physicochemical conditions. The environment may influence both intracellular and extracellular enzyme levels and activity. It is apparent that the chemical composition of a formula and the conditions under which it is stored determine whether substrates are available for microbial growth and whether growth of a given microorganism is possible. From the foregoing discussion, formulators must recognize that substrates may be immediately available in the product or after adaptation that may involve induction/derepression of genes responsible for the production of requisite enzymes.

2.5.4 Adaptation of Microorganisms in Cosmetics

Microorganisms respond to changes in their environment by altering the expression of genes, by changes in enzyme activity, and by changes in enzyme quantity. Environmental conditions, including low pH and iron limitation in acidic products containing chelating agents, affect the expression of genes and enable microorganisms to survive and utilize substrates that are available.

Adaptation is a survival strategy that permits efficient utilization of materials available to microorganisms so that they can survive and grow. Adaptive mechanisms prevent wasteful utilization of cellular reserves, including overproduction of materials that are not needed by the cell. These mechanisms are dynamic—they shift in response to the immediate needs of the cell. Adaptation involves modulation of the expression of genetic information, which enables a cell to increase the production of enzymes (i.e., enzyme induction) or to cope with specific nutrient limitation (i.e., depression of enzymes needed for siderophore synthesis). Once the needs of the cell have been fulfilled, these mechanisms are turned off, and the cell channels its metabolic reserves to synthesizing enzymes required for maintenance and growth. Adaptation may enable microorganisms to utilize substrates available in cosmetic products.

2.6 OSMOREGULATION

It was noted above that decreasing the a_w increases the stress imposed on bacteria, yeasts, and molds. Microorganisms have morphological features and physiological capabilities that enable them to adapt and grow in the presence of high concentrations of water-soluble solutes. The term *osmoregulation* refers to genetic and physiological changes that occur to permit microorganisms to adapt and grow under a_w conditions that are the result of intracellular water loss and a reduction in cell turgor. Osmoregulation is associated with the build-up of internal solute content and of intracellular osmolarity, so that cell turgor is restored; consequently, osmoregulation is a survival strategy that enables growing microorganisms to modulate their intracellular milieu in response to osmotic changes in their environment (5,46–48). The moisture requirements of microorganisms, mechanisms of osmoregulation, and the effect of osmoregulation on the survival of microorganisms in cosmetics are discussed in this section and in Chapter 3.

2.6.1 Moisture Requirements of Microorganisms

Water is a requirement for growth of all known living organisms, from bacteria to humans. As would be expected, different microorganism have different minimum water requirements. The water requirement of microorganisms may be expressed in terms of available water, or a_W. The a_W is the vapor pressure of the solution (solutes in water) divided by the vapor pressure of the solvent (water). The a_W of pure water would be 1.00, and a 1.0 M solution of an ideal solute would have an a_W of 0.9823. If a solute, such as glucose, is dissolved in water, it will lower the water vapor pressure. One mole of an ideal solute dissolved in one kilogram of water will depress the vapor water by 1.77%, and the solution will have an aqueous vapor pressure that is 98.23% of the vapor pressure of pure water (without solutes) at the same temperature. The atmosphere above this solution will have a relative humidity (R.H.) of 98.23%, and the solution will have an a_W of 0.9823. At equilibrium, the a_W is a numerical expression equal to the R.H./100% (5,49).

Use of low a_w in self-preserving products may be facilitated with an understanding of mechanisms used by microorganisms to deal with low a_w environments. Several generalizations may be made about the moisture requirements of microorganisms as follows:

Each microorganism has a characteristic optimal a_w and its own range of a_w for growth for a given set of environmental conditions.

Factors affecting the a_w requirements of microorganisms include the kind of solutes present, the nutritional value of the substrates available, temperature, O_2 supply, pH, and the presence of inhibitors.

An unfavorable a_w will decrease the growth rate and the maximum yield of cells.

The lag time before initiation of growth or germination of spores is increased as the a_w becomes more unfavorable.

Bacteria generally have higher a_w requirements than yeasts, and yeasts generally have higher a_w requirements than molds.

Different classes of microorganisms have different tolerance to low a_w, and this determines whether a particular bacterium, yeast, or mold is able to grow in a given environment (culture medium, natural habitat, cosmetic product, etc.). The reported lower limits of a_w for several microorganisms of interest to the cosmetic industry are presented in Table 2.2.

2.6.2 Osmoregulation by Gram-Negative Bacteria

Osmotic tolerance of *E. coli* depends on the accumulation of K^+ ions and osmoprotectants. The primary response to low a_w in *E. coli* is the accumulation of K^+ ions, in which the intracellular concentration of K^+ ions is nearly proportional to the osmolarity of the growth medium, and a concomitant increase in the concen-

TABLE 2.2 Lowest a_w Values Permitting Growth of Organisms

Group of microorganisms	Minimal a_w value
Many bacteria	0.91
Pseudomonas	0.97
Escherichia coli	0.95
Bacillus subtilis	0.945
Staphylococcus aureus	0.86
Halophilic bacteria	0.75
Many yeasts	0.88
Osmophilic yeasts	0.60
Many molds	0.80
Xerophilic fungi	0.65

Source: Adapted from Ref. 49.

tration of glutamate in the cytoplasm (48). A secondary response to osmotic stress is the accumulation of large amounts of compatible solutes—osmotically active solutes including amino acids, derivatives of amino acids, and carbohydrates that are noninhibitory to enzyme function or cellular metabolism at high concentration—which can be substituted for potassium glutamate. Trehalose and glycine betaine (*N,N,N*-trimethyl-glycine) are the best-studied osmoprotectants in *E. coli* (48).

Stressful conditions in natural aquatic environments, such as cold temperatures, lack of nutrients, low a_w, and UV radiation, injure bacteria so that the numbers recovered decline and/or disappear. Unfavorable conditions may cause microorganisms to enter a *dormant state* in which they are viable but nonculturable (50). Dormant cells may survive for long periods in seawater, and it is known that the survival of *E. coli* is dependent on organic matter.

Gram-negative eubacteria, such as *E. coli*, require an external supply of betaines (glycine betaine, proline betaine, or γ-butyrobetaine) or choline, which is a precursor of glycine betaine. Ghoul et al. (51) demonstrated that *E. coli* resisted saline stress by accumulating and concentrating osmoprotective substances such as glycine betaine from sediments. In addition, *E. coli* can obtain a moderate level of osmotolerance by endogenous synthesis of glutamate and trehalose when betaines are unavailable (52). This is an active area of current research, and numerous reports have examined the role in heat-shock protein DnaK in osmotic adaptation and the requirement for betaine, proline, glutamate, and trehalose in maintenance of viability at high osmolarity in gram-negative bacteria including *E. coli*, *Serratia marcescens*, and *S. typhimurium* (53–60).

Bacterial capsules may play a role in osmoregulation. It is believed that all strains of *P. aeruginosa* carry the genes for alginate production (*alg* genes) even though these genes may be silent (21). The *alg* genes are specifically activated by environmental signals such as high osmolarity (21,61,62). Several workers have reported that the alginate capsule of *P. aeruginosa* is believed to protect cells from desiccation (14,21,63).

2.6.3 Osmoregulation by Gram-Positive Bacteria

Gram-positive bacteria accumulate K^+ ions, neutral amino acids, proline, and gamma-aminobutyrate in response to osmotic stress (5). Whatmore et al. (64) reported that *Bacillus subtilis* accumulated K^+ ions and proline in response to osmotic upshock. Their findings indicated that K^+ ions accumulate rapidly (<1 hr) and that K^+ ions serve in some manner as the signal for increased proline synthesis, which required 5–9 hours. They also observed that *B. subtilis* accumulated glycine betaine from the medium and that this repressed endogenous synthesis of proline. Similarly, *Lactobacillus lactis* accumulates high concentrations of proline or betaine when grown under conditions of high osmotic strength. Molenaar et al. (65) reported that intracellular pools of these osmolytes are created by specific transport systems and that proline uptake is inhibited by betaine.

S. aureus is one of the most osmotolerant bacteria because it can grow at an a_w of 0.86 (5,49,66). Anderson and Witter (67) reported that *S. aureus* synthesizes glutamine (at the expense of glutamic acid and aspartic acids) and that proline is taken up by *S. aureus*. Killham and Firestone (68) found that salt stress stimulated accumulation of potassium salts, neutral amino acids, proline, glutamine, and alanine in *Streptomyces griseus*. Their work demonstrated that under the conditions of the study, tolerance to high salinity is associated with intracellular accumulation of free neutral amino acids and, under conditions of extreme salt stress, with selective concentration of K^+ ions.

2.6.4 Osmoregulation by Yeast and Mold

Yeasts, molds, and algae accumulate polyols, including glycerol and trehalose, in response to osmotic stress. Brewster et al. (69) noted that *Saccharomyces cerevisiae* responds to decreases in a_w by increasing glycerol synthesis and decreasing glycerol permeability, which enables accumulation of cytoplasmic glycerol up to molar concentrations. Blomberg and Adler (70) reported that salt stress induced cytoplasmic glycerol-3-phosphate dehydrogenase and that salt-induced enzyme activities were as much as 30-fold higher than in the uninduced state. This enzyme is involved with glycerol biosynthesis in yeast.

2.6.5 Effect of Osmoregulation on the Survival of Microorganisms in Cosmetics

The osmotic tolerance of microorganisms depends on the survival strategies available to a particular microorganism and the composition of the growth medium (or cosmetic product). The a_w and physicochemical composition of a product determines whether microbial growth is possible. Although determination of moisture content is useful for understanding the composition of the product, the moisture content may tell very little about the chemical or microbial dynamics of the system—unless the moisture available for microbial growth is determined. For example, a 25% NaCl solution has an a_w of about 0.92, whereas a 25% sucrose solution has an a_w of about 0.98. Although both solutions have the same moisture content, their ability to support microbial growth is different. Many bacteria, yeasts, and molds are able to grow in the presence of 25% sucrose, but only the more salt-tolerant bacteria are able to grow in the presence of 25% NaCl (5,49).

Although the moisture requirements for any given microorganism depend on several factors including nutrient availability, the type of solute used to decrease the a_w, the pH, surfactants, and other physicochemical factors, the a_w requirements for *Pseudomonas* species generally are higher (minimum $a_w = 0.97$) than most other organisms of interest to the cosmetic industry (5,49). This suggests that decreasing the a_w of a product may be an effective means of helping control the growth of pseudomonads.

Low a_w by itself may not be sufficient to kill microorganisms. For example, an a_w of 0.90 will not permit *E. coli* to grow, but viable cells may persist for extended periods in a dormant state. The presence of potassium salts and compatible solutes may enable microorganisms to persist and grow in low a_w products. Cosmetic products that contain botanicals, proteins, glycerol (or other polyols used as humectants and moisturizers), or betaines may alter the survival of microorganisms at low a_w. L-Proline and other amino acids may be available to microorganisms in cosmetic products that contain proteins, protein hydrolysates, or plant extracts. Alternately, the action of microbial proteases on proteins may release amino acids (including proline) that can be transported into microbial cells to enhance their osmotolerance. It is likely that reduction and/or elimination of potassium salts and compatible solutes may make it more difficult for microorganisms to persist in low a_w systems. Frazier and Westhoff (49) pointed out that microorganisms may be able to change the available moisture in their environment by release of metabolic water or by changing substrates so as to free water. It seems likely that internalization of compatible solutes and/or substrates used for energy may also raise the a_w. This may be a problem if only a_w is used to control microbial growth in self-preserving products.

2.7 SIDEROPHORES

Iron is an essential nutrient because it is involved in many physiological processes, including electron transport proteins and as a cofactor for enzymes. The availability of iron is limited by its low solubility at neutral and alkaline pH values. This necessitates specialized iron transport systems for bacteria to obtain iron for maintenance and growth (5).

Some microorganisms have high-affinity systems that utilize siderophores, which are small (MW = 500–1000 daltons) iron chelators that have very high association constants for their complexes with iron. Siderophores act as extracellular solubilizing agents for iron because they bind specifically to iron and make it available to cells (71). Siderophores and iron-regulated membrane proteins (IRMPs) are induced by conditions of iron limitation (72). IRMPs are membrane receptors and transport systems that recognize iron-siderophore complexes and transport them into the cell.

2.7.1 Iron Acquisition Systems in *E. coli*

There are four known iron uptake systems in *E. coli* that use siderophores. These include enterobactin and aerobactin, which are siderophores produced by *E. coli*, and ferrichrome and coprogen, which are fungal siderophores (71). Each of these systems requires a specific outer membrane receptor protein and other membrane proteins that are involved with transport of iron or the iron-siderophore complex into the cell (71).

Rutz et al. (73) noted that the native iron-acquisition system of *E. coli* contains enzymes for the synthesis and secretion of enterobactin, an outer membrane (OM) receptor protein (FepA) that specifically absorbs ferric enterobactin, and several other cell envelope proteins that act in the transport and release of ferric enterobactin. Related members of the family Enterobacteriaceae including *Klebsiella*, *Salmonella*, and *Shigella* spp., produce and transport enterobactin. Rutz and coworkers found that bacteria closely and distantly related to *E. coli* contained an IRMP that is antigenically related to *E. coli* FepA. They studied the uptake of [^{59}Fe]enterobactin and found that iron stress augmented the rate of enterobactin accumulation 20- to 40-fold. Among bacteria closely related to *E. coli*, only *Shigella flexneri* transported ferric enterobactin at low levels when subjected to iron stress. In addition, distantly related bacteria (*P. aeruginosa*, *Vibrio cholerae*, and *Haemophilus influenzae*) did not accumulate ferric enterobactin after growth in iron-deficient media, even though these bacteria contained an iron-regulated OM protein that reacted with anti- FepA monoclonal antibody (74). Other work has demonstrated that the uptake of [^{59}Fe]enterobactin by *P. aeruginosa* was augmented in medium supplemented with ferric enterobactin (74).

E. coli can transport iron by use of a process mediated by citrate (75). In addition, enterobacteria in the genera *Proteus*, *Providencia*, and *Morganella* were reported to use α-keto acids as siderophores. Drechsel et al. (76) found that members of the tribe Proteeae exhibited siderophore activity with α-keto acids possessing an aromatic or heteroaromatic side chain, such as phenylpyruvic acid and indolylpyruvic acid, which are formed by deamination of phenylalanine and tryptophan, respectively.

2.7.2 Iron-Acquisition Systems in *P. aeruginosa*

P. aeruginosa produces two siderophores: pyochelin and pyoverdin (5,72,78). Pyochelin binds to iron to yield an orange to wine red chromophore, depending on the pH. Pyochelin is believed to be synthesized from salicylic acid and cysteine (77). Pyoverdin is the intensely fluorescent, yellow-green compound that is characteristic of *P. aeruginosa*. The copper-colored appearance of aged culture media following growth of *P. aeruginosa* is a distinguishing feature of this bacterium. It is believed that pyoverdin from *P. aeruginosa* is a component of this pigment complex. Not all fluorescent pseudomonads produce growth media with this appearance (77). Although the fluorescent pseudomonads produce compounds similar to pyoverdin, it is believed that different species produce different pyoverdins.

2.7.3 Shared Pyochelins and Cooperative Growth

Ankenbauer et al. (78) demonstrated that both synthetic and natural pyochelins promoted growth of *P. aeruginosa* in iron-depleted media and functioned to mediate iron transport to the same levels. Furthermore, Sokol (79) found that *P. cepacia* was able to accumulate Fe-pyochelin for growth, whether it was supplied exogenously or produced by *P. cepacia*. The discussion of the enterobacteria (above) indicated that many bacteria are able to transport iron using siderophores from other species and strains. However, the observation by Sokol is important because many strains of *P. cepacia* do not produce pyochelins—yet they may be able to use those produced and excreted by *P. aeruginosa* and/or other pseudomonads (5). This type of cooperative growth would enable siderophore-deficient species to grow by using siderophores produced by other bacteria.

In addition to directly promoting growth in iron-limiting environments, siderophores may indirectly affect the production of virulence factors (e.g., enterotoxin A, alkaline protease, and elastase) because the synthesis of these compounds is under iron regulation in the genus *Pseudomonas* (78). Iron limitation causes derepression of genes involved in the synthesis of siderophores, IRMPs, and enzymes involved with the biosynthesis of these virulence factors. This increases the levels of these virulence factors.

The *fur* (ferric uptake regulation) gene encodes a 17 kDa polypeptide that is rich in histidine (71). The Fur protein requires Fe^{+2} or certain other divalent metal ions as cofactors to negatively regulate expression of the aerobactin operon in *E. coli* (71). Fur binds and blocks the aerobactin promoter region of DNA. The *fur* gene appears to have wide-ranging regulatory actions in microbial cells because it is involved in the production of stress proteins. The *E. coli fur* locus regulates genes involved with production of superoxide dismutase (Sod). Both *fur* and Sod gene products may act together to control the biosynthesis of enterobactin (71).

2.7.4 Effect of Siderophores on Microbial Survival in Cosmetics

The production of siderophores is a survival strategy that greatly enhances the ability of microorganisms to grow in iron-limiting environments. Iron salts generally are not used in cosmetic formulations because iron may function as a prooxidant and initiate free radical reactions that lead to autooxidation of lipids, modification of fragrances, and darkening of products. Chelating agents frequently are used to tie up trace amounts of iron and other heavy metal ions so that they are not available to act as prooxidants. Chelating agents (e.g., EDTA, hexaphosoinositol, etidronate, citric acid, and dicarboxylic acids) may also be part of the product preservative system (3,5,24). The use of chelating agents in product preservation is discussed in Chapter 9.

It is apparent that restriction of iron and use of chelating agents in formulations may provide the signal for the derepression of siderophore and IRMP synthesis in pseudomonads. A complicating factor is the finding that many bacteria are able to utilize siderophores from other species to transport iron (6,71,73,79). The type and concentration of chelating agent used and other formulation variables may determine whether contaminating microorganisms are capable of making siderophores to capture iron and other mineral nutrients required for their growth and survival.

2.8 SUMMARY

Historically, the term *adaptation* was used to describe the ability of microorganisms to grow in new environments. The material presented in this chapter reveals that microbial adaptation can involve genetic changes or changes in the level and/or activity of enzymes. These changes affect cellular metabolism and biosynthesis of cellular products including morphological features (cell envelope, EPS) and physiological capabilities (membrane ATPase, siderophores, IRMPs, lysine decarboxylase, etc.).

Survival strategies enable bacteria, yeasts, and molds to persist and grow in unfavorable environments, including cosmetic products. The survival strategies discussed in this chapter were selected because they represent mechanisms microorganisms may use to deal with conditions that may be found in self-preserving products. A discussion of other survival strategies was presented elsewhere (5).

Gram-negative bacteria have been more troublesome in cosmetic products than gram-positive bacteria, so much of the information presented was based on survival strategies of gram-negative bacteria. Although gram-positive bacteria generally have greater tolerance to low a_W than gram-negative bacteria, this is due, at least in part, to the relatively dense peptidoglycan layer of the cell envelope and, of course, the ability of some gram-positive bacteria to make endospores. Nevertheless, gram-negative bacteria do have mechanisms to deal with osmotic stress. In addition, the complex nature of the gram-negative bacterial cell envelope and EPS confers a wide range of mechanisms by which these bacteria are able to cope with various types of environmental stress including free radicals, toxic chemicals (including preservatives/antibiotics), nutrient limitation, and even desiccation.

As noted earlier, self-preserving products may be microbiostatic, not microbicidal. Although contaminating microorganisms may not be capable of growing, they may remain in the product for extended periods. Self-preserving products should be viewed differently from conventional products, which means that production practices, microbial limits for finished products, and acceptance criteria for preservative efficacy testing may not be the same as for conventional products (80).

Self-preserving products that do not contain preservatives are an emerging class of cosmetics that are intended to provide the same performance benefits as existing formulations *and* reduced irritation/sensitization potential. Many contemporary formulations for hair care, skin care, and eye care may not simply be converted to self-preserving products by formula changes alone because low-pH/low-a_w products may have unacceptable "cosmetic elegance." Development of self-preserving products may require modification of current manufacturing practices, more protective packaging, and testing to ensure that microbial survival strategies will not allow bacteria, yeasts, or molds to grow. Creativity will be required to develop self- preserving products that have desirable aesthetics, user-friendly packaging, and the ability to cope with microbial survival strategies during manufacturing and consumer use.

REFERENCES

1. D. S. Orth. In: *Handbook of Cosmetic Microbiology*. Marcel Dekker, New York, 1993, pp. 521–582.
2. D. S. Orth. The required D-value: evaluating product preservation in relation to packaging and consumer use/abuse. *Cosmet. Toiletr. 107*:39–43 (1992).
3. J. J. Kabara. Food-grade chemicals in a systems approach to cosmetic preservation. In: J. J. Kabara, ed. *Cosmetic and Drug Preservation: Principles and Practice.* Marcel Dekker, New York, 1984, pp. 339–356.
4. D. S. Orth and S. R. Milstein. Rational development of preservative systems for cosmetic products. *Cosmet. Toiletr. 104*(10):91,91,94–100,102,103 (1989).
5. D. S. Orth. In: *Handbook of Cosmetic Microbiology*. Marcel Dekker, New York, 1993, pp. 151–219.
6. D. S. Orth. In: *Handbook of Cosmetic Microbiology*. Marcel Dekker, New York, 1993, pp. 429–432.
7. P. Gacesa and N. J. Russell. The structure and properties of alginate. In: P. Gacesa and N. J. Russell, eds. *Pseudomonas Infections and Alginates: Biochemistry, Genetics and Pathology*. Chapman and Hall, London, 1990, pp. 29–49.
8. T. B. May, D. Shinabarger, R. Maharaj, J. Kato, L. Chu, J. D. DeVault, S. Roychoudhury, N. A. Zielinski, A. Berry, R. K. Rothmel, T. K. Misra, and A. M. Chakrabarty. Alginate synthesis by *Pseudomonas aeruginosa*: a key pathogenic factor in chronic pulmonary infections of cystic fibrosis patients. *Clin. Microbiol. Rev. 4*:191–206 (1991).
9. W. F. Fett, S. F. Osman, and M. F. Dunn. Characterization of exopolysaccharides produced by plant-associated fluorescent pseudomonads. *Appl. Environ. Microbiol. 55*:579–583 (1989).
10. V. Deretic, R. Dikshit, W. M. Konyecsni, A. M. Chakrabarty, and T. K. Misra. The *alg*R gene, which regulates mucoidy in *Pseudomonas aeruginosa*, belongs to a class of environmentally responsive genes. *J. Bacteriol. 171*:1278–1283 (1989).
11. S. F. Osman and W. F. Fett. Structure of an acidic exopolysaccharide of *Pseudomonas aeruginosa* HT041B. *J. Bacteriol. 171*:1760–1762 (1989).
12. D. P. Krieg, J. A. Bass, and S. J. Mattingly. Aeration selects for mucoid phenotype of *Pseudomonas aeruginosa. J. Clin. Microbiol. 24*:986–990 (1986).

13. J. R. W. Govan. Characteristics of mucoid *Pseudomonas aeruginosa in vitro* and *in vivo*. In: P. Gacesa and N. J. Russell, eds. *Pseudomonas Infections and Alginates: Biochemistry, Genetics and Pathology*. Chapman and Hall, London, 1990, pp. 50–75.
14. J. W. Costerton, M. R. W. Brown, J. Lam, K. Lam, and D. M. G. Cochrane. The microcolony mode of growth *in vivo*—an ecological perspective. In: P. Gacesa and N. J. Russell, eds. *Pseudomonas Infections and Alginates: Biochemistry, Genetics and Pathology*. Chapman and Hall, London, 1990, pp. 76–94.
15. R. L. Anderson. Iodophor antiseptics: intrinsic microbial contamination with resistant bacteria. *Infect. Control Hosp. Epidemiol. 10*(10):443–446 (1989).
16. J. W. Costerton, R. T. Irvin, and K.-J. Cheng. The bacterial glycocalyx in nature and disease. *Ann. Rev. Microbiol. 35*:299–324 (1981).
17. J. W. Costerton, G. G. Geesey, and K.-J. Cheng. How bacteria stick. *Sci. Am. 238*:86–95 (1978).
18. A. M. Emmerson. The role of the skin in nosocomial infection: a review. *J. Chemother. 1*(suppl. 1):12–18 (1989).
19. S. E. Smith and J. A. Simpson. The contribution of *Pseudomonas aeruginosa* alginate to evasion of host defense. In: P. Gacesa and N. J. Russell, eds. *Pseudomonas Infections and Alginates: Biochemistry, Genetics and Pathology*. Chapman and Hall, London, 1990, pp. 135–159.
20. L. P. T. M. Zevenhuizen and A. G. Ebbink. Exocellular pyruvate-containing galacto-glucan of *Achromobacter* spp. *Arch. Microbiol. 96*:75–82 (1974).
21. D. E. Ohman and J. B. Goldberg. Genetics of alginate biosynthesis in *Pseudomonas aeruginosa*. In: P. Gacesa and N. J. Russell, eds. *Pseudomonas Infections and Alginates: Biochemistry, Genetics and Pathology*. Chapman and Hall, London, 1990, pp. 206–220.
22. U. B. Sleytr, P. Messner, D. E. Minnikin, J. E. Heckels, M. Virji, and R. B. B. Russell. Structure of bacteria and their envelopes. In: I. Hancock and I. Poxton, eds. *Bacterial Cell Surface Techniques*. John Wiley & Sons, New York, 1988, pp. 1–31.
23. E. Jawetz, J. L. Melnick, E. A. Adelberg, G. F. Brooks, J. S. Butel, and L. N. Ornston (Eds.). *Medical Microbiology*, 18th ed. Appleton & Lange, Norwalk, CT, 1989, pp. 14–23.
24. S. P. Denyer, W. B. Hugo, and V. D. Harding. Synergy in preservative combination. *Int. J. Pharm. 25*:245–253 (1985).
25. T. A. Dahl, W. R. Midden, and P. E. Hartman. Comparison of killing of gram-negative and gram-positive bacteria by pure singlet oxygen. *J. Bacteriol. 171*:2188–2194 (1989).
26. L. C. McDonald, H. P. Fleming, and H. M. Hassan. Acid tolerance of *Leuconostoc mesenteroides* and *Lactobacillus plantarum*. *Appl. Environ. Microbiol. 56*:2120–2124 (1990).
27. J. W. Foster. Beyond pH homeostasis: the acid tolerance response of Salmonellae. *ASM News 58*(5):266–270 (1992).
28. G. R. Bender, S. V. W. Sutton, and R. E. Marquis. Acid tolerance, proton permeabilities, and membrane ATPases of oral streptococci. *Infect. Immun. 53*:331–338 (1986).
29. H. Kobayashi, T. Suzuki, N. Kinoshita, and T. Unemoto. Amplification of the *Streptococcus faecalis* proton- translocating ATPase by a decrease in cytoplasmic pH. *J. Bacteriol. 158*:1157–1160 (1990).
30. J. L. Slonczewski. pH-regulated genes in enteric bacteria. *ASM News 58*(3):140–144 (1992)

31. D. S. Orth. Microbiological considerations in cosmetic formula development and evaluation. II. The dynamic role of microorganisms on the skin and in aqueous test systems, and the role of the acid mantle in the etiology of dry skin. *Cosmet. Toiletr. 104*(5):51–56,58,60,62,64 (1989).
32. R. E. Marquis, G. R. Bender, D. R. Murray, and A. Wong. Arginine deiminase system and bacterial adaptation to acid environments. *Appl. Environ. Microbiol. 53*:198–200 (1987).
33. A. Casiano-Colon and R. E. Marquis. Role of the arginine deiminase system in protecting oral bacteria and an enzymatic basis for acid tolerance. *Appl. Environ. Microbiol. 54*:1318–1324 (1988).
34. C. Vander Wauven, A. Pierard, M. Kley-Raymann, and D. Haas. *Pseudomonas aeruginosa* mutants affected in anaerobic growth on arginine: evidence for a four-gene cluster encoding the arginine deiminase pathway. *J. Bacteriol. 160*:928–934 (1984).
35. K. Blomqvist, M. Nikkola, P. Lehtovaara, M. Suihko, U. Airaksinen, K. B. Straby, J. K. C. Knowles, and M. E. Penttila. Characterization of the genes of the 2,3-butanediol operons from *Klebsiella aerogenes. J. Bacteriol. 175*:1392–1404 (1993).
36. T. D. K. Brown, C. R. S. Periera, and F. C. Størmer. Studies of the acetate kinase-phosphotransacetylase and the butanediol-forming systems in *Aerobacter aerogenes. J. Bacteriol. 112*:1106–1111 (1972).
37. F. C. Størmer. Evidence for induction of the 2,3-butanediol-forming enzymes in *Aerobacter aerogenes. FEBS Lett. 2*:36–38 (1968).
38. L. Johansen, K. Bryn, and F. C. Størmer. Physical and biochemical role of the butanediol pathway in *Aerobacter* (*Enterobacter*) *aerogenes*. J. Bacteriol. 123:1124–1130 (1975).
39. R. J. Magee and N. Kosaric. The microbial production of 2,3- butanediol. *Adv. Appl. Microbiol. 32*:89–161 (1987).
40. D. S. Orth. Principles of preservative efficacy testing. *Cosmet. Toiletr. 96*(3): 43,44,48–52 (1981).
41. D. S. Orth and C. M. Lutes. Adaptation of bacteria to cosmetic preservatives. *Cosmet. Toiletr. 100*(2):57–59,63,64 (1985).
42. E. Levy. Insights into microbial adaptation to cosmetic and pharmaceutical products. *Cosmet. Toiletr. 102*(12):69–74 (1987).
43. P. I. Bowman and S. M. Lindstrom. Resistance of *Pseudomonas* strains to imidazolidinyl urea. *J. Soc. Cosmet. Chem. 36*:413–423 (1985).
44. J. Close and P. A. Nielsen. Resistance of a strain of *Pseudomonas cepacia* to esters of *p*-hydroxybenzoic acid. *Appl. Environ. Microbiol. 31*:713–722 (1976).
45. W. C. Noble and J. A. Savin. Steroid cream contaminated with *Pseudomonas aeruginosa. Lancet 1*:347–349 (1966).
46. C. S. Dattanada and J. Gowrishankar. Osmoregulation in *Escherichia coli*: complementation analysis and gene-protein relationships in the *proU* locus. *J. Bacteriol. 171*:1915–1922 (1922).
47. L. N. Csonka. Physiological and genetic responses of bacteria to osmotic stress. *Microbiol. Rev. 53*:121–147 (1989).
48. M. Jebbar, R. Talibart, K. Gloux, T. Bernard, and C. Blanco. Osmoprotection of *Escherichia coli* by ectoine: uptake and accumulation characteristics. *J. Bacteriol. 174*:5027–4035 (1992).

49. W. C. Frazier and D. C. Westhoff. *Food Microbiology*, 3rd ed. McGraw-Hill, New York, 1978, pp. 3–16,82.
50. D. B. Roszak, D. J. Grimes, and R. R. Colwell. Viable but nonrecoverable stage of *Salmonella enteriditis* in aquatic systems. *Can. J. Microbiol. 30*:334–338 (1984).
51. M. Ghoul, T. Bernard, and M. Cormier. Evidence that *Escherichia coli* accumulates glycine betaine from marine sediments. *Appl. Environ. Microbiol 56*:551–554 (1990).
52. O. B. Styrvold and A. R. Strøm. Synthesis, accumulation, and excretion of trehalose in osmotically stressed *Escherichia coli* strains: influence of amber suppressors and function of the periplasmic trehalase. *J. Bacteriol. 173*:1187–1192 (1991).
53. M. Sugiura and M. Kisumi. Proline-hyperproducing strains of *Serratia marcescens*: enhancement of proline analog-mediated growth inhibition by increasing osmotic stress. *Appl. Environ. Microbiol. 49*:782–786 (1985).
54. M. J. Gauthier and D. Le Rudulier. Survival in seawater of *Escherichia coli* cells grown in marine sediments containing glycine betaine. *Appl. Environ. Microbiol. 56*:2915–2918 (1990).
55. M. J. Gauthier, G. N. Flatau, D. Le Rudulier, R. L. Clemént, and M. C. Combarro. Intracellular accumulation of potassium and glutamate specifically enhances survival of *Escherichia coli* in seawater. *Appl. Environ. Microbiol. 57*:272–276 (1991).
56. J. Meury and M. Kohiyama. Role of heat shock protein DnaK in osmotic adaptation of *Escherichia coli. J. Bacteriol. 173*:4404–4410 (1991).
57. P. M. Munro, M. J. Gauthier, V. A. Breittmayer, and J. Bongiovanni. Influence of osmoregulation processes on starvation survival of *Escherichia coli* in seawater. *Appl. Environ. Microbiol. 55*:2017–2024 (1989).
58. J. D. Fowler and C. R. Robertson. Metabolic behavior of immobilized aggregates of *Escherichia coli* under conditions of varying mechanical stress. *Appl. Environ. Microbiol. 57*:93–101 (1991).
59. W. G. Roth, M. P. Leckie, and D. N. Dietzler. Restoration of colony-forming activity in osmotically stressed *Escherichia coli* by betaine. *Appl. Environ. Microbiol. 54*:3142-3146 (1988).
60. B. Marthi and B. Lighthart. Effects of betaine on enumeration of airborne bacteria. *Appl. Environ. Microbiol. 56*:1286–1289 (1990).
61. S. Fujiwara, N. A. Zielinski, and A. M. Chakrabarty. Enhancer-like activity of AlgR1-binding site in alginate gene activation: positional, orientational, and sequence specificity. *J. Bacteriol. 175*:5452–5459 (1993).
62. P. H. Gilligan. Microbiology of airway disease in patients with cystic fibrosis. *Clin. Microbiol. Rev. 4*:35–51 (1991).
63. N. J. Russell and P. Gacesa. Future prospects. In: P. Gacesa and N. J. Russell, eds. *Pseudomonas Infections and Alginates: Biochemistry, Genetics and Pathology*. Chapman and Hall, London, 1990, pp. 221–227.
64. A. M. Whatmore, J. A. Chudek, and R. H. Reed. The effects of osmotic upshock on the intracellular solute pools of *Bacillus subtilis. J. Gen. Microbiol. 136*:2527–2535 (1990).
65. D. Molenaar, A. Hagting, H. Alkema, A. J. M. Driessen, and W. N. Konings. Characteristics and osmoregulatory roles of uptake systems for proline and glycine betaine in *Lactococcus lactis. J. Bacteriol. 175*:5438–5444 (1993).
66. J. C. Measures. Role of amino acids in osmoregulation of nonhalophilic bacteria. *Nature* (London) *257*:398–400 (1975).

67. C. B. Anderson and L. D. Witter. Glutamine and proline accumulation by *Staphylococcus aureus* with reduction in water activity. *Appl. Environ. Microbiol. 43*:1501–1503 (1982).
68. K. Killham and M. K. Firestone. Salt stress control of intracellular solutes in streptocomycetes indigenous to saline soils. *Appl. Environ. Microbiol. 47*:301–306 (1984).
69. J. L. Brewster, T. de Valoir, N. D. Dwyer, E. Winter, and M. C. Gustin. An osmosensing signal transduction pathway in yeast. *Science 259*:1760–1763 (1993).
70. A. Blomberg and L. Adler. Roles of glycerol and glycerol-3-phosphate dehydrogenase (NAD^+) in acquired osmotolerance of *Saccharomyces cerevisiae. J. Bacteriol. 171*:1087–1092 (1989).
71. J. H. Crosa. Genetics and molecular biology of siderophore-mediated iron transport in bacteria. *Microbiol. Rev. 53*:517–530 (1989).
72. D. M. G. Cochrane, H. Anwar, M. R. W. Brown, K. Lam, and J. W. Costerton. Iron deprivation in vivo: surface protein antigens of *Pseudomonas aeruginosa* in infection. *Antibiot. Chemother. 39*:125–135 (1987).
73. J. M. Rutz, T. Abdullah, S. P. Singh, V. I. Kalve, and P. E. Klebba. Evolution of the ferric enterobactin receptor in gram-negative bacteria. *J. Bacteriol. 173*:5964–5974 (1991)
74. K. Poole, L. Young, and S. Neshat. Enterobactin-mediated iron transport in *Pseudomonas aeruginosa. J. Bacteriol. 172*:6991–6996 (1990).
75. U. Pressler, H. Staudenmaier, L. Zimmerman, and V. Braun. Genetics of the iron dicitrate transport system of *Escherichia coli. J. Bacteriol. 170*:2716–2624 (1988).
76. H. Drechsel, A. Thieken, R. Reissbrodt, G. Jung, and G. Winkelmann. α-Keto acids are novel siderophores in the general *Proteus*, *Providencia*, and *Morganella* are produced by amino acid deaminases. *J. Bacteriol. 175*:2727–2733 (1993).
77. C. D. Cox. Iron transport and serum resistance in *Pseudomonas aeruginosa. Antibiot. Chemother. 36*:1–12 (1985).
78. R. G. Ankenbauer, T. Toyokuni, A. Staley, K. L. Rinehart, Jr., and C. D. Cox. Synthesis and biological activity of pyochelin, a siderophore of *Pseudomonas aeruginosa. J. Bacteriol. 170*:5344–5351 (1988).
79. P. A. Sokol. Production and utilization of pyochelin by clinical isolates of *Pseudomonas cepacia. J. Clin. Microbiol. 23*:560–562 (1986).
80. D. S. Orth. In: *Handbook of Cosmetic Microbiology.* Marcel Dekker, New York, 1993, pp. 521–582.

3

Water Activity and Self-Preserving Formulas

Davin C. Enigl
HACCP Validations—sm
Sacramento, California

Kent M. Sorrells
National Steak and Poultry
Owasso, Oklahoma

By nature of their composition, certain drug and cosmetic formulations lend themselves to microbial contamination. Cosmetic microbiology began in the early 1900s with concern for visible mold growth. The importance of contaminated products was confirmed by the addition of cosmetics to the Food and Dairy Law of 1906. This law was revised with creation of the Food, Drug and Cosmetic Act of 1938 and enforced by the establishment of the U.S. Food and Drug Administration. In the preceding years, concern for microbiological safety, especially gram-negative bacteria, was first apparent in a Swedish report (1) and showed a need for preservation systems in nonsterile drug and cosmetic products (2). While the addition of biocides to formulated products was one solution to the problem, their presence has caused dermatological problems.

Preservation of foodstuffs from the earliest times has been accomplished by reducing or eliminating water from the product. The food industry has used reduced water activity (a_w) formulas for years to control the growth of microorganisms. This chapter outlines some principles for controlling growth of spoilage microorganisms by reducing the available amount of water in a cosmetic or pharmaceutical product.

3.1 WATER ACTIVITY (a_w) OR EQUILIBRIUM RELATIVE HUMIDITY

The a_w can be defined as the ratio of the vapor pressure of a product (P) to that of pure water (P_0) at a constant temperature (3):

$$a_w = \frac{P}{P_0}$$

With pure water (no capillary or absorptive forces), the value would be one. Other expressions of a_w can be in terms of:

1. Equilibrium Relative Humidity (ERH)

$$\text{ERH}(\%) = a_w \times 100$$

2. Osmotic Pressure

$$\text{Osmotic Pressure} = \frac{-RT \ln a_w}{\bar{v}}$$

where R is the gas constant, T is the absolute temperature, ln a_w is the natural logarithm of a_w value, and $\bar{v}$ is the partial molar volume of water. ERH refers to the headspace above the product after equilibrium is reached, while a_w refers to the product itself.

When products are formulated with various solutes, a_w can be expressed in the terms of Raoult's law, which states:

$$a_w = \frac{P}{P_0} = \frac{n_1}{n_1 + n_2}$$

where P is the vapor pressure of a solution, P_0 is the vapor pressure of pure water, n_1 is the number of moles of solvent, and n_2 is the number of moles of solute.

The a_w of some solutes such as sodium chloride, sucrose, glucose, invert sugar, and glucose syrup and their solutions are presented in Tables 3.1 and 3.2. The use of amino acids for potential lowering of a_w has also been reported (4). The effect of anhydrous glycerin on the a_w is shown in Figure 3.1.

Computer programs are available to predict a_w for some food products and could be adapted for cosmetics and drugs. The Flour Milling and Baking Research Association, Chorleywood, United Kingdom, offers "ERH CALC" for MS-DOS compatible computers. An a_w prediction program for the Macintosh is available from Ted Labuza and Kathy Nelson at the University of Minnesota Department of Food Science and Nutrition (5).

TABLE 3.1 a_w of Various Concentrations of Sodium Chloride and Sucrose Solutions at 25°C

	Sodium chloride		Sucrose	
a_w	Molality	% (w/w)	Molality	% (w/w)
1.000	0.0	0.0	0.0	0.0
0.995	0.150	0.88	0.272	8.52
0.990	0.300	1.74	0.534	15.45
0.980	0.607	3.43	1.03	26.07
0.960	1.20	6.57	1.92	39.66
0.940	1.77	9.38	2.72	48.22
0.920	2.31	11.90	3.48	54.36
0.900	2.83	14.18	4.11	58.45
0.880	3.32	16.28	4.93	62.77
0.860	3.80	18.18	5.58	65.63
0.840	4.26	19.94		
0.820	4.71	21.59		
0.800	5.15	23.13		
0.753	6.16	26.5		

Molality = number of gram moles per kilogram of water.
Source: Adapted from Ref. 3.

TABLE 3.2 a_w of Glucose Solutions at 25°C

	Glucose		Invert sugar	Glucose syrup
a_w	Molality	% (w/w)	% (w/w)	% (w/w)
0.995	0.259	4.45	2.05	1.57
0.990	0.542	8.90	4.11	3.15
0.980	1.04	15.74	8.22	6.30
0.960	2.21	28.51	16.43	12.59
0.940	3.38	37.93	24.65	18.89
0.920	4.31	43.72	32.87	25.19
0.900	5.24	48.54	41.09	31.49
0.880	6.27	53.05	49.30	37.78
0.860	7.81	58.45	57.52	44.08
0.850	9.00	61.84	61.63	47.23

Glucose syrup dextrose equivalent value was 55.0.
Molality = number of gram moles per kilogram of water.
Source: Adapted from Ref. 3.

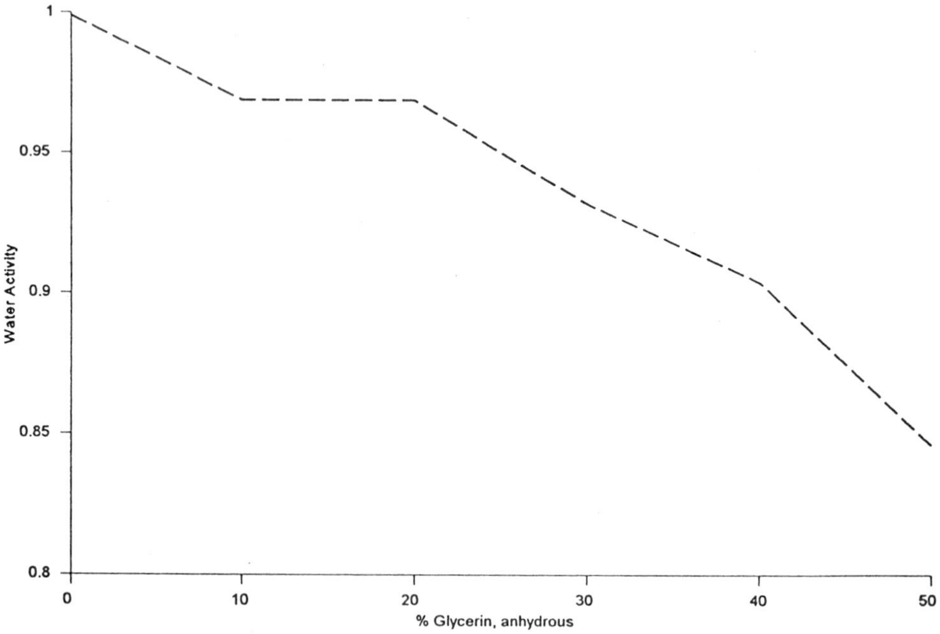

FIGURE 3.1 a_w of glycerin added to water. (D. C. Enigl, unpublished.)

3.2 a_w BASICS

a_w-based preservative systems are common and widely used in food (6–8) and solid dosage form drugs (9). The available water influences microbial activity, flavor, color, texture, and mechanical properties of cosmetics and pharmaceuticals (10).

Microorganisms require the presence of available water for growth and metabolism. This available water is best measured by a_w. The term a_w has been accepted as predicting inhibition of growth and metabolism by solutes or desiccation. a_w is not the moisture content. Moisture content cannot even make a rough estimate of a_w. Microorganisms differ in their response to a_w. A reduction in a_w generally results in a decrease in growth rate (increase in generation time) (11–13), enzyme activity (11,14,15), sporulation (6,11), and toxin production (11). Generally, yeasts and molds can grow at lower a_w than bacteria. Minimum a_w allowing bacterial growth is listed in Table 3.3. Table 3.4 shows the minimum a_w for growth of selected yeasts and molds. The minimum a_w for growth for most bacteria is 0.90. *Staphylococcus aureus*, an exception, can grow at an a_w of 0.86 or above. Salt-loving (halophilic) bacteria grow at a very low a_w of 0.75 but, have not been reported as a spoilage problem in cosmetics or drugs (16,17). Osmophilic (xerophilic, "dry-loving") yeasts can grow down to an a_w of 0.75, but most yeasts require at

TABLE 3.3 Minimum a_w Permitting Growth of Selected Bacteria

Bacterium	a_w
Bacillus cereus	0.95
B. megaterium	0.95
B. stearothermophilus	0.93
B. subtilis	0.90
Clostridium botulinum type A	0.95
C. botulinum type B	0.94
C. botulinum type E	0.97
C. perfringens	0.95
Escherichia coli	0.95
Halobacterium halobium	0.75
Micrococcus halodenitrificans	0.86
M. lysodeikticus	0.93
Pseudomonas fluorescens	0.97
Salmonella spp.	0.95
Staphylococcus aureus	0.86

Source: Adapted from Ref. 26.

least 0.87 and molds at least 0.70. The minimum a_w values listed in Tables 3.3 and 3.4 are under ideal conditions and can increase or decrease depending on pH, temperature, nutrient availability, and other factors in the formulation (emulsion, surfactant, water mobility due to syneresis) and packaging (anaerobic atmosphere). If the product is stored at higher temperatures, the a_w will go up and could allow growth if the product is preserved by a_w alone.

3.2.1 Range of Growth

Most bacteria grow rapidly at a_w levels between 0.995 and 0.980. A reduction in a_w from this range can adversely affect bacterial growth. Lowered a_w causes an increase in the lag phase of growth, decrease in growth rate, and decrease in total cell yield (13). At very low a_w, it can be assumed that the lag phase becomes infinite, i.e., no growth (13). The growth of most bacteria is confined to an a_w above 0.90. As stated above, yeasts and molds can grow at a much lower a_w with a limiting range above 0.70 (3). Gram-positive bacteria are more tolerant to low a_w than gram-negative bacteria, especially *S. aureus*. The solute used to adjust a_w has been shown to affect growth rates differently (11). Sucrose and sodium chloride were shown to affect the growth rate of *Pseudomonas fluorescens* more than glycerol when used to adjust the a_w of glucose minimal medium (18). Nutrients will increase the range of a_w values at which microorganisms can grow (19).

Table 3.4 Minimum a_w Permitting Growth of Selected Fungi

Organism	a_w
Fungus	
Alternaria citri	0.84
Aspergillus candidus	0.75
A. flavus	0.78
A. fumigatus	0.82
A. niger	0.77
A. restrictus	0.75
A. sydowii	0.78
A. versicolor	0.78
Botrytis cinerea	0.93
Chrysosporium fastidum	0.69
Emericella nigulans	0.78
Eremascus albus	0.70
Eurotium amstelodami	0.70
E. chevallieri	0.71
E. repens	0.71
E. rubrum	0.70
Mucor plumbeus	0.93
Paecilomyces variotii	0.84
Penicillium chrysogenum	0.79
P. expansum	0.83
P. viridicatum	0.81
Rhyzopus nigricans	0.93
Stachybotrys atra	0.94
Wallemia sebi	0.75
Xeromyces bisporus	0.61
Yeast	
Debaryomyces hansenii	0.83
Saccharomyces bailii	0.80
S. cerevisiae	0.90
Zygosaccharomyces rouxii	0.62

Source: Adapted from Refs. 6, 26.

3.2.2 Influence on Spore Germination and Sporulation

Spores are either bacterial or fungal. Bacterial spores have an *increased* heat resistance with lower a_w. Therefore, care must be taken if heat sterilization is involved. The heat process may not be adequate if a product is reformulated with a lower a_w (20).

Most *Bacillus* spp. spores germinate at an a_w of 0.94–0.89 (21). The apparent limit for cell multiplication, i.e., growth, is 0.90–0.91 (3). The limits for growth, sporulation, and germination for *Clostridium* spp. are similar to those for *Bacillus*. Solute type has been shown to affect growth versus germination (12,13). For more information on bacterial spores and a_w, see Refs. 6 and 22.

3.3 MEASUREMENT OF a_w

The measurement of a_w is usually performed in the laboratory by commercially available hair, chilled-mirror, electric hygrometer, or direct vapor pressure measurement equipment. Comparison of different methods of a_w measurement has been reported by Labuza (23) and Troller and Christian (24). Reviews of a_w instrumentation have been reported by Rodel et al. (25) and Troller et al. (26).

3.3.1 Calibration

Instrument calibration uses saturated salt solutions of known a_w. A list of a_w values of many saturated salt solutions suitable for the calibration of electronic hydrometers is shown in Table 3.5. Reagent-grade salts and deionized or distilled water should be used in the preparation of standard salt solutions. The salt is added to a sealable container and water added in small amounts, followed by stirring until free liquid is absorbed. The amount of undissolved salt (slush) should be at least 25% of the total volume in the container. The standard salt solutions should be allowed to equilibrate to the measurement temperature before use.

Calibration procedures involve the establishment of a standard curve, adjustment for the set point(s), or a combination of these. Standard curves are constructed using at least five different salts of known a_w range to be measured. Standard curves are needed for instruments that do not read ERH or a_w directly or have no adjustable set point. Readings are obtained after equilibrium has been reached and plotted against the known a_w values. Calibration plots should not be extrapolated.

Instruments with adjustable set points are adjusted to the a_w of the specific salt standard. To improve accuracy, standardize the instrument using two salt solutions nearest to the expected a_w range of the products. After screening the samples, recalibration may be done in a tighter range and the products tested again.

3.3.2 Samples

The product sample is placed in the sample chamber, sensor attached, and allowed to equilibrate with respect to temperature and a_w. Maintaining samples at a constant temperature is very important (27). A constant-temperature water circulation jacket is sometimes used, or the sensor is sealed in a constant-temperature air incubator. Equilibrium is assumed to be reached when two consecutive readings differ by less than 0.01 a_w units for direct readout equipment or when a plateau is reached on

TABLE 3.5 Equilibrium Relative Humidities of Some Saturated Solutions

	Percent Equilibrium Relative Humidity ($a_w \times 100$)		
	20°C	25°C	30°C
Salt			
LiBr	6.61	6.37	6.16
LiCl	11.31	11.30	11.28
$MgCl_2$	33.07	32.78	32.44
K_2CO_3	43.16	43.16	43.17
$Mg(NO_3)_2$	54.38	52.89	51.40
NaBr	59.14	57.57	56.03
Na_2NO_3	75.36	74.25	73.14
NaCl	75.47	75.29	75.09
KBr	81.67	80.89	80.27
KCl	85.11	84.34	83.62
KNO_3	94.62	93.58	92.31
K_2SO_4	97.59	97.30	97.00
Crystalline products[a]			
Mannitol		0.977	
Maltose		0.952	
Lactose-monohydrate		0.931	
Lactose-mixed forms		0.990	
Dextrose-monohydrate		0.891	
Sucrose		0.844	
Sorbitol		0.725	
Fructose		0.634	

[a]Adapted from Ref. 5.
Source: Adapted from Ref. 26.

recording equipment. After equilibrium and reading, a desiccant should be placed in the chamber to lower the a_w or the chamber opened for at least 10 minutes. This lets the instrument get back to ambient environmental conditions (28).

The a_w is found from the digital readout, a recorder's plot, or a standard curve. Direct readout equipment is the easiest to use and does not require a standard curve. Duplicate samples should be measured for a_w, then averaged. The FDA recommends reading at 15-, 30-, 60-, and 120- (further intervals: 60 minutes) minute intervals to ensure that equilibrium has been reached (29). The *Federal Register* of March 16, 1979, incorporated an a_w specification for food beyond which susceptibility to proliferation of molds or bacteria is considered unlikely (30). No such specifications have been included regarding cosmetics or pharmaceuticals.

3.3.3 Formulations and a_w

The information referred to above allows prediction of the a_w of a simple solution of known composition. However, in mixed compositions of cosmetics and pharmaceuticals, the a_w is much more complicated. To obtain the a_w of a certain formulation, it is necessary to measure a_w values at a range of water levels and plot percent (%) water versus a_w to obtain a sorption isotherm (Fig. 3.2). The factors in formulations that reduce water vapor pressure (a_w) include adsorption of water molecules onto surfaces, capillary forces, and the solution of components (solutes). In the isotherm, it is generally accepted that the steep lower section represents the adsorption of water to form a monolayer of molecules on the adsorption sites of solid materials. With the addition of more water, the isotherm rises steeply again as solvents dissolve and capillary spaces are filled (3).

Methods that measure total moisture are important in developing the isotherm graph. Isotherm graphs relate how the water is bound. Will the system get enough energy to break the bound water out of the product structure into the gas phase? a_w measures only active, unbound, "free" water. Moisture methods, such as "Karl Fisher," measure both bound and unbound (total) water (R. Albert,

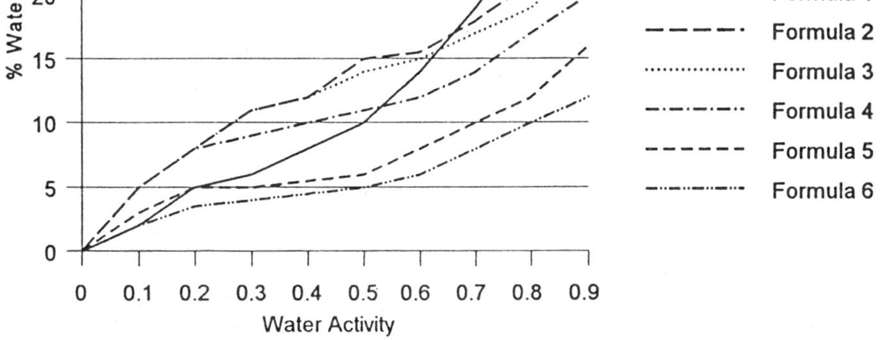

FIGURE 3.2 Example of water vapor adsorption isotherms of some different hypothetical formulas. (From Ref. 3.)

personal communication). At the microbial level, about 0.2–1.0 μm, the amount of unbound water could be high even when an a_w instrument shows a low "average" macroscopic a_w (such as an emulsion). Perhaps we need a microscopic a_w probe to more accurately measure a_w of w/o emulsions because the water droplets are merely suspended in the oil phase.

Isotherms can be obtained by addition of water to dry ingredients or by drying a wet preparation. Each method results in a different isotherm (Fig. 3.3). The difference between the two curves is called hysteresis and the divergent isotherms, the hysteresis loop (25). If the a_w of cosmetics and pharmaceuticals is produced by drying, the desorption isotherm is important and hysteresis is a problem because the a_w remains high for any given moisture level. However, most cosmetics and pharmaceuticals are formulated by adding water to different ingredients with varying a_w-lowering abilities and the adsorption isotherm becomes important. The visible symptom of a problem with hysteresis is the formation of droplets of water coming out of the surface of the product as it desorbs water. This is sometimes called "sweating" or "leaking." The water in the droplets can have a very high a_w. This high a_w may be above the minimum required for growth. Therefore, the product may not only fall apart from an esthetic and chemical/physical point of

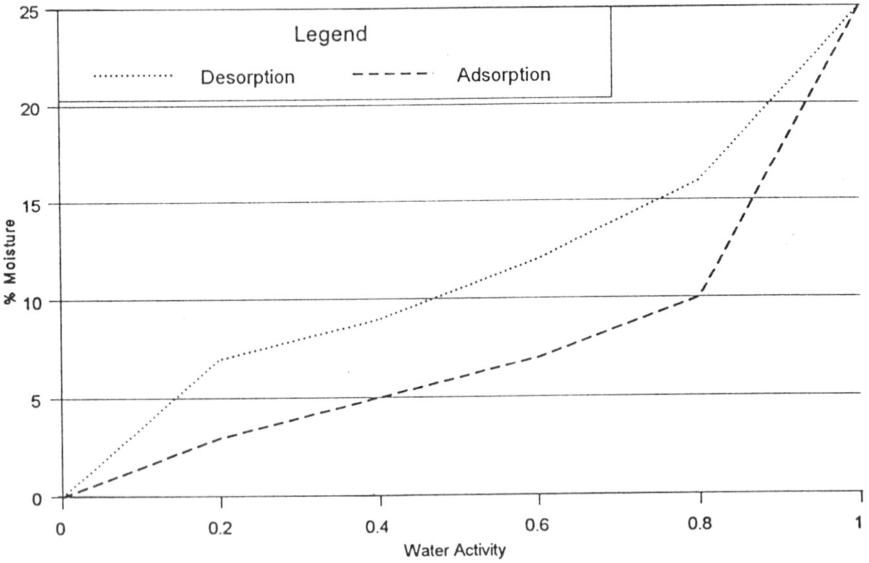

FIGURE 3.3 Graphic representation of sorption isotherms for a product when water is added to dry ingredients (lower curved line from lower left going up to upper right) and when water is removed by desiccation (upper curved line going from right down to lower left). (Adapted from Ref. 3.)

view, but also may no longer be adequately preserved. Formulations showing marked hysteresis at high levels of a_w allow more rapid microbial growth by desorption than by adsorption (11,31).

Some examples of preservative systems that use a_W are antiperspirants that use up to 25% aluminum chlorohydrate (or zirconium complexes) along with acidic pH, anhydrous powders, oils, bar soaps, anhydrous sunscreens, and complexion bars. Additional formulas may be found in Resnik and Chirife (32). The range for these formulas encompasses a_W values from about 0.57 to 0.97 at different temperatures.

David Bone published formulas for calculating a_w in a new way (33). These formulas calculate the amount of solute needed to tailor-make a series of products with specific a_w. Formulas with an a_w of 0.90, for example, would require 6.17 molal and for an a_w of 0.85 would require 9.80 molal of sucrose, glycerin, sorbitol, or lactose. These molalities are theoretical and can only be applied to ideal solutions. Table 3.6 lists critical factors relating to how a_w might change in the formulation from the theoretical calculations (34). Also, yeasts and molds are more tolerant of glycerin than sucrose, i.e., more glycerin is needed to have the same stability as sucrose (34).

To formulate their salad dressing products, Meyer et al. (35) used the empirical formula $a_w = 0.97 - 2.91 \times 10^{-3}$ (% sucrose) – 0.01 (% NaCl), where % sucrose and % NaCl are the total weight percent of each component in the final product. This formula can be used for finding a starting point with prototype cosmetic and pharmaceutical products.

Methylcellulose, gum, and starch, for example, do not decrease a_w at use levels. Water is still unbound and is available to microorganisms. Adding 1–2% of these components to a formula does not help lower bound water. The effect of other ingredients on a_w may be unknown until the formula is tested with an a_w instrument (see Tables 3.7 to 3.10).

Table 3.6 Critical Factors Relating to Formulating a_W-Based Preserved Products

1.	Differences in the a_W of input mix ingredients
2.	Adsorption isotherms for the input ingredients
3.	Order of addition of components
4.	Method and time of mixing
5.	Physico-chemical nature of input ingredients
6.	Processing method
7.	Processing rate
8.	Processing conditions
9.	Time interval between mixing and processing
10.	Temperature of input ingredients
11.	Presence or absence of surface active ingredients

Source: Adapted from Ref. 33.

Table 3.7 a_w of Some Surfactant-Based Cosmetic Products[a]

Product	a_w
Shower and bath gel	0.975
Shampoo 1	0.965
Antidandruff 1	0.977
Antidandruff 2	0.978
Conditioner 1	0.959
Conditioner 2	0.939
Antidandruff conditioner	0.939
Skin antiseptic	0.976
Antiacne wash	0.882
Liquid soap	0.910
Facial cleanser	0.920
Makeup remover	0.978

[a]Rotronic-DT water activity equipment was provided courtesy of David McDougall, McDougall Air Systems, Irvine, CA.
Source: D. C. Enigl, unpublished.

Table 3.8 a_w of Some Emulsion-Type Products[a]

Product	a_w
Hand cream 1	0.840
Hand cream 2	0.857
Facial moisturizer 1	0.972
Facial moisturizer 2	0.970
Facial moisturizer 3	0.988
Hand lotion 1	0.920
Hand lotion 2	0.980
Cream 1 (light)	0.969
Cream 2 (heavy)	0.976
Cream 3	0.970
Cream 4	0.980
Body lotion	0.976

[a]Rotronic-DT water activity equipment was provided courtesy of David McDougall, McDougall Air Systems, Irvine, CA.
Source: D. C. Enigl, unpublished.

3.3.4 The Required D-Value and Formulation

A D-value is the time (usually in expressed in hours) that a preservative or preservative system takes to kill 90% of the microorganisms inoculated into a

TABLE 3.9 a_w of Some Raw Materials[a]

Product	a_w
Surfactant 1	<0.4
Surfactant 2	0.830
Surfactant 3	0.995
Ammonium lauryl sulfate	0.980
Polysorbate 80	<0.4
Sodium lauryl ether sulfate	0.988
Oleyl betaine	0.921
Tego betaine	0.995
Anhydrous lactose	<0.4
Lactose monohydrate	<0.4
Magnesium stearate	<0.4

[a]Rotronic-DT water activity equipment was provided courtesy of David McDougall, McDougall Air Systems, Irvine, CA.
Source: D. C. Enigl, unpublished.

TABLE 3.10 a_w of Some Contaminated Products from In-Home Use Tests with Different Package Designs[a]

Product	a_w after in-home use	a_w before in-home use	a_w difference after use
Shampoo 1 with pump	0.997	0.965	+0.032
Shampoo 2 with flip top	0.986	0.965	+0.021
Shampoo 3 in tube	0.972	0.965	+0.007
Shower and bath gel with pump	0.990	0.975	+0.015

[a]Rotronic-DT water activity equipment was provided courtesy of David McDougall, McDougall Air Systems, Irvine, CA.
Source: D. C. Enigl, unpublished.

product. Because bacteria, yeasts, and molds usually die logarithmically, the time of 90% kill correlates to a one log reduction. Therefore, if 1000 bacteria (3 logs) are expected to be contaminating the product on a daily basis, the D-value must be 8 hours (24 hours divided by 3 logs) or less for the product to self-preserve. If the D-value is greater than 8 hours and the same level of contamination continues, the product will increase in microbial count, even if no growth occurs. This is due to the preservative system not being able to cope with the repeated contamination of a multidose drug or cosmetic. The 1000 microbes per 24 hours contamination rate is very likely and far from worst-case. Worst-case is probably more like 1,000,000 (6 logs) per day, making the D-value for self-preservation 4 hours or less. This is why we cannot agree with the USP (≤112 hour D-value) or the CTFA (≤56-hour D-value) acceptance requirements.

Using D-values is preferable to the USP and CTFA criteria because D-values are quantitative and the other methods are qualitative. Therefore, D-values will tell how close the formulator is to having an adequate preservative system and how much more work is ahead.

Customer use/abuse and package design are major factors contributing to the D-value needed to preserve the product. Orth et al. (36) used the Required D-value (RDV) expressed as follows:

$$\text{RDV} = \frac{\text{Maximum acceptable D-value for target microbes}}{\text{(Packaging factor) (Consumer use/abuse factor)}}$$

The "target microorganisms" should be selected from those presenting the greatest risk of contamination or adaptation to the preservative system (37).

The packaging factor is either 1 = low risk (single-use container), 5 = moderate risk (multiple-use container where water contact is unlikely or container is completely water-tight and waterproof), or 10 = high risk (multiple-use, near water or soiled finger contamination potential, container is not waterproof, eye shadow/mascara with reusable applicator).

The consumer use/abuse factor is defined as 1 = low risk (single-use system), 5 = moderate risk (finger touching inside, use in bathroom or kitchen sink), 10 = high risk (using in shower, bathtub, finger or mouth saliva contact potential, possible intentional product dilution, and use over a period of several days). See Chapter 10 for a discussion of the role of packaging in product preservation.

A rapid method for estimation of D-values, such as the method of Orth and Enigl (38), can be performed to learn how well the product is preserved. Keep in mind that the USP acceptable D-value is ≤112 hours and the CTFA is ≤56 hours. Orth does not consider the USP or CTFA guidelines to be adequate and recommends ≤4 hours for disease-producing microorganisms and ≤28 hours for spoilage/nonpathogens (39). This is wise and is on the "safe" side for preservation. Increasing the preservation to get low D-values in the 4-hour range could increase skin irritation of cosmetics (especially near the eye). Orth et al. (40) are aware of this, which is why they (and others, including the authors) advocate testing novel/unconventional means of preservation for synergistic combinations to reduce preservative components (Tables 3.11 to 3.15).

3.4 EFFECTS OF WATER ON PRESERVATION

3.4.1 Use of a_w in Preservation

Cosmetic and pharmaceutical formulators of aqueous products have not used a_w to the extent used in the food industry or the solid-dosage form (compressed tablet)

TABLE 3.11 Conventional Preservatives That Could Be Tested by D-Value Criteria for Reduction in Concentration When Used in Reduced a_w Based Systems

Preservative	% Conventional concentration	Possible changes for low-a_w systems
Methyl paraben	0.3	0.2, add butylene glycol
Propyl paraben	0.2	0.1, add butylene glycol
Butyl paraben	0.1	0.05, add butylene glycol
Alcohol, ethyl	15–20	5–10, add fungicide
Salicylic acid	0.1–0.5	0.1–0.2
Propylene glycol	15	10–20, add fungicide
EDTA	0.1–0.5	0.05–0.2, add to others on this list.
Diazolidinyl urea	0.05–0.3	0.0 (not accepted in Japan)
DMDM hydantoin	0.15–0.4	0.0 (provisionally accepted in Japan)
Imidazolidinyl urea	0.05–0.5	0.0 (provisionally accepted in Japan)
Potassium sorbate	0.025–0.2	0.05, add 0.05 sodium benzoate, acidify
Sodium benzoate	0.1–0.2	0.05, add 0.05 potassium sorbate, acidify
Phenoxyethanol	0.5–2.0	0.2–1.0, add EDTA. antioxidants, add to glycerin, glycol or alcohol.

pharmaceutical industry. The formulators in the food industry plan for a_w-based preservative systems at the beginning of formulation. *At the beginning of product planning*, a list of bacteria, yeasts, and molds versus a_w range like the ones shown in Table 3.3 and 3.4 are reviewed.

Formulation chemists may consult a microbiologist for bacteria or a mycologist for yeasts and molds and determine what microorganisms are likely to contaminate or spoil a specific product. For instance, it is unlikely a halophilic (salt-loving) bacterium will be encountered in cosmetic or pharmaceutical products (16,17). This class of bacteria grows at extremely low a_w and can be disregarded because they are unlikely to be encountered. Since preservative effects overlap, several classes of microorganisms might still be unable to grow for other reasons (see Tables 3.13 and 3.14). The product need not be specifically protected from certain classes of microorganisms (i.e., halophilic, thermophilic, thermoduric, xerophilic, or anaerobic spore-forming bacteria). Some of these classes may only need to be held at low levels and their viable spores killed only as they germinate. Initial assessment of core requirements will make formulation simpler and more cost effective. Challenge tests should be initiated with standard test strains and microorganisms likely to spoil the product.

Preservation systems should be constructed during the initial stages of formu-

TABLE 3.12 Properties versus a_W of the Ideal Preservative

Property[a]	Comparison to a_W
1. Broad spectrum of antimicrobial activity: bacteria, yeasts, and molds	Very good for bacteria but not as good for xerophilic yeasts and molds
2. Effective and stable over time and wide range of pH and temperature	Stable over pH, but good vapor barrier is needed; increased temperature raises a_W
3. Compatible with other ingredients	Depends if solutes are compatible
4. Not affected by physical properties of the product	Usually not a factor, unless extremely low a_W
5. High aqueous vs. oil solubility (partition coefficient)	Not applicable, since solutes are usually highly water soluble
6. Quick acting to prevent adaptation	A big problem: a_W does not directly decrease D-value, Only extends lag phase and provides a hostile environment (i.e., stress)
7. Nontoxic by ingestion, nonsensitizing, nonirritating	a_W-lowering methods can be very mild and nontoxic (e.g., sugars, glycerin)
8. Comply with government regulations	Most solutes are generally recognized as safe (GRAS)
9. Cost effective	Very cost effective,usually needed in the product anyway (e.g., glycerin)
10. No injury repair or regrowth of microbes	See comment 6 above

[a]Adapted from Ref. 63.

formulation and not as an afterthought. Package integrity and antimicrobial properties of "natural" ingredients are important early on. These factors may eliminate the need for conventional preservatives. A good preservation system should provide a barrier between the contaminating microorganisms in the environment outside and the cGMP manufactured product inside the container. A preservative system is not a substitute for poor or unsanitary manufacturing practices.

One has to balance the conventional preservative system versus the "ideal preservative" itself (Table 3.12). Reasons to do this include (1) cost containment, (2) reduction of allergic reactions, (3) reduction of irritation, (4) ease of compounding, and (5) satisfying requests for preservative-free products. How can we optimize the preservative system?

Each product will have a different resistance profile to different microorganisms, but *S. aureus*, *P. aeruginosa*, *Escherichia coli*, *Aspergillus niger*, and *Can-*

Table 3.13 Components Contributing to a Preservative System Compared to a_W-Based Systems

Component		Comparison to a_W
1.	Conventional preservatives and antibiotics	a_W can easily be combined with conventional preservatives
2.	Acids and alkalis	Widely used in the food industry (e.g. low pH)
3.	Alcohols (ethyl, isopropyl, benzyl)	Compatible with many solutes, e.g., glycerin
4.	Cationic detergent surfactants (cetyl pyridinium chloride)	a_W may not even be needed in these type products; sometimes already adequately preserved
5.	Anionic detergents surfactants (sodium lauryl sulfate)	Can be used advantageously
6.	Esters (glyceryl monolaurate)	Unknown, probably would work well
7.	Humectants (glycerin, propylene glycol, sorbitol)	This is an a_W-based system
8.	Aqueous solutes (sugar, dextrins, salts)	This is an a_W-based system
9.	Phenolic antioxidants (BHA, BHT, TBHQ)	Highly recommended for a_W-based system
10.	Chelating agents (tetrasodium EDTA)	Highly recommended for a_W-based system
11.	Glycols (propylene glycol, butylene glycol)	This is an a_W-based system
12.	Colors (dyes)	Novel, needs research
13.	Fragrances	Phenoxyethanol recommended
14.	Compatible combinations of 1–13 above	

Table 3.14 Physical and Chemical Requirements for Microbial Growth

Physical requirements
- Temperature range: minimum/maximum
- The pH of the growth medium
- Hydrostatic pressure
- Freedom from harmful radiation, UV, gamma, etc

Chemical requirements
- Water, range of a_W dehydration.
- Nutrients (carbon, nitrogen, phosphorus, sulfur sources).
- Minerals: zinc, etc.
- Oxygen or lack of (proper oxidation-reduction potential)
- Organic growth factors
- Freedom from toxic chemicals: copper, enzymes, lytic compounds, free radicals

dida albicans are usually important and are required to be tested by USP. Additional cultures, such as *Bacillus subtilis*, should be tested to obtain data on how spore formers will respond to the preservative system. *Burkholderia (Pseudomonas) cepacia* or other waterborne pseudomonads are important in eye products. *S. epidermidis* has been known to survive where *S. aureus* may be killed. Recent evidence of disease from *S. epidermidis* makes this important (41).

Some other ideas for the formulator wishing to use low a_w include the use of glycerin, acids, cationic surfactants, and antimicrobial fragrances. Glycerin is added as a liquid without adding water and can act both as a solvent and a humectant. Dry sugar requires adding water. Glycerin will not increase the a_w, while sugar needs water to function and could become a potential nutrient source. Another advantage of glycerin is that it widely used and an old "safe" formula ingredient having moisturizing effects on human skin. A disadvantage of glycerin is that in low concentrations it also could be a nutrient (see Table 3.15). Microorganisms are more tolerant to glycerin than to NaCl. The formulator needs to add more glycerin to get the same a_w (and molar concentration) compared with smaller molecular weight humectants (Table 3.1).

As can be seen, approximately 35% glycerin was required to decrease the a_w to 0.92, where most bacteria, but not yeasts or molds, are inhibited (i.e., have an infinitely long lag phase and will not grow). The a_w was decreased to 0.85 with the addition of 50% glycerin. In practice, these large amounts of glycerin would not be absolutely required since the log phase of some bacteria may be increased by addition of as little as 5% glycerin (Table 3.15) and addition of other "hostile" formula components (Table 3.13). Therefore, an intermediate amount such as 20–30% glycerin is recommended. This amount creates a hostile environment based on suboptimal a_w (see Ref. 42).

Products such as antidandruff shampoos and many hair conditioners are acidic or have cationic surfactants and may not need any conventional preservative (see Chapter 6). Many cationic surfactants are antimicrobial (e.g., Tergitol 7 patented by Union Carbide), kill gram-positive bacteria, even spores as they germinate, in very low concentrations if combined with low a_w (as low as 0.5% Tergitol 7). If the a_w is 0.85 or below, the product may need no further addition of bacterial preservatives and may only need an antifungal because of xerophilic molds (*Eurotium*, *Wallamia*). To predict microbial stability, antimicrobial effectiveness tests (AET), sometimes called preservative efficacy tests (PET), can be performed. We recommend an AET/PET "D-value" (see Sec. 3.3.4) test on bacteria to decide if additional bactericides are needed. If the product is safe from abuse, add only the preservatives needed to pass the acceptance criteria of the linear regression method.

Some products such as surfactant-based cleansing washes, foaming facial cleansers, and liquid-type shampoos are not easily preserved and may require formaldehyde donors (some are provisionally acceptable in Japan). Liquid hand

TABLE 3.15 Conventional Preservatives and Selected Other Components' Relationship to a_w-Based Preservative Systems

Conventional preservatives possibly compatible with a_w-based preservative systems:
- Methyl paraben
- Propyl paraben
- Butyl paraben
- Methylchloroisothiazolinone and methylisothiazolinone
- Imidfazolidinyl urea
- Quaternary ammonium (benzalkonium chloride 0.1%)

Emulsifiers with antimicrobial effect in a_w-based preservative systems:
- Lipophilic acids (lauric acid, myristic acid)
- Nonionic surfactsants have little effect
- Polysorbate 20 and 80 below critical micelle concentration (CMC) may enhance antimicrobial activity; however, above CMC interferes with preservatives and is used to neutralize some preservatives
- Glycerin monolaurate use at 0.5–2%
- Glyceryl stearate has no known antimicrobial effect

Some humectants used:
- Glycerin is a substrate for growth <5.0% but lowers a_w at higher concentrations
- Sorbitol is a substrate for growth <5.0% but lowers a_w at higher concentrations
- Sugars, amino acids, salts, dextrin (starches and gums do not work)

Phenolic antioxidants recommended:
- BHT
- BHA
- TBHQ

Chelating agents recommended:
- EDTA

Fragrances recommended:
- Phenoxyethanol

soap, hand cream, antidandruff shampoos, and conditioners may contain a cationic surfactant and kill microorganisms quickly. Generally, most emulsion-based creams have a high a_w of 0.97, and this could be dropped to 0.91–0.92, to stop the growth of most bacteria of concern, except *S. aureus*. At that point, only a fungicide and a gram-positive bactericide may be needed. This strategy might *not* produce a truly self-preserving product, just a biostatic product. A broad-spectrum antimicrobial would be required if a biocidal product is desired. See Table 3.8 for a_w of some emulsions and creams [e.g., night cream, facial moisturizer, body lotion, hand lotion, hand cream (higher viscosity than a lotion), sunscreens, daily-type UV skin protectants, and SPF 15 cream]. The easiest to preserve are nonaqueous products if the consumer keeps from adding water. For instance, some sunblocks have virtually no water.

A fragrance, phenoxyethanol, along with benzalkonium chloride might be all that is needed to preserve low-a_w products. This could be the basis for a preservative system that would be acceptable internationally. A big problem is the restrictions placed on preservatives by Japan. A goal of the authors is to produce a "universal" preservative system that is acceptable internationally (see Chapter 8).

See Table 3.7 for a survey of some clear surfactant-based soaps and conditioners. Note that each product is a multipreservative system including lowered a_w, pH, waterproof package design, chelator, glycerin, cationic surfactant, citric acid, other acid, fragrance, and sometimes an antimicrobial pigment or herb (e.g., cloves).

See Table 3.9 for a_w of some raw materials, such as Polysorbate 80, surfactants, sodium lauryl ether sulfate, sodium lauryl sulfate, ammonium lauryl sulfate, tego betaine, and oleyl betaine.

3.4.2 Other Considerations

Greater spoilage risk exists during water desorption (drying of wet ingredients) than during water absorption (adding water to dry ingredients), as discussed previously. This should not be a problem since most aqueous cosmetics and pharmaceuticals are compounded by adding water, not desiccation. Wet granulations and wet-coated tablets such as furosemide, triamterene, verapamil hydrochloride, and cyclobenzaprine hydrochloride might involve drying during manufacture.

These products should be tested for viable microorganisms with a method such as the USP Microbial Limit Tests (43). Some microbiologists imply stability assurance with some kind of challenge test (44). The Cosmetic, Toiletry and Fragrance Association (CTFA) or United States Pharmacopeia (USP) Antimicrobials Preservatives—Effectiveness test (PET/AET) method can be employed (45). However, we suggest using the D-value calculation. Even "anhydrous" solid dosage forms have been known to absorb moisture from the air and grow mold (46).

Increasing nutrients in a formula and fluctuations in temperature can provide for a less hostile environment for microorganisms that results in better growth at extreme a_w values or with conventional preservatives. It may be desirable to include protein in a shampoo from a sales aspect, however, protein increases the range for growth and neutralizes many conventional preservatives (47). The initial D-value may be acceptable, yet the presence of protein may result in a shorter shelf life evident only during prolonged stability studies. Competition with bacteria (if present) keeps the numbers of yeasts and molds low at a_w values at between 0.999 and 0.980 (20). Different solutes affect microbes differently, narrowing or widening the range of a_w values for growth and usually selecting for different populations of organisms. Antioxidants such as BHA, BHT, and TBHQ have also been used to potentiate preservative action and decrease D-values. Although usually beneficial, antioxidants can adversely affect the color of the product, turning some from white to pink or red (see Chapter 7 for an in-depth review).

3.5 EXAMPLES OF a_w TO CONTROL MICROORGANISMS

Preservation should be thought of as a system, not just adding a preservative. a_w-based systems are usually part of a multiple antimicrobial system. Combinations of solutes/humectants and conventional preservatives are used in many intermediate-moisture food formulations (5). a_w is rarely used by itself, especially in cosmetics where the consumer is notorious for adding water to extract the last bit of product out of the container.

A hostile environment consists of denying a microorganism the physical and chemical requirements for growth or survival (Table 3.14). Combinations of sublethal factors will increase the hostility of the environment and increase lag phase. If the environment is hostile enough, the lag phase will be extended to infinity. For practical purposes this provides at least three years of shelf stability.

Microbiologists should work with the product developer/formulator at the early stages, building a good stability factor into the product. This depends on the initial bioburden and possible consumer abuse due to the addition of contaminants to the product during use. It is necessary that package design and storage will be protective (48,49) (see Chapter 10). a_w-based systems need waterproof package design (Table 3.10) and labels should announce storage cautions, e.g., "Do Not Add Water to Product," or "Do Not Dilute," "Do Not Use Under Water or in Bath or Shower." A major problem is consumer compliance with these instructions.

3.6 USE OF a_w IN REDUCING PRESERVATIVE LEVELS IN FORMULAS

In Germany it was widely reported that a certain preservative was a sensitizer and caused irritation of the skin. While these reports may be media "hype," the issues raised cannot be ignored (50,51). Also, formaldehyde and formaldehyde donors sensitized people in Japan after World War II when formaldehyde was cheap and used in concentrations well over that needed to preserve products.

See Table 3.11 for preservatives that can be tested for use at reduced rates with a_w-based preservative systems. Mold spoilage of low- to medium- (<15%) alcohol products has occurred at the interface of the product and the container, where alcohol concentration is lower because evaporation of the alcohol leaves behind unevaporated water. This watery area of the container has a higher a_w than the liquid of the product and is an area of possible microbial spoilage. Products susceptible to this type of spoilage include syrups, therapeutic solutions, skin ointment vehicles, and mouthwash. On the other hand, addition of 4–7% alcohol is recommended for chlorhexidine gluconate hand wash and oral rinse solutions to prevent growth of gram-negative bacteria such as *P. aeruginosa.*

3.7 CAUTIONS TO CONSIDER WHEN USING a_w

3.7.1 Microorganisms and Complaint Samples

Complaint samples often show an increase in a_w, which may enable microorganisms to produce putrid odor, off-color and cloudy appearance. The suspected cause of these symptoms is often contaminated water that gets into the product. As a precaution, one manufacturer added parabens to an anhydrous body oil. Anhydrous products normally would not need additional preservation because water is absent. However, additional preservatives in an a_w-based system can easily be justified based on the RDV of Orth et al. (36). Table 3.10 shows some in-home use products that had an increase in water presumably caused by the consumer.

Package integrity is important. The package must be waterproof if the product is likely to be used in or near water. Package materials need a multilayer moisture barrier such as Suran® with an inner coating of polyethylene. The main body of the package may be polypropylene. Nylon lets oxygen into the package, and another barrier should be added if nylon must be used. Nylon is sometimes used for the cap, which can lead to oxidation of the product surface and degradation of the preservative system.

The formulator must take into account consumer complaints from spoiled products. To predict microbial stability, several AETs and PETs have been devised. The most common PET required by regulatory agencies in the United States is the Antimicrobial Preservatives—Effectiveness test of the USP (45). The USP has the easiest acceptance criteria to pass from a regulatory perspective, but products that can pass only USP criteria probably are the least safe from a microbial stability standpoint. The RDV is a safer method to predict potential spoilage in the "real" world. There are great differences in acceptance criteria used by the USP and the linear regression methods. The latter method uses D-values, which provide a quantitative assessment of the speed of microbicidal activity of a formula for test organisms. Also, the USP does not require "self-preservation" for fungi.

a_w is interconnected with lowering the D-value but not by itself. Lowering the a_w increases the lag phase of the microbial growth curve (20). Microbiologists call this an "extended lag phase" (ELP). This means that low a_w does *not* necessarily decrease the D-value. Low a_w is a hostile environment for any living organism (including humans) and will increase stress on the microorganisms, making them vulnerable to attack from other factors in the preservative system (42). Therefore, lowering a_w alone may make ingredients such as conventional preservatives more effective, but may not by itself decrease the D-value.

When performing an AET/PET on low-a_w products, ELP can be important since the AET/PET USP method uses an incubation time for the total plate count (TPC) of only 48–72 hours (43) and the ELP may be 5–7 days for bacteria and 14–45 days for yeasts and molds. It is interesting that later in the same reference (52), USP 23 requires several weeks of incubation for sterility tests, where ELP may also occur. Modifications found in the USP Microbial Limits Preparatory Test must also be used in AET/PET plate counts.

ELP will vary with extent of injury. There will most likely be a range of injury. Some cells will not be injured at all, while a great number may be injured slightly or severely. Of course many cells will have an infinite lag phase and can be considered dead. As many as 90% of the total "final" count (noninjured plus injured) may be evident at 48 hours, however, it is known that as much as a 300% increase in counts can occur in some products when using improved culture media, neutralizers, injury-recovery techniques, broth preenrichment, agitation, and extended incubation times (53,54).

The incubation time must be increased if ELP is suspected. It is important to use a humidified incubator or plastic bag with moist paper towels to prevent drying during the prolonged incubation time. Rapid AET/PET methods still can be used because they are intended as screening methods. However, we recommend increasing incubation time—a minimal inconvenience considering the great savings in technician time and materials (38).

3.7.2 Sources of Contamination Other Than the Consumer

Raw materials may be a source of contamination. Thus, it is important to use low-bioburden material to prevent possible carryover into the finished product. The product may also become contaminated during processing: in the bulk tank, blending tank during mixing, transfer tanks, lines (especially in dead legs), pumps if not disassembled and only cleaned in place (CIP), and primary storage tanks. Standing water after CIP is a big problem. High microbial counts will confirm a "flushout" effect if standing water is left in the production equipment over a weekend or holiday. Cleaning crews find it difficult to remove o-rings without damage. Often cleaning crews will not remove and clean the o-rings and the surface behind the o-rings. They will only wash the top and then air-dry or compressed air-dry only the top surface. This practice leaves a wet, nutrient-rich area for microorganisms to breed and adapt to the diluted preservative systems. Chemical sanitizers are neutralized by unclean surfaces, product, moisture, and some cleaning solutions. Unless heat is used to penetrate behind the o-rings and into the microsized cracks and stainless steel polishing grooves, the next batch of product may get inoculated with high levels of bacteria, yeasts, and molds. Spoilage probability is greatly increased. Leaving processing equipment wet is one of the quickest ways to destroy a product by microbial spoilage. Rotating the chemical sanitizer or product preservative system is no help. Rotation only hides the problem and eventually makes things worse. Microorganisms adapt to rotated sanitizers. If rotation of chemical sanitizers continues without proper cleaning or standing water is left, the contaminating microorganisms may adapt to all the chemical sanitizers. This is particularly a problem with a_W-based preservative systems since standing water contaminates the product, possibly with up to 10,000,000 microorganisms. Wet equipment can also dilute and destroy the self-preserving character of preservative systems. Standing water increases the a_W of the product itself, providing a more hospitable environment which then makes other preservatives less effective. Standing water behind o-rings or within damaged o-rings increases the likelihood of microbial adaptation to the conventional preservatives. Microbial adaptation leads to the development of "superbugs" capable of invading and colonizing other nondiluted products (see Chapter 2).

Truck transport has been implicated in *P. aeruginosa* contamination due to condensation of water in the upper part of the transport vessel and faulty or contaminated air filters incapable of trapping bacteria and fungi during transport. This happens especially in farmland (e.g., tomato farms in central California) and cities with high levels of airborne contamination.

Methods for plate counts also need a_w adjustment. Molds such as *Wallemia sebi* cannot be detected using the usual high-a_w media such as potato dextrose agar (PDA) or sabouraud dextrose agar (SDA) because *low* a_w is required for growth (55,56). The a_w of the culture medium must be adjusted. The use of Dichloran 18% glycerin agar has been suggested by Hocking and Pitt (55). This medium contains 18% glycerin, which lowers the a_w of the medium from conventional PDA/SDA at 0.999 to an a_w of 0.950. The adjustment may also apply to mold such as *Eurotium amstelodami* and other xerophilic molds (Tables 3.3 and 3.4). Even yeasts grow better with slightly lower a_w. Albert et al. (55) found reduction of a_w from 1.000 (moisture content, 51.7–65.3%) to 0.930 (moisture content 28.1–41.7%) dramatically increased fungal fruiting body development. While decreasing the risk of bacterial spoilage, lower a_w may contribute to a greater danger from yeast and mold spoilage. This aspect should be remembered by those whose say dry solid dosage forms are microbiologically stable without hard supporting data from challenge studies using properly adjusted low-a_w culture media.

Enigl (unpublished) found that addition of 20% sucrose to PDA or dichloran rose bengal chloramphenicol agar increased total plate counts of yeast in low-a_w food products compared with conventional PDA. Multicomponent preservative systems can be degraded due to low water because of the oil/water partition coefficients of raw materials. If the partition coefficient is less than unity, the preservative is increased in the aqueous phase, where the greatest antimicrobial action takes place. Unfortunately, a decrease in aqueous phase preservative concentration occurs if the partition coefficient is greater than unity (58). Therefore, for low-a_w–based systems, the partition coefficient should be less than unity.

A preservative system can be weakened when one component counteracts another. Lecithin and Polysorbate 80 can neutralize lipophilic preservatives (e.g., parabens, chlorhexidine gluconate) in the presence of water. Formaldehyde donors in water or at elevated temperatures are less soluble and may gas-off, thereby weakening the preservative system. Ironically, a D-value series during a stability study may show an abnormal *increase* in preservative effectiveness with age (lower D-value than at the beginning of the study). This happens when preservatives gas-off. If packaging allows release of formaldehyde during the gas-off, preservative concentration suddenly may decrease, resulting in D-values of several hundred hours (Enigl, unpublished).

Mascaras can be anhydrous at the time of manufacture. This is the ultimate in low a_W. The problem is that the consumer can (probably will) introduce water into the mascara container and brush. To test this type of problem/product, special preservative testing methods are required (59). A direct filter membrane test by the FDA has been developed for mascaras and could be used for other dry solid dosage forms (60).

Problems with test methods also occur when measuring unbound water in emulsions. The a_w of oil-in-water (o/w) emulsions may be measured with some degree of confidence. Water- in-oil (w/o) emulsions may not be preserved reliably by a_w alone because of the difficulty in accurate a_w measurement. It is almost impossible to measure a_w in w/o emulsions since the oil "hides the water" from a_w-measuring equipment.

3.7.3 Recovery Systems

AET/PET methods require injured cell–recovery systems. Injured cells cannot grow, unless they are repaired, leading to low TACs on plates lacking a suitable recovery system and falsely low D-values. Injured cells have repaired and grown to spoil products and have caused infections in susceptible consumers. This is called the "Phoenix phenomenon" (61). A culture medium such as R2A recovers injured bacteria better from deionized water than tryptic soy agar (TSA) or tryptic soy agar with soy lecithin and polysorbate 80 (TSALT). Activated charcoal, pyruvate, catalase, lecithin, and polysorbate 20/80 have been used to recover injured microbes or stop preservatives from carrying over into the culture medium (44,58). Low-nutrient culture media will recover injured bacteria better than high-nutrient media (62,63).

A method to test the tolerance of a product to preservative dilution and neutralization can be performed. Use a 1:1 dilution (or another applicable dilution ratio) of the product with water to simulate in-use conditions.

3.8 SUMMARY

The food industry has used a_w for many years to preserve and increase the shelf life of foodstuffs. The food industry has used a_w as the main starting point to satisfy consumer requests for "preservative-free" products and reduce the cost of conventional preservatives. There is no technical reason why self-preserving formulas cannot be used as a "new system" for cosmetics, OTC drugs, and pharmaceuticals. It must be remembered, however, that foodstuffs are meant to be ingested, whereas topical products are applied to the skin. The latter often require different microbiological standards.

The important thing to remember about a_w is that it alone may not reduce microbial numbers. Self-preserving products relying on a_w alone may not be able to "cleanse" themselves as quickly as products preserved with conventional preservatives. The use of a_w could be important in postuse contamination where growth would be inhibited and controlled.

The combination of low a_w and conventional preservatives fits into the concept of reducing preservatives whenever possible. In many cases it is possible to reduce the preservative levels in formulas with reduced a_w and still maintain product stability. This means that the products could be used by people who are sensitive to and easily irritated by higher preservative levels.

As with any different technology, it is important to conduct experiments relating to the efficacy of the system. All formulas should be measured for a_w and challenged with appropriate microorganisms. The RDV should then be used to predict the microbial stability of the formula.

While many concerns must be addressed before using reduced a_w-based preservative system(s), the authors feel the benefits are worthwhile. Multiple-factor preservative systems can be enhanced and/or reduced by starting the formulation process based on low a_w. Hopefully this chapter has stimulated some ideas on a_w and self-preserving formulas and has supplied sufficient information to enable formulators to apply the concept in future formulation work.

REFERENCES

1. L. O. Kallings, O. Ringertz, and L. Silverstolpe. Microbiological contamination of medical preparations. *Acta Pharm. Suecica. 3*: (1966).
2. J. C. Curry. A history of cosmetic microbiology. *Cosmet. Technol.* 5(2):24,26,29,30.
3. J. H. B. Christian. Reduced water activity. In: *Microbial Ecology of Foods*. J. H. Skinner et al., eds. Academic Press, New York, 1980, pp. 70–91.
4. J. Chirfe, C. Ferro Fontan, and O. C. Scorza. A study of the water activity lowering behavior of some amino acids. *J. Food Technol. 15*:383–387 (1980).
5. S. Hegenbart. Exploring dimensions in intermediate moisture foods. *Food Prod. Design 3*(4):32–54 (1993).
6. J. M. Jay. In: *Modern Food Microbiology*. D. Van Nostrand Co. New York, 1970, pp. 1–6, 30–32, 165–169.
7. G. A. Wistreich and M. D. Lechtman. *Microbiology*, 4th ed. Macmillan, New York 1984, pp. 882–883.

8. P. M. Foegeding and F. F. Busta. Chemical food preservatives. In: *Disinfection, Sterilization, and Preservation*, 4th ed. S. S. Block, ed. Lea and Febiger, Philadelphia, 1991, pp. 802–832.
9. R. A. Fassihi. Preservation of medicines against microbial contamination. In: *Disinfection, Sterilization, and Preservation*, 4th ed. S. S. Block, ed. Lea and Febiger, Philadelphia, 1991, pp. 871–886.
10. Decagon Devices, Inc. Monograph for CX-1 Water Activity System. Pullman, WA, 1992, pp. 1–8.
11. W. H. Sperber. Influence of water activity on foodborne bacteria: a review. *J. Food Protect. 46*(2):142–150 (1983).
12. J. A. Troller. The water relation of food-borne bacterial pathogens: a review. *J. Milk Food Technol. 36*(5):276–288 (1973).
13. W. J. Scott. Water relations of food spoilage microorganisms. *Adv. Food Res.* 7:83–127 (1957).
14. B. A. Prior and C. P. Kenyan. Water relations of glucose-catabolizing enzymes of *Pseudomonas fluorescens. J. Appl. Bacteriol. 48*:211–222 (1980).
15. L. W. Acker. Water activity and enzyme activity. *Food Technol. 23*:27–40 (1969).
16. J. G. Holt and N. R. Krieg (Eds.). *Bergey's Manual of Systematic Bacteriology*. Vol. 1. Williams and Wilkins, Baltimore, 1984, pp. 261–267.
17. J. G. Holt, J. T. Staley, M. P. Bryant, and N. Pfenning (Eds.). *Bergey's Manual of Systematic Bacteriology*. Vol. 3. Williams and Wilkins, Baltimore, 1989, pp. 2216–2231.
18. B. A. Prior. The effect of water activity on the growth and respiration of *Pseudomonas fluorescens. J. Appl. Bacteriol. 44*:97–106 (1978).
19. E. O. Morris. Effect of environment on microorganisms. In: *Recent Advances in Food Science*. J. Hawthorn and J. M. Leitch, eds. Butterworths, London, 1962, pp. 24–36.
20. L. R. Beuchat. Microbial stability as affected by water activity. *Cereal Foods World 26*(7):345–349 (1981).
21. B. J. Marshall, D. F. Ohye, and J. H. B. Christian. Tolerance of bacteria to high concentrations of NaCl and glycerol in the growth medium. *Appl. Microbiol. 21*:363–364 (1971).
22. H. Pivnick. Curing salts and related materials. In: *Microbial Ecology of Foods*. Academic Press, New York, 1980, pp. 136–159.
23. T. V. Labuza, K. Acott, S. R. Tatini, R. Y. Lee, J. Flink, and W. McCall. Water activity determination: a collaborative study of different methods. *J. Food Sci. 41*:910–917 (1976).
24. J. A. Troller and J. H. B. Christian. *Water Activity and Food*. Academic Press, New York, 1978.
25. V. Rodel, K. Krispien, and L. Leistner. Measuring water activity (a_w value) of meat and meat products. *Fleischwirtschaft 59*(6):849–851 (1979).
26. J. A. Troller, D. T. Bernard, and V. W. Scott. Measurement of water activity. In: *Compendium of Methods for the Microbiological Examination of Foods*. American Public Health Association, Washington, DC, 1984, pp. 124–134.
27. Novasina AG, Zurich, Switzerland. Technical Communication Thermoconstanter for Water Activity Measurement. Product Information, 1992, p. TC-1.
28. Rotronic Instrument Corp., Huntington, NY. Technical Communication: Operating Instructions for DT and WA Measuring Station, 1992, pp. 1–3.

29. L. Stoloff. Calibration of water activity measuring instruments and devices: Collaborative study. *J. Assoc. Off. Anal. Chem. 61*(5): 1166–1167 (1978).
30. FDA regulations. *Fed. Reg. 44*(53):16209–16230 (1979).
31. K. M. Acott and T. P. Labuza. Microbial growth response to water sorption preparation. *J. Food Technol. 10*:603–611 (1975).
32. S. L. Resnik and J. Chirife. Proposed theoretical water activity values at various temperatures for selected solutions to be used as reference sources in the range of microbial growth. *J. Food Protect. 51*(5):419–423 (1988).
33. D. P. Bone. Water activity—its chemistry and applications. *Food Product Devel.* (August/September):81–94 (1969).
34. K. J. Horner and G. D. Anagnostopoulos. Combined effects of water activity, pH, and temperature on the growth and spoilage potential of fungi. *J. Appl. Bacteriol. 36*:427–436 (1973).
35. R. S. Meyer, M. A. Grant, L. O. Luedecke, and H. K. Leung. Effects of pH and water activity on microbiological stability of salad dressing. *J. Food Protect. 52*(7):477–479 (1989).
36. D. S. Orth, R. F. Barlow, and L. A. Gregory. The required D- value: evaluating product preservation in relation to packaging and consumer use/abuse. *Cosmet. Toiletr. 107*(12):39–43. (1992).
37. D. S. Orth and C. M. Lutes. Adaptation of bacteria to cosmetic preservatives. *Cosmet. Toiletr. 100*(2):57–59, 63–64 (1985).
38. D. S. Orth and D. C. Enigl. Preservative efficacy testing by a rapid screening method for estimation of D-values. *J. Soc. Cosmet. Chem. 44*:329–336 (1993).
39. D. S. Orth. Standardizing preservative efficacy test data. *Cosmet. Toiletr. 106* (3):45–51 (1991).
40. D. S. Orth, C. M. Lutes Anderson, D. K. Smith, and S. R. Milstein. Synergism of preservative system components: use of the survival curve slope method to demonstrate anti-*Pseudomonas* synergy of methyl paraben and acrylic acid homopolymer/copolymers in vitro. *J. Soc. Cosmet. Chem. 40*:347–365 (1989).
41. M. S. Cooper (Ed.). Non-sterile products: evolving pathogens, *Staphylococcus epidermidis. Microbiol. Update Pharm. Med. Devices, Cosmet. 12*:3 (1994).
42. V. N. Scott. Interaction of factors to control microbial spoilage of refrigerated foods. *J. Food Protect. 52*(6):431–435 (1989).
43. Microbiological Tests: Microbial Limit Tests <61>. United States Pharmacopeia 23, United States Pharmacopeial Convention, Rockford, MD, 1995, pp. 1681–1686.
44. M. S. Cooper (Ed.). Non-sterile products: *Pseudomonas. Microbiol. Update Pharm. Med. Devices, Cosmet. 12*:3 (1994).
45. Microbiological Tests, Antimicrobial Preservatives—Effectiveness <51>. United States Pharmacopeia 23, United States Pharmacopeial Convention, Rockford, MD, 1995, p. 1681.
46. M. S. Cooper (Ed.). Microbiological control topics for drugs, cosmetics, and foods: dry dosage forms. *Microbiol. Update Pharm. Med. Devices Cosmet. 10*:2 (1992).
47. D. S. Orth. Linear regression method for rapid determination of cosmetic preservative efficacy. *J. Soc. Cosmet. Chem. 30*:321–331 (1979).
48. M. S. Cooper (Ed.). Container sterility integrity. *Microbiol. Update Pharm. Med. Devices Cosmet. 10*:3 (1992).

49. T. J. McCarthy. Formulated factors affecting the activity of preservatives: packaging, deterioration-light. In: *Cosmetic and Drug Preservation: Principles and Practice.* J. J. Kabara, ed. Marcel Dekker, New York, 1984, pp. 378-381.
50. E. Schopf and A. Baumgartner. Contact allergies to preservatives in cosmetic preparations. *J. Appl. Cosmetol. 8*:39–49 (1990).
51. A. C. de Groot and J. W. Weyland. Contact allergy to methyldibromoglutaronitrile in the cosmetics preservative Euxyl K 400. *Am. J. Contact Dermatitis* 2:31–32 (1991).
52. Microbiological Tests, Sterility Tests <71>. United States Pharmacopeia 23. United States Pharmacopeial Convention, Rockville, MD, 1995, pp. 1686–1690.
53. D. C. Enigl, A. D. King, and T. Torok. *Talaromyces trachyspermus*, a heat-resistant mold isolated from fruit juice. *J. Food Protect. 56*:1039–1042 (1993).
54. D. A. Power and P. J. McCuen. *Manual of BBL Products and Laboratory Procedures.* 6th ed. BCYE Agar Base, 1988, pp. 176–177.
55. A. D. Hocking and J. I. Pitt. Dichloran-glycerin medium for enumeration of xerophilic fungi from low moisture foods. *Appl. Environ. Microbiol. 39*:488–492 (1980).
56. L. R. Beuchat and M. Soledad Tapia De Daza. Evaluation of chemicals for restricting colony spreading by a xerophilic mold, *Eurotium amstelodami* on dichloran-18% glycerin agar. *Appl. Environ. Microbiol. 58*:2093–2095 (1992).
57. R. A. Albert, B. Huebner, and L. W. Davis. Role of Water Activity in the Spoilage of Alfalfa Hay. *J. Dairy Sci. 72*:2573–2581 (1988).
58. H. S. Bean, S. M. Heman-Ackah, and J. Thomas. The activity of antibacterials in two-phase systems. *J. Soc. Cosmet. Chem. 16*:15–30(1965).
59. D. G. Ahern, J. Sanghvi, and G. J. Haller. Mascara contamination: in-use and laboratory studies. *J. Soc. Cosmet. Chem. 29*:127–131 (1978).
60. A. T. Tran, A. D. Hitchins, and S. W. Collier. Direct contact membrane method for evaluating preservative efficacy in solid cosmetics. *Int. J. Cosmet. Sci. 12*:175–183 (1990).
61. D. S. Orth, C. M. Lutes, S. R. Milstein, and J. J. Allinger. Determination of shampoo preservative stability and apparent activation energies by the linear regression method of preservative efficacy testing. *J. Soc. Cosmet. Chem. 38*:309–319 (1987).
62. *Standard Methods for the Examination of Water and Waste Water*, 18th ed. American Public Health Association, Washington, DC, 1992, pp. 33–34.
63. D. S. Orth. Principles of preservation. In: *Guide to Microbiological Control in Pharmaceuticals.* S. P. Denyer and R. M. Baird, eds. Ellis Horwood, London, 1990, pp. 241–250.

4

The Role of Surfactants in Self-Preserving Cosmetic Formulas

Oreste Cozzoli
Stazione Sperimentale per le Industrie degli Oli e dei Grassi
Milan, Italy

Surfactants are chemical compounds that possess surface activity. They are classified as anionic, nonionic, cationic, or amphoteric based on their net charge in solution and chemical structure. When dissolved in water, surfactants lower the surface tension (i.e., interfacial tension) by positive adsorption at the liquid/vapor surface or at the water/oil interface. The primary use of these compounds is as detergents, foaming or wetting agents, solubilizers, or dispersants. Surfactants also act as co-emulsifiers because they are partly water soluble and partly oil soluble. This hydrophilic:lipophilic balance (HLB) is a function of specific structure and temperature.

When present at concentrations above the critical micelle concentration (CMC) in aqueous solution, surfactants can form micelles. Preservatives absorbed or enclosed in micelles have decreased preservative efficacy due to a reduction in effective concentration.

Abe (1) reported on the effect of surfactants on bactericidal and fungicidal action. At concentrations below the CMC, surfactants can help solubilize preservatives and make them more effective. Linfield et al. (2) found that nonionic detergents of the polyoxyethylene ester series exhibited a potentiating effect of certain germicidal agents in deodorant soaps at relatively low levels. Molnar (3) reported that some surfactants increase the antimicrobial activity of bromsalan and other ring-halogenated substances.

Such studies demonstrate that the influence of the surfactant and vehicle are very important in any technical assessment of the antimicrobial potency of a compound. In practice, the formulator needs to determine empirically the effects of surfactants in each system. While many surfactants are known for their antimicrobial effects, their role in helping to reduce or eliminate the use of preservatives has received little attention. This chapter presents information on the antimicrobial properties of various classes of surfactants to provide the reader with an understanding of the multifunctional properties of surfactants and their role in preservative systems.

4.1 ANIONIC SURFACTANTS

Surfactants are classed as anionic when they ionize in aqueous solution to produce negatively charged organic ions that have surface activity.

4.1.1 Fatty Acid Soaps

Soaps are among the earliest examples of self-preserving products. Their antimicrobial properties depend on alkalinity and low water content (i.e., water activity; see also Chapter 3). Much of the literature on the antimicrobial properties of fatty acids in acid pH systems can be found in an informative monograph prepared by Kabara (4) and in a review of antimicrobial effects of anionic surfactants by Kastner (5). The reader is referred to Chapter 5 for additional information on the antimicrobial action of fatty acids.

4.1.2 Other Anionic Surfactants

Generally, anionics show only weak antibacterial effects (6). Voss (7) reported that the effect of alkyl (C_{12}) benzene sulfonate (ABS) on gram-negative bacteria is negligible, while some effects were seen in the case of gram- positive bacteria. The effect of ABS against *Staphylococcus aureus* is intensified by divalent inorganic cations (8). Oba and Tomiyama (9) also reported inhibition of *S. aureus* by ABS at a concentration of 2.5 μg/ml at pH < 5 and 20 μg/ml at pH > 7. By way of comparison, cetyltrimethylammonium bromide (CTAB) is active at concentrations of 0.25–5 μg/ml at pH >7 and 0.5–1.0 μg/ml at pH $= 6$.

Damm (10) reported that alkali salts of palmitic acid killed *Escherichia coli* at a concentration of 0.72%. In addition, sodium dodecyl sulfate (SDS), sodium dodecyl aminopropionic acid, sodium ricinoleate, bile acids, and alkylaryl sulfonates are effective against gram-positive bacteria. SDS and desoxycholate were effective against influenza A, smallpox, and psittacosis viruses, with virus-inactivating concentrations being in the concentration range that causes hemolysis.

Schaller et al. (11) reported that sodium lauryl ether sulfate inhibited *Microsporum gypseum*, *Aspergillus niger*, and *Epidermophyton floccosum*. The data in Table 4.1 show the low bactericidal activity of various anionics in comparison with other types of surfactants.

The sulfated partial ester of a modified undecylenic acid (Grillosan DS 7911) is an anionic surfactant reported to have outstanding dermatological, bactericidal, and fungicidal properties (12). It is claimed to be active against a broad range of bacteria and yeasts (*Candida* and *Pityrosporum ovale*), which makes it suited for use in antidandruff shampoos, feminine hygiene products, and foot care products. The disodium undecylenic acid monoethanol amidosulfosuccinate (Rewocid SBU 185) (13) is an undecylenic acid derivative with the following formula:

$$CH_2{=}CH{-}(CH_2)_8{-}\overset{\overset{\displaystyle O}{\|}}{C}{-}NH{-}CH_2{-}CH_2{-}O{-}\overset{\overset{\displaystyle O}{\|}}{C}{-}CH_2{-}\underset{\underset{\displaystyle SO_3Na}{|}}{CH}{-}COONa$$

This compound combines the skin-substantive properties of the sulfosuccinate with the antimicrobial effect of undecylenic acid. Anionic surfactants are used in liquid soaps, (antidandruff) shampoos, body washes, (acne) facial cleansers, and many other types of cosmetics and over-the-counter (OTC) drug products. Anionics exert chaotropic (i.e., destabilizing) effects on microbial cell membranes when present in concentrations above the CMC. This class of surfactants is more effective against gram-positive bacteria than against gram-negative bacteria. The general resistance of gram-negative organisms is due at least in part to the protection afforded by the outer membrane (see Sec. 2.2). Anionic surfactants should be considered part of a product preservative system because they contribute to the antimicrobial action of the formula.

4.2 NONIONIC SURFACTANTS

Surfactants are classified as nonionic when they do not ionize in aqueous solution. Nonionics frequently are used to help solubilize other ingredients in formulations. When formulating products, it is important to note that nonionic surfactants may have a potentiating effect on antimicrobial action at concentrations below the CMC and they may have an inhibiting effect on biocidal activity at concentrations above their CMC. Schmolka (14) proposed that antimicrobials become encapsulated above the CMC of nonionics, which makes less of the biocide available to kill microorganisms. To help avoid this, he recommended keeping the ratio of nonionic to antimicrobial less than 4:1. Schmolka explained that a low interfacial tension exists at low nonionic concentrations and that this may make microorganisms more susceptible to the biocide.

TABLE 4.1 Minimum Effective Concentration of Some Surfactants (in ppm)

	S. aureus	*E. coli*	*Ps. aeruginosa*	*S. typhosa*	*P. ovale*	*C. albicans*	*A. niger*
Anionics							
Standapol SHC 101	1467	>8000	>8000	>8000	>8000	>8000	4000
Stepan DS-60	111	>8000	>8000	>8000	2445	1200	400
Ivory soap	1511	>8000	>8000	5689	1867	4000	1867
Neutral toilet bar	2000	>8000	>8000	>8000	800	1867	800
Amphoterics							
Monateric LA	453	1511	>8000	1022	236	>8000	285
Miranol CS	>8000	>8000	>8000	>8000	>8000	>8000	>8000
Miranol HS	>8000	>8000	>8000	>8000	1333	>8000	1911
Nonionic with some cationic characteristics							
Monazoline C	80	187	>8000	107	13	164	20
True cationics							
Bromat	<5	667	1778	516	<5	22	18
Mytab	<5	178	1378	156	<5	26	13
Nonionic							
Emulphogene BC720	253	>8000	>8000	>8000	>8000	>8000	>8000

Standapol SHC 101 = anionic condensate of di-basic half-ester; Stepan DS-60 = dodecylbenzene sodium sulfonate; Ivory soap = sodium oleostearate; Neutral toilet bar = principally acyl isethionate and stearic acid; Monateric LA = 2-Cu-1-(ethyl-beta-oxypropanoic acid) imidazoline; Miranol CS = sulfonated imidazolinium compound (coconut acid derivative); Miranol HS = sulfonated imidazolinium compound (lauric acid derivative); Monazoline C = 1-hydroxyethyl, 2 C9-17 imidazoline; Bromat = cetyl trimethyl ammonium bromide; Mytab = myristyl trimethyl ammonium bromide: Emulphogene BC720 = tridecyl-oxypoly(ethyleneoxy)ethanol.
Source: Ref. 3

4.2.1 Monoesters of Fatty Acids

The antimicrobial activity of monoesters of fatty acids was reviewed by Kabara (4). Kabara et al. (15) noted that some hydrophilic group is necessary for biological activity. Although esters of fatty acids with monohydric alcohols are inactive against microorganisms, esters of α-hydroxy fatty acids with monohydric alcohols and esters of fatty acids with polyhydric alcohols are reported to have antimicrobial activity (15,16). These data suggest that one or more free hydroxyl groups is necessary for biological activity of the ester. In fact, Kabara et al. (17) found that esters of glycerol and fatty acids were more active than the corresponding fatty acids. The structure-function relationship for saturated and unsaturated esters follows the antimicrobial activity of their respective fatty acids; consequently, lauric acid (C_{12}) and palmitoleic acid ($C_{16:1}$) form the most active saturated and unsaturated fatty acid monoesters (16,18). (For a detailed discussion, see Chapter 5.)

Beuchat (19) and Shibasaki (20) reported that glyceryl monolaurin was the most active monoglyceride—even in the presence of other complexing agents. Antimicrobial activity has been reported for esters of more complex polyhydric alcohols. Kato and Arima (21) reported that the sucrose ester of lauric acid was active against gram-negative bacteria; however, other workers have found that these and other sucrose esters are primarily active against gram-positive bacteria and fungi (16,22). In contrast to the data obtained with glycerol esters, the diester of sucrose was more active than the sucrose monoester. Sucrose dicaprylate possessed the highest antimicrobial activity of the sucrose esters, but is less active than monolaurin (19,23,24). The monoesters of glycerol and the diesters of sucrose are reported to have higher antimicrobial activity than their corresponding free fatty acids, but also compare favorably in activity with commonly used preservatives such as the parabens, sorbic acid, and dehydroacetic acid (23).

Another class of esters has been developed from a group of compounds called polyglycerol esters. As was observed with sucrose esters, the medium-chain fatty acids appear to have the most antimicrobial activity. The fatty acid moiety has been found to determine the overall biocidal activity, regardless of whether the polyol was tri-, hexa-, or decapolyglycerol (16).

4.2.2 Polyethylene Glycol Monoesters of Alkyldihydroxybenzoic Acids

Uchibori et al. (25) prepared tri- and tetraethylene glycol monoesters of pentyl-, hexyl-, heptyl-, and octyldihydroxybenzoic acids by the reaction of tri- and tetraethylene glycol monochloride with alkyldihydroxybenzoic acids obtained from resorcinol by acylation and then reduction to the alkylresorcinol, followed by carboxylation. Antimicrobial activities were ascertained by determining the minimum inhibitory concentration (MIC). Antibacterial activity of the tetraethylene glycol monoesters was increased markedly with an increase in the number of carbon atoms of their alkyl group. Uchibori and coworkers reported that the monoester of octyldihydroxybenzoic acid was as active as a common cationic surfactant. Heptyldihydroxybenzoic acid was reported to be the most active triethylene glycol monoester. All of the monoesters studied, except tetraethylene glycol monoester of pentyldihydroxybenzoic acid, were highly effective against *Microsporum gypseum* and *Trichophyton interdigitale.*

4.2.3 Alkanolamides

Bistline et al. (26) studied several fatty acid amides and concluded that among the diethanolamides, only the *N,N*-bis-(2-hydroxyethyl)lauramide showed significant antimicrobial activity (MIC of 100 ppm against *S. aureus*). The *N,N*-bis(2-hydroxyethyl)myristamide had a MIC of 1000 ppm against *S. aureus*. Undecylenic acid monoethanolamide (Steinazid U 185) is another alkanolamide with reported fungicidal and bacteriostatic activity. The manufacturer of this compound suggested its use as a deodorizing additive for soaps, creams, and shampoos.

Many N-substituted amides of long-chain fatty acids have been shown to possess antimicrobial activity (27–29). Kenney (30) reviewed the work of Bailey et al. (31) on the antimicrobial properties of fatty ester-amides. These workers prepared 27 ester-amide derivatives of diethanolamine or *N*- methylethanolamine, each containing one or more fatty acyl moieties. The structure for the amide acyl compounds was $RCON(CH_2CH_2OR_1)_2$, and the structure for the ester acyl derivatives was $RCON(CH_3)CH_2CH_2OR_1$. The amide acyl moieties included lauroyl, palmitoyl, oleoyl, erucoyl, benzoyl, and *p*-toluoyl groupings, and the ester moieties included the lauroyl, palmitoyl, oleoyl, erucoyl, benzoyl, *p*-toluoyl, and trimethylacetyl and furoyl groupings. These compounds were screened for antimicrobial activity by measuring zones of inhibition around filter paper discs saturated with the test compound on agar seeded with *S. aureus, E. coli, Aspergillus flavus* or other *Aspergillus* spp., or a yeast (*Candida albicans* or *Torula* spp.). The MICs indicated that the benzoyl oleoyl amide ester had good activity against *S. aureus* and moderate activity against *E. coli* and *A. flavus.* Their data indicated that the erucoyl benzoyl ester of *N*-methylethanolamine had good activity against both bacteria and *Torula* spp. These workers observed that some of the compounds tested could serve a dual role as both plasticizer and antimicrobial agent in some product applications. Kenney noted that finding relationships between surface active agents and their bactericidal activity brings cosmetic formulators closer to being able to formulate self-preserving compositions. It is likely that fatty ester-amides such as these may contribute to the preservative systems of self-preserving products.

4.2.4 Polyoxyethylenic Derivatives

Although the antibacterial action of polyoxyethylenic derivatives has been reported (32–35a), it is also known that nonionics interfere with preservatives (14,36,37). It is likely that solubilization of preservatives in the micelle causes a decrease in concentration of the compound in the aqueous phase, with a corresponding loss of antimicrobial potency. It is also possible that hydrogen bonds form between phenolic preservatives and the polyoxyethylene chain of the surfactant. Evans (38) reported that the distribution constant for *p*-hydroxybenzoic acid between water and the micelles of a polyoxyethylene octylphenol ether favored the micelles by a factor of about 3000:1. The apparent loss of antimicrobial activity that occurs in the presence of nonionics requires a greater amount of preservative to be used to obtain the desired antimicrobial activity. The use of higher concentrations of preservatives increases costs and the likelihood of sensitization and/or irritation. This is a good reason for attempting to formulate self-preserving products with surfactants (other than nonionics) whenever possible.

4.2.5 Sugar Esters

In his review on fatty acid derivatives in cosmetics and toiletries, Johnson (39) noted that the use of dimethylformamide (DMF) led to fears of residual DMF in the final product. The use of sucrose esters in cosmetics, pharmaceuticals, and foods was quite rare until Croda produced these materials by a patented process that involved no toxic solvent (USP 3.480.616) and marketed them under the Crodesta® trademark (40).

Sucrose has three hydroxymethyl groups available for esterification. By selectively esterifying one, two, or three groups, a wide range of lipophilic or hydrophilic surfactants may be formed. The HLB range is from 3 to 15 and is obtained without ethylene oxide. The absence of ethylene oxide, its inverse solubility characteristics, compatibility with antimicrobial preservatives and antioxidants, and low irritation properties are cited as beneficial and unique characteristics of the sucrose esters (40). Lauric, stearic, and acetylated stearic acids, in addition to tallow fatty acids, may be used in sucrose esters. Sucrose esters are reported to have bacteriostatic action toward gram-positive bacteria, and to a lesser extent towards gram-negative bacteria (41). Although antimicrobial activity is claimed over a wide pH range, it is best at pH $\leq$ 6.0. Lerk and Sucker (42) studied the effect of sucrose laurate on *E. coli*, *P. aeruginosa*, *S. aureus*, and other test organisms and found that two oral liquid preparations could only be partially preserved with 2.86% sucrose laurate. They found that adequate preservation was obtained with the micellar solution of 2.86% sucrose laurate and 0.02% propyl 4-hydroxybenzoate.

4.2.6 Alkyl Glycosides or Alkyl Glycopyranosides

Matsumura et al. (43) synthesized *n*-alkyl and α- and β-glucopyranosides, α-D-mannopyranosides, and β-D-galactopyranosides with alkyl chains having 8–12 carbon atoms. When examined by the agar dilution method, all alkyl glycosides studied showed a broad spectrum of antimicrobial activity against *S. aureus*, *B. subtilis*,

Sarcina lutea, E. coli, Salmonella typhi, Pseudomonas aeruginosa, C. albicans, Saccharomyces cerevisiae, T. interdigitale, M. gypseum, Penicillium chrysogenum, and *A. niger*. The alkyl glycosides with *n*-dodecyl groups were particularly effective against gram-positive bacteria and the fungi tested. The shorter-alkyl-chain-length glycosides were less active than the *n*-dodecyl derivatives; however they were still considered to be active.

It was found that the activities between anomeric isomers of *n*-dodecyl glucosides were different. The α-isomers of *n*-dodecyl glucosides showed better antimicrobial activities against bacteria than the β-isomers, although the α- isomers were less effective against fungi than the β-isomers. The antimicrobial properties were influenced by the structure of the glycopyranosyl residue of the *n*-alkyl glycoside (mannopyranosyl > glucopyranosyl > galactopyranosyl group) in addition to a dependence on the alkyl chain length. When compared with conventional ethoxylated nonionics, it was found that ethyoxylated dodecanol and tetradecanol showed activity against *S. aureus* and *B. subtilis*, but no activity toward the gram-negative bacteria. In addition, their antimicrobial activity generally was less effective against both gram-positive and gram-negative bacteria than that of *n*-dodecyl glucoside and mannoside.

4.2.7 Amino Group–Containing Oligosaccharides

Amino group–containing oligosaccharides with the general formula ANHR (A = oligosaccharide residue having 2–6 glycoside bonds; R = C_{6-24} aliphatic cyclic or alicyclic hydrocarbyl) have been reported to have surfactant activity and good bactericidal and fungicidal activity—even in the presence of proteins (44). *N*-dodecyl-maltosylamine was prepared by refluxing maltose monohydrate with *n*-dodecylamine in methanol for 30 minutes. This compound was reported to have MICs of 25, 10, and 50 μg/ml against *S. aureus*, *B. subtilis*, and *C. albicans*, respectively.

4.2.8 Glucosamine Derivatives

Matsumura et al. (43,45) synthesized methyl 2-acylamino-2-deoxy-D-glucopyranosides, *n*-alkyl 2-amino-2-deoxy-D-glucopyranoside hydrochlorides, and *n*-alkyl 2-acetylamino-2-deoxy-D-glucopyranosides and screened them for antimicrobial activity. The MICs for the compounds tested are given in Tables 4.2 and 4.3. Methyl *N*-acyl glucosaminides (MeαNCn, MeβNCn) and *n*-alkyl glucosaminide hydrochlorides (CnαN, CnβN) showed a broad sprectrum of antimicrobial activity and much stronger antimicrobial activity than the corresponding *n*-alkyl glucopyranosides. The glucosamine derivatives containing a C_{12} alkyl chain had the highest activity. Data such as these suggest that use of these surfactants in product formulas would increase the activity of the preservative system.

TABLE 4.2 Antimicrobial Activity of Glucosamine Derivatives Containing Alkyl Chain and Alkyl Glucopyranosides (I)[a]

Organisms	Compounds [MIC (μg/ml)[b]]									
	MeαNC8	Meα NC10	Meα NC12	Meα NC14	MeβNC10	MeβNC12	MeβNC14	C8βGlc[c]	C10βGlc[c]	C12βGlc[c]
S. aureus	>400[d]	>400	100	200	400	50	100	400	100	10
B. subtilis	>400	>400	100	400	400	50	>400	400	100	25
S. lutea	10	10	10	10	10	10	10	100	50	10
E. coli	10	10	10	10	10	10	10	400	200	400
S. typhi	400	>400	400	200	400	50	>400	400	400	400
P. aeruginosa	10	5	5	5	5	25	10	400	200	200
C. albicans	>400	>400	200	100	100	100	>400	>400	200	400
S. cerevisiae	>400	>400	50	200	200	50	>400	>400	100	400
T. interdigitale	50	100	50	50	50	50	25	400	200	200
M. gypseum	50	100	50	50	50	50	25	400	200	200
P. chrysogenum	50	100	50	50	50	50	25	400	400	200
A. niger	>400	>400	>400	200	200	400	>400	>400	400	200

[a]Control always produced growth of the m croorganism.
[b]MIC, minimum inhibitory concentration.
[c]C8βGlc:n-octyl β-D-glucoside, C10βGlc:n-decyl β-D-glucoside, C12βGlc:n-dodecyl β-D-glucoside.
[d]No inhibition, maximum concentration tes:ed listed.

MeαNC8	methyl 2-octanoylamino-2-deoxy-α-D-glucopiranoside
MeαNC10	methyl 2-decanoylamino-2-deoxy-α-D-glucopiranoside
MeαNC12	methyl 2-dodecanoylamino-2-deoxy-α-D-glucopiranoside
MeαNC14	methyl 2-tetradecanoylamino-2-deoxy-α-D-glucopiranoside
MeβNC10	methyl 2-decanoylamino-2-deoxy-β-D-glucopiranoside
MeβNC12	methyl 2-dodecanoylamino-2-deoxy-β-D-glucopiranoside
MeβNC14	methyl 2-tetradecanoylamino-2-deoxy-β-D-glucopiranoside

Source: Ref. 45.

Table 4.3 Antimicrobial Activity of Glucosamine Derivatives (II)[a]

	Compounds [MIC (μg/ml)[b]]											
Organisms	C8αNac	C10αNAc	C12αNAc	C8βNAc	C10βNAc	C12βNAc	C8αN	C10αN	C12αN	C8βN	C10βN	C12βN
S. aureus	400	200	>400	>400[c]	>400	>400	400	100	10	400	50	10
B. subtilis	>400	200	>400	>400	>400	>400	>400	100	10	400	100	10
S. lutea	>400	200	>400	>400	>400	>400	400	50	10	400	100	10
E. coli	>400	>400	>400	>400	>400	>400	400	50	10	400	50	25
S. typhi	>400	>400	>400	>400	>400	>400	400	100	10	>400	200	25
P. aeruginosa	400	>400	>400	>400	400	400	400	200	50	>400	200	50
C. albicans	>400	>400	>400	>400	>400	>400	>400	400	200	>400	400	400
S. cerevisiae	>400	>400	>400	>400	>400	>400	>400	100	10	400	100	25
T. interdigitale	400	400	>400	400	>400	400	400	50	25	400	100	25
M. gypseum	400	400	>400	400	>400	400	400	100	25	400	200	50
P. chrysogenum	>400	400	>400	400	>400	400	>400	400	200	>400	400	200
A. niger	>400	400	>400	400	>400	>400	>400	400	400	>400	>400	400

C8αNAc	n-octyl 2-acetylamino-2-deoxy-α-D-glucopyranoside
C10αNAc	n-decyl 2-acetylamino-2-deoxy-α-D-glucopyranoside
C12αNAc	n-dodecyl 2-acetylamino-2-deoxy-α-D-glucopyranoside
C8βNAc	n-octyl 2-acetylamino-2-deoxy-β-D-glucopyranoside
C10βNAc	n-decyl 2-acetylamino-2-deoxy-β-D-glucopyranoside
C12βNAc	n-dodecyl 2-acetylamino-2-deoxy-β-D-glucopyranoside
C8αN	n-octyl 2-amino-2-deoxy-α-D-glucopyranoside hydrochloride
C10αN	n-decyl 2-amino-2-deoxy-α-D-glucopyranoside hydrochloride
C12αN	n-dodecyl 2-amino-2-deoxy-α-D-glucopyranoside hydrochloride
C8βN	n-octyl 2-amino-2-deoxy-β-D-glucopyranoside hydrochloride
C10βN	n-decyl 2-amino-2-deoxy-β-D-glucopyranoside hydrochloride
C12βN	n-dodecyl 2-amino-2-deoxy-β-D-glucopyranoside hydrochloride

[a]Control always produced growth of the microorganism
[b]MIC, minimum inhibitory concentration
[c]No inhibition, maximum concentration tested listed
Source: Ref. 45.

4.2.9 Diethanolamides of *N*-Lauroyl Dipeptides

Another class of interesting compounds is represented by diethanolamides of *N*-lauroyldipeptides [R-NH-CH(R_1)-CO-NH-CH-(R_1)-CO-N-($CH_2CH_2OH)_2$, where R = lauroyl, R_1 = H(glycine), isobutyl (leucine), benzyl (phenylalanine), thiomethylene (cysteine) or -CH_2-CH_2-CH_2-N^{α} (proline)]. Mhaskar and Lakshminarayana (46) examined the structure/activity relationships to determine the correlation of the amino acid structures with antibacterial properties against *S. aureus* and *E. coli.* The results of their study are shown in Table 4.4. The leucine derivative showed larger zones of inhibition against both test organisms than the other diethanolamides of *N*-lauroyl dipeptides. In contrast, the cysteine derivative did not produce any zones of inhibition in this study.

4.3 CATIONIC SURFACTANTS

Cationic surfactants ionize in aqueous solution to produce positively charged organic ions that are responsible for surface activity. The antimicrobial action of cationic surfactants is well known for benzalkonium chloride (BAC) and cetyl pyridinium chloride (CPC). Their use often reduces and/or eliminates the need for additional preservatives in some formulations (32,39).

Jungermann (47) reported on the antimicrobial activity of fatty nitrogen derivatives (i.e., both quaternary ammonium salts in which one, two, or three alkyl groups were fatty and amine derivatives). Reck and Harwood (48) found that unsaturation in the fatty alkyl chain does not affect the antimicrobial activity. Although these compounds were not as active as BAC against gram-positive bacteria, their work indicated that some amines and amine acetates, such as dodecylamine and dodecylamine acetate, had antibacterial activity.

TABLE 4.4 Antibacterial Activity[a] of Diethanolamides of N-Lauroyl Dipeptides[b]

Acylated dipeptide moiety	*Staphylococcus aureus* diameter (cm)	*Escherichia coli* diameter (cm)
C_{12}-gly-gly	1.1	0.8
C_{12}-leu-leu	1.5	1.5
C_{12}-phe-phe	0.8	0.7
C_{12}-pro-pro	1.1	—
C_{12}-cys-cys	—	—
C_{12}-gly[c]	1	1.5
Streptomycin (control)	2.2	2.1

[a]Determined as zone of inhibition of bacterial growth on nutrient agar plate.
[b]Tested as 1% solution in methanol.
[c]For comparative studies.
Source: Ref. 46.

4.3.1 Quaternary Ammonium Salts

Several cosmetic and OTC drug products have used cationic surfactants for antimicrobial activity (49–51). Domagt (50) identified the bactericidal effect of quaternary ammonium compounds (quats) in 1935. Quats such as cetyl trimethyl-ammonium bromide (cetrimide; CTAB), CPC, and BAC have been used as active antimicrobials in hard-surface disinfectants, in mouthwashes and toothpastes, and as components of the preservative system of cosmetic products (i.e., contact lens wetting solutions, emulsions, conditioners). Harry (51) reported that quats are used as antiseptics at concentrations ranging from 50 to 500 ppm, with higher concentrations being necessary for antiseptic action in creams. Linfield et al. (2) presented data on preservation of cleansing cream and skin lotion that contained stearyl dimethyl amine oxide and other quats. Aqueous solutions of quats may be effective hair conditioners and have antistatic properties. It is possible that some conditioners have self-preserving formulas due to the low pH and the quaternary ammonium compounds used in their formulas.

Advantages of quats are their broad spectrum of antimicrobial activity, activity over the range of pH values found in cosmetic formulations, low toxicity at use concentrations, high aqueous solubility, high surface activity (excellent detergency), continued bacteriostatic action at very low concentrations (5–20 ppm), good uptake, and skin tolerance. In addition, they are colorless and odorless (52,53). Disadvantages of quats include their inactivation by anionics (including soaps and free fatty acids), cotton fibers, and some proteins.

Although nonionic surfactants frequently cause a decrease in the antimicrobial activity of phenolic preservatives and formaldehyde donors (32), the antimicrobial action of some quats may be increased in the presence of nonionic surfactants (54). Table 4.5 shows the influence of anionic and nonionic surfactants

TABLE 4.5 Minimum Bactericidal Concentration (in µg/ml) for Two Antimicrobial Agents in the Presence of Various Detergents[a]

Antimicrobial agents	Sec. alkane sulfonate (anionic)	Ether sulfate (anionic)	Nonylphenol ethoxylate (nonionic)	Fatty alcohol ethoxylate (nonionic)	Cocamine oxide (nonionic; acid cat.)
Benzalkonium chloride	1000	2000	60	500	500
Didecyldimonium chloride	1000	2000	60	60	1000

[a]Test conditions: *Escherichia coli*, 10^7 organisms/ml; contact time, 1 hr; pH 6.
Source: Ref. 54.

the activity of BAC and didecyldimonium chloride. Although the anionic surfactants interfered with bactericidal action against *E. coli* in this system (i.e., MICs were 1000–2000 ppm), it is seen that the antibacterial action of both quats was good in the presence of nonylphenol ethoxylate (i.e., MICs were 60 ppm).

Cocodimethyl benzyl ammonium chloride (Noramium DA50) is marketed for its microbicidal activity against a broad range of pathogenic bacteria, fungi, and viruses (50). It was reported (55) that "hard" quats (i.e., those with a very short alkyl chain) were used widely for cleaning and disinfection of surfaces, but some of these compounds have been reported to be toxic. Quats with longer alkyl chains were believed to be less toxic than quats with shorter alkyl chains because they are absorbed less readily.

This led to the development of "soft" surfactants which, subsequent to their desired effect, hydrolyzed to release nontoxic/non–surface active by-products. Bodor (56) developed quats that had substantial antibacterial action in nature, yet remained relatively nontoxic due to the ability of such agents to degrade into nontoxic by-products subsequent to exerting their antibacterial effects. These soft quats were surface active agents with antimicrobial activity, and described generically as follows:

R O
| | ||
$-N^{+}$-CH-O-C-$R_1 \cdot X^-$
|

R O
| | ||
$=N^{+}$-CH-O-C-$R_1 \cdot X^-$

|
N^-
where | represents a tertiary open-chain or cyclic aliphatic amine and |
=N represents an aromatic amine. R represents a member selected from the group consisting of a hydrogen atom, a C_1-C_{20} open-chain or cycloalkyl group, a C_1-C_{20} alkoxyalkyl group, a C_1- C_{20} acyloxyalkyl group, a C_1-C_{20} haloalkyl group, a C_1-C_{20} carboxyalkyl group, an aryl group, and a substituted aryl group, whose substituents are selected from the group consisting of a halogen atom, an O-C_{1-4} alkyl group, an O-C_{1-8} acyl group, a nitro group, a carboxyl group, and a carboethoxy group. R_1 represents various alkyl groups.

According to Green et al. (57), certain polymeric quats, especially the product made by the condensation of 1,4-dihalo-2-butene and 1,4-bis-dimethylamino-2-butene, are highly effective microbicidal agents in aqueous systems, including cosmetics and other emulsions containing nonionic emulsifiers, and do

TABLE 4.6 Antibacterial Activity of Two Polymeric Quaternary Ammonium Compounds Against *Enterobacter aerogenes*

		Number of bacteria surviving after:		
	Concentration (ppm)	30 min	60 min	180 min
Product A	20	5.900	1.400	<100
	50	3.300	450	<100
	100	1.250	350	<100
Product B	20	>300.000	>300.000	94.000
	50	>300.000	166.000	28.100
	100	>300.000	87.500	14.350

Source: Ref. 57.

not cause undue foaming. The quaternary ammonium moieties of these compounds are part of a long chain, rather than being on branches that are bonded to the long chain. These quats are reported to be effective at levels as low as about 500 ppm; however, the preferred range is about 1000–2000 ppm. Table 4.6 compares the antibacterial activity of Product A (the reaction product of 1,4-bis-dimethylamino-2-butene and 1,4-dichloro-2-butene) with Product B [the reaction product of bis(2-chloroethyl)ether and tetramethyl ethylenediamine] against *Enterobacter aerogenes.*

Suling and O'Leary (58) found that lauryl/myristyldimethylammonium acetate or cetyltrimethylammonium chloride potentiated the activity of chlorotetracycline on antibiotic-resistant strains of *E. coli*, *Proteus mirabilis*, and *Klebsiella pneumoniae.* These quats appeared to increase the uptake of tetracycline, based on inhibition of protein synthesis observed with *E. coli.* Both the betaine and the cationic surfactant proved to be effective with penicillin G on strains of staphylococci. It was suggested that these compounds inhibit formation of penicillinase. Although they were unable to demonstrate potentiation of antimicrobial activity of streptomycin sulfate with these quats, their work suggests application for increasing the effectiveness of some classes of antimicrobial compounds.

Several quats used as components of product preservative systems or as disinfectants are listed in Table 4.7. Baydar and Johnston (59) investigated the synthesis and the antimicrobial properties of quaternary ammonium salts of triglycerides and reported on two compounds:

1. Propane 1,2,3-triyltris (carbonyloxy(heptadec-8-1,11-diyl)oxycarbonyl-methylene-*N*-*N*-dimethyl-*N*-octadecylammonium chloride)
2. Propane 1,2,3-triyltris (carbonyloxy(heptadec-8-1,11-diyl)oxycarbonyl-methylene-*N*-*N*-dimethyl-*N*-dodecylammonium chloride)

TABLE 4.7 Brief Monographs for the Antimicrobial Quaternary Ammonium Salts

			Ref.
1.	Chemical name	Lauryl trimethyl ammonium chloride	54
	Structural formula	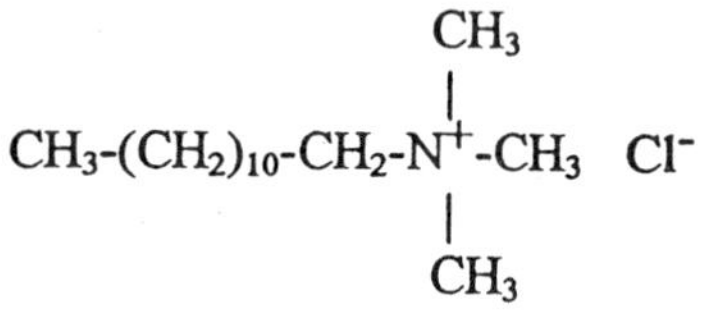	54
	Nonproprietary or registered names	Laurtrimonium chloride	54
	CAS number	112-00-5	54
	Applications		
	Antimicrobial activity	Alkyl as dodecyl	78
2.	Chemical name	Cetyl trimethyl ammonium chloride	54
	Structural formula	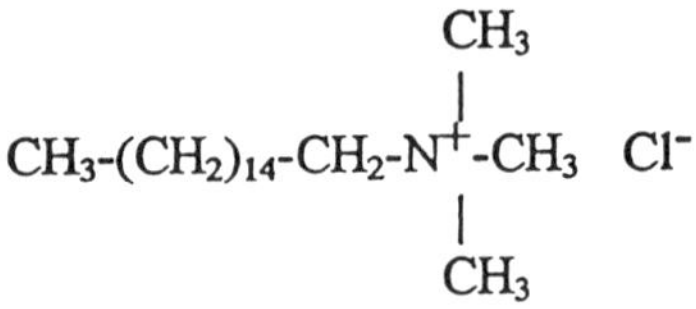	54
	Nonproprietary or registered names	Cetrimonium chloride	54
	CAS number	112-02-7	54
	Applications	Oral hygiene	54
	Antimicrobial activity	Alkyl as hexadecyl	78
3.	Chemical name	Stearyl trimethyl ammonium chloride	54
	Structural formula	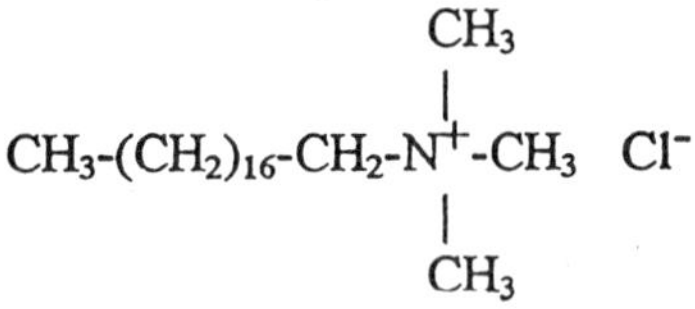	54
	Nonproprietary or registered names	Steartrimonium chloride	54
	CAS number	112-03-8	54
	Applications	Hair products; disinfectants	130
	Antimicrobial activity		130
4.	Chemical name	Cetyl trimethyl ammonium bromide	54

	Structural formula	$\begin{array}{c} CH_3 \\ \vert \\ CH_3\text{-}(CH_2)_{14}\text{-}CH_2\text{-}N^{+}\text{-}CH_3 \quad Br^- \\ \vert \\ CH_3 \end{array}$	54
	Nonproprietary registered names	Cetrimonium bromide	54
	CAS number	57-09-0	54
	Applications	Oral hygiene	54
	Antimicrobial activity		14
5.	Chemical name	Myristyl dimethyl benzyl ammonium chloride	54
	Structural formula	$\begin{array}{c} CH_3 \\ \vert \\ CH_3\text{-}(CH_2)_{13}\text{-}N^{+}\text{-}CH_2\text{-}C_6H_5 \quad Cl^- \\ \vert \\ CH_3 \end{array}$	54
	Nonproprietary or registered names	Myristalkonium chloride	54
	CAS number	139-08-2	54
	Applications		
	Antimicrobial activity		
6.	Chemical name	Dodecyl dimethyl dichlorobenzyl ammonium chloride	121
	Structural formula	$\begin{array}{c} CH_3 \\ \vert \\ CH_3\text{-}(CH_2)_{11}\text{-}N^{+}\text{-}CH_2\text{-}C_6H_5Cl_2 \quad Cl^- \\ \vert \\ CH_3 \end{array}$	121
	Nonproprietary or registered names	Preventol RB	121
	CAS number	102-30-7	122
	Applications	Disinfectant addition to cosmetic products	121
	Antimicrobial activity		121
7.	Chemical name	Alkyl dimethyl benzyl ammonium chloride	54

TABLE 4.7 Continued

			Ref.
	Structural formula	$R\text{-}N^{+}(CH_3)_2\text{-}CH_2\text{-}C_6H_5\ Cl^-$ $\quad R = C_{12}H_{25}, C_{14}H_{29}, C_{16}H_{33}$	54
	Nonproprietary or registered names	Benzalkonium chloride	54
	CAS number	8001-54-5	54
	Applications	Oral hygiene, eye preparations	54, 14
	Antimicrobial activity		47, 14, 123
8.	Chemical name	Diisobutyl phenoxy ethoxy ethyl dimethyl benzyl ammonium chloride	54
	Structural formula	$CH_3\text{-}C(CH_3)_2\text{-}CH_2\text{-}C(CH_3)_2\text{-}C_6H_4\text{-}O\text{-}CH_2\text{-}CH_2\text{-}O\text{-}CH_2\text{-}CH_2\text{-}N^{+}(CH_3)_2\text{-}CH_2\text{-}C_6H_5\ Cl^-$	54
	Nonproprietary or registered names	Benzethonium chloride	54
	CAS number	121-54-0	54
	Applications	Oral hygiene	54
	Antimicrobial activity		47, 14
9.	Chemical name	1-Hexadecyl pyridinium chloride	54
	Structural formula	$C_5H_5N^{+}\text{-}CH_2\text{-}(CH_2)_{14}\text{-}CH_3\ Cl^-$	54
	Nonproprietary or registered names	Cetylpyridinium chloride	54
	CAS number	123-03-5	54
	Applications	Oral hygiene	54
	Antimicrobial activity		47
10.	Chemical name	Didecyl dimethyl ammonium chloride	54
	Structural formula	$CH_3\text{-}(CH_2)_9\text{-}N^{+}(CH_3)_2\text{-}(CH_2)_9\text{-}CH_3\ Cl^-$	54

	Nonproprietary or registered names	Didecyldimonium chloride	54
	CAS number	7173-51-5	54
	Applications	Antimicrobial scrub	54
	Antimicrobial activity		54, 123, 124
11.	Chemical name	Phenoxyethyl dimethyl dodecyl ammonium bromide	
	Structural formula	$C_6H_5\text{-}O\text{-}CH_2CH_2\text{-}\underset{CH_3}{\overset{CH_3}{N^+}}\text{-}(CH_2)_{11}\text{-}CH_3\quad Br^-$	54
	Nonproprietary or registered names	Domiphen bromide	54
	CAS number	538-71-6	54
	Applications	Oral hygiene	54
	Antimicrobial activity		54
12.	Chemical name	Dodecyl dimethyl ethylbenzyl ammonium chloride	125
	Structural formula	$C_{12}H_{25}\text{-}\underset{CH_3}{\overset{CH_3}{N^+}}\text{-}CH_2\text{-}C_6H_4\text{-}C_2H_5\quad Cl^-$	125
	Nonproprietary or registered names	Quaternium-14	125
	CAS number		125
	Applications	Shampoo	
	Antimicrobial activity		
13.	Chemical name	Cis-isomer 1-(3-chloroallyl)-3,5,7-triaza-1-azoniaadamantane chloride	125
	Structural formula	$(CH_2\text{-}N\text{-}CH_2)_3\text{-}N^+\text{-}CH_2\text{-}CH{=}CHCl\ Cl^-$	125
	Nonproprietary or registered names	Quaternium-15	125
	CAS number	4080-31-3	14
	Applications	Shampoo, hair care products, lotions	125, 14
	Antimicrobial activity		14

TABLE **4.7** Continued

			Ref.
14.	Chemical name	Lanolinamidopropyl dimethylethyl ammonium ethosulfate	125
	Structural formula	CH_3 | $R\text{-}CO\text{-}NH\text{-}CH_2\text{-}CH_2\text{-}N^{+}\text{-}C_2H_5\ ^{-}OSO_3C_2H_5$ | CH_3	125
	Nonproprietary or registered names	Quaternium-33	125
	CAS number		
	Applications	Shampoo	125
	Antimicrobial activity		
15.	Chemical name	Alkyl dimethyl benzyl ammonium saccharinate	131
	Structural formula	CH_3 | $R\text{-}N^{+}\text{-}CH_2\text{-}C_6H_5\ ^{-}NCOSO_2C_6H_4$ | CH_3	
	Nonproprietary or registered names	Steinaquat QA 100; Onyxide 3300	131, 126
	CAS number		
	Applications	Deodorizing soaps; hair preparations	131, 126
	Antimicrobial activity		
16.	Chemical name	N (lauryl colaminoformylmethyl) pyridinium chloride	127
	Structural formula		127
		O ‖ O ‖ $CH_3(CH_2)_{10}\text{-}C\text{-}O\text{-}CH_2CH_2\text{-}NH\text{-}C\text{-}CH_2\text{-}{}^{+}NC_5H_5\ \ Cl^{-}$	
	Nonproprietary or registered names	Emcol E 607	127
	CAS number		
	Applications	Creams, deodorants	127
	Antimicrobial activity		127

whose structural formulas are as follows:

$$
\begin{array}{l}
\qquad\qquad\qquad\qquad\qquad OCOCH_2{}^{+}NCH_3CH_3R' \ \ Cl^{-} \\
\qquad\qquad\qquad\qquad\qquad | \\
|\text{-}OCO(CH_2)_7CH{=}CHCH_2CH(CH_2)_5CH_3 \\
| \\
| \\
| \qquad\qquad\qquad\qquad\quad OCOCH_2{}^{+}NCH_3CH_3R' \ \ Cl^{-} \\
| \qquad\qquad\qquad\qquad\quad | \\
|\text{-}OCO(CH_2)_7CH{=}CHCH_2CH(CH_2)_5CH_3 \\
| \\
| \\
| \\
|\text{-}OCO(CH_2)_7CH{=}CHCH_2CH(CH_2)_5CH3 \\
\qquad\qquad\qquad\qquad\qquad | \\
\qquad\qquad\qquad\qquad\qquad OCOCH_2{}^{+}NCH_3CH_3R' \ \ Cl^{-}
\end{array}
$$

where R′ for compound 1 is $(CH_2)_{17}CH_3$ and R′ for compound 2 is $(CH_2)_{11}CH_3$.

The presence of the quaternary group in the triglyceride molecule was found to modify its properties, making it comparable with a variety of solvents that enable it to be used in cosmetic formulations. Baydar and Johnston compared the minimum lethal concentration (MLC) for the C-12 and C-18 triglyceride quaternaries with BAC and found that the C-12 quat had comparable antibacterial activity with BAC against *S. aureus*, somewhat lower activity against *C. albicans* and *P. aeruginosa*, and slightly higher activity against *Aspergillus wentii* (Table 4.8).

Schlossman and McCarthy (60) described lanolin quats and prepared two products:

$$
\begin{array}{l}
\qquad\quad O \qquad\qquad\qquad\qquad\quad C_2H_5 \\
\qquad\quad \| \qquad\qquad\qquad\qquad\quad\ | \\
LAN\text{-}C\text{-}NH\text{-}CH_2CH_2CH_2\text{-}N^{+}\text{-}R \ \ X^{-} \\
\qquad\qquad\qquad\qquad\qquad\qquad\ | \\
\qquad\qquad\qquad\qquad\qquad\quad C_2H_5
\end{array}
$$

one of which has $R = CH_2\text{-} C_6H_5$ and $X = Cl$, the other $R = C_2H_5$ and $X = SO_4C_2H_5$.

These products were active against both gram-positive and gram-negative bacteria, they were compatible with anionic surfactants, and they possessed low toxicity when compared to stearalkonium chloride. Schlossman and McCarthy noted that compounds such as these could be used in a wide range of cosmetics. Quaternium-33 (and) Ethyl Hexanediol (Lanoquat 1756; Emery Industries) (61) is a lanolin derivative similar to the products above, but having the quaternary nitrogen represented as follows:

$$
\begin{array}{l}
\ CH_3 \\
\quad | \\
\text{-}N^{+}\text{-}C_2H_5 \\
\quad | \\
\ CH_3
\end{array}
$$

Table 4.8 Minimum Lethal Concentration (MLC) Values

Challenge organism	Quat	MLC ppm Mq^{-1}	Benzalkonium chloride
S. aureus	C-12	60	50
S. aureus	C-18	500	
C. albicans	C-12	250	100
C. albicans	C-18	2000	
P. aeruginosa	C-12	2000	900
P. aeruginosa	C-18	—	
P. rettgeri	C-12	2000	900
P. rettgeri	C-18	—	
A. wentii	C-12	250	400
A. wentii	C-18	—	

Source: Ref. 59.

Zone of inhibition tests indicated no decrease in the bactericidal activity using low pH shampoos containing 1.5% and 3.0% active solutions of quaternium-33 (and) ethyl hexanediol, with and without 3.2% active TEA lauryl sulfate. Their tests indicated that 1% active quaternary solution of the product was more than twice as effective against *S. aureus* as 1% active solutions of stearalkonium chloride or quaternium-7.

4.3.2 Aminimides

The aminimides are tertiary amides joined to a quaternary ammonium nitrogen, $R\text{-}CO\text{-}N^{-}\text{-}N^{+}\text{-}\ R_1R_2R_3$, where R-CO is an aliphatic acyl radical (i.e., lauroyl, myristoyl, palmitoyl, or stearoyl), and R_1, R_2, and R_3 are small organic groups, such as methyl or hydroxypropyl. In these compounds, the quaternary nitrogen is so strongly cationic that the amido nitrogen is rendered weakly anionic (54). Kabara et al. (62) studied the antimicrobial activity of acyl aminimides with R derived from C_{12} through $C_{18:1}$ fatty acids, in which R_1 and $R_3 = CH_3$, and $R_2 = CH_3$, CH_2CH_2OH, or $CH_2CHOHCH_3$. Kabara and coworkers concluded that these bipolar compounds have antimicrobial activity against both gram-positive bacteria and yeasts and that their activity against gram-negative bacteria is low or absent. The acyl derivatives tested showed maximum activity at chain lengths of C_{14} and C_{16}, and unsaturation was an important factor contributing to antimicrobial activity.

Kabara and Haitsma (63) screened the in vitro antimicrobial activity of amine acylimide aminimides in which the acyl portion of the imide was kept constant as either an acetimide or a methacrylimide group and the chain length of the tertiary amine was a variable. They found that among the acetimide amines, the acetimide trimethylamine, which is the simplest derivative, exhibited no activity. Antimicrobial activity was observed when the chain length of the alkyl portion was increased from C_{12} to C_{18}, with maximum activity observed with a C_{16} chain length (63,64).

4.4 AMPHOTERIC SURFACTANTS

Surfactants are classified as ampholytic when they have two or more functional groups which, depending on the conditions of the medium, can be ionized in aqueous solution and give the compound the characteristics of either an anionic or a cationic surfactant. Surfactants are amphoteric only if the charge on the hydrophobe changes as a function of pH. These surfactants carry a negative charge in basic media and a positive charge in acidic media. At intermediate pH values, the surfactant is a "zwitterion" because it contains both negatively and positively charged groups. The two main types of amphoteric surfactants are the cyclic imidazolines and substituted betaines (39).

4.4.1 Alkyl Betaines

The betaines have not shown the general antimicrobial effectiveness as other surface-active quats; however, Kenney (65) proposed a variety of surface-active betaines and amine oxides for controlling body odor. The compositions consisted essentially of aqueous or nonaqueous mixtures of an acyl-*N*-betaine and an alkyl-*N*,*N*- dimethylamine oxide, an alkyl-*N*,*N*-dihydroxyethylamine oxide, or an acylamino t-amine oxide, preferably adjusted to a pH of 4.7–5.2 because microorganisms responsible for underarm odor are readily inhibited at pH 5.5 or less.

It is possible that there is a synergistic effect between betaines and other products. According to Moore (66), mixtures of long-chain betaines and quats had increased activity against gram-negative organisms; however, there was a loss of activity against gram-positive bacteria. Moore also reported a considerable increase in the activity of phenols when solubilized with surfactant betaines. Michaels (67) reported a disinfecting shampoo containing lauryl betaine together with lauramine oxide. This illustrates the potential application of amphoteric surfactants for enhancing antimicrobial activity in cosmetic formulations.

4.4.2 Alkyl Amidopropylbetaines

Alkyl amidopropyl dimethylaminoacetic acid betaine (TEGO Betain L 7; Goldschmidt) is an alkyl amidopropyl betaine that has fungistatic and bacteriostatic properties (68). The formulas, antimicrobial activities, and product applications of betaines are presented in Table 4.9 (69). Although betaines may not be sufficient for preservation of formulas, combination with other compounds with antimicrobial properties may result in a synergism that increases the efficacy of the preservative system.

Table 4.9 Characteristics of Some Surfactants

Chemical name	Formula	Commercial name	Activity against	Use	Ref.
Alkyl amidopropyl dimethylaminoacetic acid betaine	$C_{11-17}H_{23-35}$-CONH-$(CH_2)_3$-$N^+(CH_3)_2$-CH_2COO^-	TEGO-Betain L 7	Bacteria Fungi	Shampoo Shower and bath preparation Liquid soap Intimate hygiene solution	68, 69
Alkyl diethylenediamino glicine Di(alkyl diaminoethyl) glicine	R-NH-C_2H_4-NH-C_2H_4-NH-CH_2-COOH (R-NH-C_2H_4-NH-$C_2H_4)_2$-N-CH_2-COOH	TEGO 103 G	Bacteria Fungi Bacilli TBC Virus	Antisepsis of human skin Antisepsis of hands	128
Dodecyl aminoethyl glicine	$C_{12}H_{25}$-NH-C_2H_4-NH-CH_2-COOH	TEGO 103 S	Bacteria Fungi Yeast	Antisepsis of hands	128

4.4.3 Alkyl Aminopropionates and Alkyl Iminodipropionates

Aminopropionates and alkyl iminodipropionates are mild yet effective detergents used in shampoos and germicidal systems (70). Several of the aminopropionates and alkyl iminodiproprionates sold under the trade name Deriphat are shown with their structural formulas and physical forms in Table 4.10. These compounds are claimed to possess bactericidal properties (71–77). Hueck et al. (78) examined the antibacterial and antifungal activities of several of these compounds. Although the compounds evaluated did not appear to be effective against the bacteria tested, they were about as effective as BAC and formaldehyde against fungi (Table 4.11).

Dudzinski and Wakeman (79) reported powerful antimicrobial action of amphoteric surfactants containing a tertiary amino group attached to the 2-carbon of a long-chain alkyl group and a carboxyethyl group substituent on the nitrogen atom, with the following general structure:

$$
\begin{array}{cccccc}
 & & R' & & O & \\
 & & | & & \| & \\
R- & CH- & N- & CH_2CH_2 & C- & OM \\
 & | & & & & \\
 & CH_3 & & & &
\end{array}
$$

where R is a straight-chain alkyl having from 6 to 16 carbon atoms, R′ is a lower alkyl (i.e., methyl or ethyl), and M is either hydrogen, an alkali metal, or an alkylamine group (i.e., mono-, di-, or triethanol or propanol amine). These workers reported MIC values of 10 ppm for *S. aureus* and *Streptococcus faecalis*.

4.4.4 Alkyl Imidazolines

Imidazoline derivatives are multifunctional ingredients because they possess both surfactant and antimicrobial properties (80–82). Schmitz and Harris (81) outlined the evolution of ampholytic surfactants, discussed the relation of surface activity to disinfection, and noted the increased bactericidal activity of ampholytic surfactants resulting from an increase in the number of N groups in the molecules. Molnar (3) studied the influence of various surfactants on the activity of bromsalan and other ring-halogenated compounds. He found that both carboxyl and sulfontate surfactants and imidazoline derivatives were effective in increasing the antimicrobial activity of ring-halogenated compounds.

4.4.5 Amino Acid Derivatives

N-Acylamine acid derivatives exhibit surfactant activity, antibacterial activity, and mildness to the skin (83). Fieser et al. (84) synthesized *N*-acylamino acids for use as fat-emulsifying agents. Kameda et al. (85) reported on the antimicrobial action

TABLE 4.10 Commercially Available Propionate Derivatives

Product	Chemical structure	Formula	Physical form	Typical pH
Deriphat 151	Sodium-*N*-coco-β-aminopropionate	$RNHCH_2CH_2COONa$	Flake 97% active	12
Deriphat 151C	*N*-coco-β-Aminopropionic acid	$RNHCH_2CH_2COOH$	42% solution in water	7
Deriphat 170C	*N*-Lauryl/myristil-β-amino-propionic acid	$RNHCH_2CH_2COOH$	50% solution in water	6
Deriphat 154	Disodium-*N*-tallow-β-imino-dipropionate	$RN(CH_2CH_2COONa)_2$	Flake 97% active	12
Deriphat 160	Disodium-*N*-lauryl-β-imino-dipropionate	$RN(CH_2CH_2COONa)_2$	Flake 97% active	12
Deriphat 160C	Partial sodium salt of *N*-lauryl-β-iminodipropionic acid	$RN(CH_2CH_2COOH)(CH_2CH_2COONa)$	30% solution in water	7

Source: Ref. 129.

Table 4.11 Inhibiting Concentrations (in ppm) of Deriphats for Some Bacteria and Fungi in Comparison with Those of Benzalkonium Chloride and Formaldehyde

	Bacteria							
	Gram-negative		Gram-positive		Fungi			
Compound	*Escherichia coli*	*Pseudomonas fluorescens*	*Bacillus subtilis*	*Staphylococcus aureus*	*Aspergillus niger*	*Chaetomium globosum*	*Myrothecium verrucaria*	*Trichoderma viride*
Benzalkonium chloride	200	300	3	4	60	10	40	80
Formaldehyde	30	30	30	30	100	30	100	1000
Deriphat 151	10000	3000	30	100	100	60	100	100
Deriphat 151C	>4500	>4500	45	100	140	15	140	140
Deriphat 170C	1500	500	50	150	150	50	100	150
Deriphat 154	>10000	10000	100	>10000	100	30	100	100
Deriphat 160	10000	3000	300	10000	300	100	300	300
Deriphat 160C	>3000	900	300	900	300	90	300	300

Source: Ref. 78.

of *N*-lauroyl-DL-phenylalanine. Fosdick et al. (86) found that sodium *N*-acyl sarcosinate had strong anticaries action in addition to its known antimicrobial properties. Infante et al. (87) investigated the physiochemical and antimicrobial properties of N-acyl arginine esters of common fatty acids.

Long-chain N^{α}-acyl diaminoacid derivatives are known to have antimicrobial activity (88,89). Particularly, Yoshida et al. (89) prepared N^{α}-acyl(C_{10-16})-L-arginine ester salts and compared their surface and antimicrobial properties with those of benzyldimethyl[2-[2-(p-1,1,3,3-tetramethylbutylphenoxy)ethoxy]-ethyl] ammonium chloride (I). Among the salts, the lauroyl derivatives showed the best surface and antimicrobial activities and were comparable with I. Infante et al. (90,91) found that some structural modifications of long-chain N^{α}-acyl-L-arginine and N^{α}-acyl-L-lysine homologs affected the antimicrobial properties of these compounds, although only those molecules with blocked carboxyl groups (i.e., cationic molecules) showed antimicrobial activity.

Infante et al. (92) prepared N^{α}-acylamino acids that contained a tetralkylammonium moiety (i.e., cationic). The sodium salt of these tetralkylammonium derivatives can be considered as zwitterionic, rather than ampholytic compounds (66). The antibacterial activity of these compounds is shown in Table 4.12. Note that N^{α}-lauroyl-L-lysine ethyl ester hydrochloride (LLECI) had MICs of ≤ 128 μg/ml for all test organisms and that N^{α}-lauroyl-L-arginine methyl ester hydrochloride (LAMCI) and N^{α}-lauroyl-L-arginine ethyl ester hydrochloride (LAECI) had MICs of ≤ 128 μg/ml for all test organisms except *Serratia marcescens.* Infante and coworkers concluded that (1) the activity was inhibited when the carboxyl group of the amino acid is unprotected, (2) the amphoteric derivatives were inactive below 128 μg/ml, (3) all the cationic derivatives studied showed antimicrobial activity, (4) the effectiveness of the cationic derivatives decreases when they contain the tetralkylammonium group instead of the amine or guanidine groups, (5) the iodide anion does not deactivate the molecule, and (6) there is no direct relationship between the cationic character of the molecule and the increment of antimicrobial activity.

Mhaskar et al. (93) investigated the effect of structural variation in the fatty acid chain on the antibacterial activity of *N*-acyl-L-leucines. Their studies revealed that, in general, the acid form of the *N*-acyl leucines exhibited greater activity than the methyl ester form and that the compounds tested had greater activity against gram-positive bacteria than against gram-negative bacteria. The stearoloyl- and dihydrosterculoyl- acyl moieties had MICs of 0.06 and 0.04 μg/ml against *S. aureus*, respectively, but they were markedly less effective against *E. coli* (Table 4.13). Mhaskar and coworkers believed that the presence of a cyclopropane group, a hydroxy group, or unsaturation in the acyl chain increased the antibacterial action of the *N*-acyl leucines studied. The activity of the dihydrosterculoyl moiety was believed to highlight the potential antibacterial nature of the cyclopropane ring. They also noted that shifting the hydroxyl group toward the amide linkage resulted in diminished antibacterial activity. This work is significant in that it demonstrates

TABLE 4.12 MIC Values (μg/ml) of the N^{α}-Lauroyl Diamino Acids Synthesized

	GRAM +											GRAM –							
COMP	C.al	S.a.	S.e.	S.f.	C.ag	S.1.	B.s.	B.m.	B.c.	M.1.	M.a.	E.c.	K.p.	C.f.	S.m.	B.b.	P.a.	S.d.	A.c.
LLMCl	32	32	16	32	4	8	16	16	64	4	16	16	32	32	128	16	64	32	32
LLECl	8	8	4	8	2	4	8	8	16	8	16	8	16	64	32	64	128	128	8
LLSCl	>128	>128	>128	>128	>128	>128	>128	>128	>128	>128	>128	>128	>128	>128	>128	>128	>128	>128	>128
LLMqI	—	128	32	64	32	64	64	64	64	32	64	>128	64	>128	>128	>128	—	>128	64
LLEqI	128	128	32	128	>128	64	64	64	—	32	64	>128	>128	>128	>128	—	—	>128	32
LLEqCl	—	128	32	64	32	32	64	64	64	>128	32	64	>128	>128	>128	—	>128	128	32
LLSqI	>128	>128	>128	>128	>128	>128	>128	>128	>128	>128	>128	>128	>128	>128	>128	>128	>128	>128	>128
LAMCl	8	8	4	16	2	8	8	16	16	8	8	16	16	32	>128	16	64	32	16
LAECl	16	8	2	8	2	8	8	8	16	4	4	16	16	32	128	16	64	32	8
LASCl	>128	>128	>128	>128	>128	>128	>128	>128	>128	>128	>128	>128	>128	>128	>128	>128	>128	>128	>128

COMP.: Compounds; LLMCl; N^{α}-lauroyl-L-lysine methyl ester hydrochloride; LLECl: N^{α}-lauroyl-L-lysine ethyl ester hydrochloride; LLSCl: sodium salt of N^{α}-lauroyl-L-lysine hydrochloride; LLMqI: $N^{\varepsilon}N^{\varepsilon}N^{\varepsilon}$-trimethyl N^{α}-lauroyl-L-lysine methyl ester iodide salt; LLEqI: $N^{\varepsilon}N^{\varepsilon}N^{\varepsilon}$-trimethyl N^{α}-lauroyl-L-lysine ethyl ester iodide salt; LLEqCl: $N^{\varepsilon}N^{\varepsilon}N^{\varepsilon}$-trimethyl N^{α}-lauroyl-L-lysine ethyl ester chloride salt; LLSqI: sodium salt of $N^{\varepsilon}N^{\varepsilon}N^{\varepsilon}$-trimethyl N^{α}-lauroyl-L-lysine iodide; LAMCl: N^{α}-lauroyl-L-arginine methyl ester hydrochloride; LAECl: N^{α}-lauroyl-L-arginine ethyl ester hydrochloride; LASCl: sodium salt of N^{α}-lauroyl-L-arginine hydrochloride.

C.al.: *Candida albicans*; S.a.: *Staphylococcus aureus*; S.e.: *Staphylococcus epidermis* ;*S.f.: Streptococcus faecalis*; C.ag.: *Corynebacterium agropinii*; S.l.: *Sarcinia lutea*; B.s.: *Bacillus subtilis*; B.m.: *Bacillus megaterium*; B.c.: *Bacillus cereus*; M.l.: *Micrococcus luteus*; M.a.: *Micrococcus auranticus*; E.c.: *Escherichia coli*; K.p.: *Klebsiella pneumoniae;* C.f.: *Citrobacter freundii*; S.m.: *Serratia marcescens*; B.b.: *Bordetella bronchiseptica*; P.a.: *Pseudomonas aeruginosa*; S.d.: *Salmonella derby*; A.c.: *Acinetoobacter calcoaceticus.*

Source: Ref. 92.

Table 4.13 Minimum Inhibitory Concentration (MIC)[a] of *N*-Acyl-L-leucines $(CH_3)_2CHCH_2CH(COOH)NHCOR$

Acyl moiety (R-)	*Staphyloccus aureus* (mg/ml)	*Escherichia coli* (mg/ml)
Azelaoyl	0.75	1.4
Stearoloyl	0.06	1.0
9,10-Epoxystearoyl	0.75	—
Dihydrosterculoyl	0.04	1.0
5-*trans*-Octadecen-7,9-diynoyl	0.27	5.0
6-Hydroxy-7,9-octadecadiynoyl	7.0	10.0
6-(R)-Hydroxy-7,9-octadecadiynoyl	7.0	10.0
Stearoyl	—	—
Oleoyl	12.0	20.0
Ricinoleoyl	1.0	—

[a]MIC was determined based on retention of clarity of the nutrient medium during inhibition of bacterial growth.
Source: Ref. 93.

that several *N*-acylamino acid derivatives have antimicrobial properties. It is likely that these compounds would contribute to the activity of product preservative systems. Other workers have reported on the antimicrobial properties of *N*-(2-hydroxy substituted phenoxy propyl) glycines and *N*,*N*-bis[2-(*N*′-acylamino)ethyl] glycines (94–97).

Several protein-based surfactants have been derived from protein hydrolysates (i.e., casein, keratin, collagen) and fatty acids. These have been reported to function as amphoteric surfactants that are nonsensitizing and nonirritating (98–101). In order to obtain protein-based amphoteric surfactants with antimicrobial properties, Molinero et al. (102) synthesized long-chain N^{α}-acyl-L-arginine dipeptides by condensation of N^{α}-lauroyl arginine and amino acids from a collagen hydrolysate. They prepared two types of condensation products, which had the following general structures:

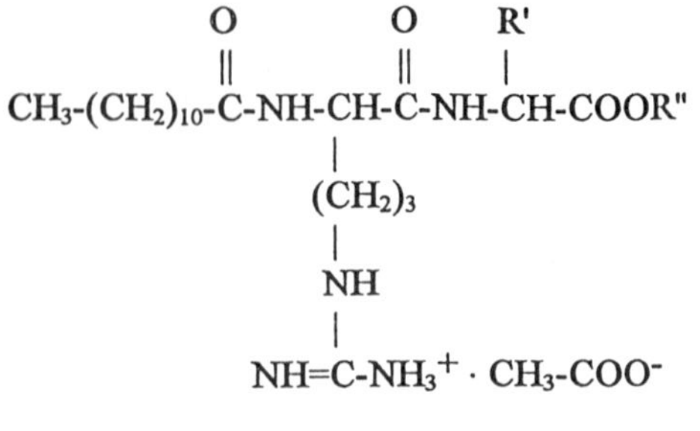

where R′ = residue of amino acid coming from collagen hydrolysate, R″ = H (condensate I), and R″ = CH_3 (condensate II).

Although these compounds had less antimicrobial activity than the methyl or ethyl ester of N^{α}- lauroyl arginine, all compounds described here may be considered as mild preservatives. In particular, condensate II was much more active against bacteria than condensate I. Molinero and coworkers concluded that this was related to the higher cationic character of the dipeptides in this condensate.

Infante et al. (103) prepared N^{α}- lauroyl-arginine dipeptides that contained neutral amino acids (glycine and phenylalanine) using L-arginine and pure amino acids (instead of amino acid mixtures from collagen). They compared the antibacterial properties of these compounds as methyl ester derivatives (cationic amphiphiles) and as free acid derivatives (amphoteric amphiphiles) to the N^{α}-lauroyl-arginine methyl ester derivative (cationic monomer) and to the N^{α}-lauroyl-arginine derivative (amphoteric monomer) and obtained the MICs shown in Table 4.14. It was found that MICs of amphiphilic dipeptides with protected carboxyl groups (i.e., cationic surfactants) were very similar to those of the amphoteric ones in which the α-carboxyl group was free, so that all derivatives were substantially active. They also found that the antimicrobial activities of the pure amphoteric dipeptide derivatives were higher than those of amphoteric dipeptide derivative mixtures obtained from collagen.

Infante and coworkers believed that the antimicrobial activity was due to the *N*-lauroyl-arginine residue, which is the common structural component in all amphiphiles under study. They thought that the amphoteric dipeptide derivatives had antimicrobial activity because of the absence of intramolecular ionic interactions leaving the guanidine group free, thereby increasing water solubility and enabling the compounds to perform as surfactants. They believed that the strong basic guanidine group could interact with polyanionic components of the bacterial cell surface, which would allow the surfactant to pass through the membrane and interfere with microbial metabolism.

Infante et al. (104) extended this work by synthesizing N^{α}-lauroyl-arginyl dipeptides that contained acidic and basic amino acids (i.e., glutamic acid and lysine) and by comparing the antibacterial activity of these compounds [as methyl ester (cationic surfactants) and free acid (amphoteric surfactants) hydrochloride salts] to the N^{α}-lauroyl-arginine methyl ester (i.e., cationic monomer) derivative and to the N^{α}-lauroyl-arginine (amphoteric monomer) derivative. The results of these studies are shown in Table 4.15. Review of the data in this table reveals that the cationic N^{α}-lauroyl arginyl dipeptides of lysine had comparable antibacterial activity against gram-positive bacteria with N^{α}-lauroyl-arginine methyl ester. While the latter compound had activity against several gram-negative bacteria, the former had little activity against the gram-negative bacteria.

Other workers have reported on phosphate-containing amphoteric surfactants (105–108). These compounds in general have been found to be less active

TABLE 4.14 Minimum Inhibitory Concentration (μg/ml) of N^{α}-Lauroyl-arginine Peptides

		Cationic derivatives			Amphoteric derivatives		
	Compounds	N^{α}-Lauroyl-arginine methyl ester	N^{α}-Lauroyl-arginyl-glycine methyl ester	N^{α}-Lauroyl-arginyl-phenylalanine methyl ester	N^{α}-Lauroyl-arginine[a]	N^{α}-Lauroyl-arginyl-glycine	N^{α}-Lauroyl-arginyl-phenylalanine
1	*Candida albicans*	32	32	16	256	32	4
1	*Staphylococcus epidermidis*	8	16	4	64	16	4
1	*Streptococcus feacalis*	16	64	32	256	64	4
1	*Corynebacterium agropyri*	8	8	4	64	8	2
1	*Bacillus subtilis*	8	16	16	128	16	4
1	*Bacillus pumilus*	—	—	—	—	16	4
1	*Micrococcus luteus*	16	32	8	64	16	2
1	*Micrococcus aurantiacus*	16	16	4	128	16	2
2	*Alcaligenes faecalis*	256	256	256	—	128	32
2	*Escherichia coli*	16	32	16	>256	16	4
2	*Klebsiella pneumoniae*	256	>256	256	256	>128	>128
2	*Citrobacter freundii*	256	>256	256	>256	>128	>128
2	*Serratia marcescens*	256	>256	256	>256	>128	>128
2	*Pseudomonas aeruginosa*	128	128	256	>256	>128	>128
2	*Salmonella typhimurium*	32	64	128	—	>128	128
2	*Bordetella bronchiseptica*	16	64	32	—	>128	32

[a]N-lauroyl-L-arginine dipeptides mixture from collagen.
1 = gram-positive bacteria; 2 = gram-negative bacteria.
Source: Ref. 103.

Table 4.15 Minimum Inhibitory Concentration (μg/ml) of Lipopeptidic Surfactants

Compounds	Cationic derivatives			Amphoteric derivatives		
	N^{α}-Lauroyl arginine methyl ester	Methyl ester of N^{α}-lauroyl-L-arginyl-L-glutamic chloride	Methyl ester of N^{α}-lauroyl-L-arginyl-L-lysine dicholoride	N^{α}-Lauroyl arginine[a]	N^{α}-Lauroyl-L-arginyl-L-glutamic chloride	N^{α}-Lauroyl-L-arginyl-L-lysine dichloride
1 *Candida albicans*	32	128	128	>128	>128	>128
1 *Staphylococcus epidermidis*	8	>128	32	>128	>128	>128
1 *Streptococcus feacalis*	16	>128	64	>128	>128	>128
1 *Corynebacterium agropyri*	8	32	32	>128	>128	32
1 *Bacillus subtilis*	8	>128	32	>128	>128	128
1 *Bacillus pumilus*	—	>128	32	—	>128	128
1 *Micrococcus luteus*	16	64	32	>128	>128	128
1 *Micrococcus aurantiacus*	16	128	32	>128	>128	128
2 *Alcaligenes faecalis*	>128	>128	>128	>128	>128	>128
2 *Escherichia coli*	16	128	32	>128	>128	32
2 *Klebsiella pneumoniae*	>128	>128	>128	>128	>128	>128
2 *Citrobacter freundii*	>128	>128	>128	>128	>128	>128
2 *Serratia marcescens*	>128	>128	>128	>128	>128	>128
2 *Pseudomonas aeruginosa*	128	>128	>128	>128	>128	>128
2 *Salmonella typhimurium*	32	>128	>128	>128	>128	>128
2 *Bordetella bronchiseptica*	16	>128	>128	>128	>128	>128

Source: Ref. 104.

than amino acid derivatives. The biomimetic phospholipids have antimicrobial activity and may be used to reduce preservative requirements of cosmetic and drug formulations (see Chapter 6).

Abe (1) prepared the β-alkoxypropionic acids, $RO(CH_2CH_2)_nCH_2CH_2COOH$, with alkoxy groups ranging from C_4H_9O to $C_{18}H_{35}O$. He reported that the β-decyloxypropionic and β-octyloxypropionic acids were more active than the other alkyloxypropionic acid groups. In general, it was found that an increase of the ether moieties increased their antibacterial activities. Oxyethylene groups enhanced the activity against *S. aureus*, but had relatively little effect against *Penicillium* species. Abe and Osanai (109) prepared the octylamine, dodecylamine, octadecylamine, and triethanolamine salts of β-dodecyloxy- and β-tetradecyloxypropionic acids, without or with one or two oxyethylene groups between the alkoxygroup and the propionic acid group, $R(OCH_2CH_2)_nOCH_2CH_2COOH{\cdot}R'NH_2$. They found that the dodecyl moiety was more potent than the tetradecyloxy moiety.

Abe et al. (110) described the preparation of several carboxylic amphoteric surfactants that have antibacterial activity against *S. aureus* and *E. coli*. These workers synthesized substituted glycine–type amphoteric surfactants containing long-chain alkoxy and methylated amino groups and found that all amphoterics studied were more effective against gram-positive cocci than against gram-negative bacilli and fungi (111). They reported that the introduction of an aminoethyl group between the alkoxyethyl and amino radicals of the substituted glycines also increased their antimicrobial activities.

In studying *N*-substituted glycine-, alanine-, and betaine-type amphoteric surfactants containing a long-chain alkylthio group, Abe and Osanai (112) found that *N*-substituted amino acid amphoterics containing an alkylthio group generally had strong antimicrobial properties and excellent surface activities, except for the alanine-type derivatives.

Seppic (113) proposed collagen-derived amphoteric surfactants with the following formula:

$$R\text{-}\overset{\displaystyle O}{\overset{\|}{C}}\text{-}NH\text{-}(CH_2)_3\text{-}\underset{\underset{\displaystyle CH_3}{|}}{\overset{\overset{\displaystyle CH_3}{|}}{N^+}}\text{-}\ [\text{ Polypeptide }]\ \text{-}COO^-,\ NaCl$$

Here, the alkyl amido propyl dimethylamino polypeptide alkyl chain is made up of C_8 to C_{18} fatty acids, and the polypeptide chain contains a collagen hydrolysate with two to five amino acids. These compounds were reported to have good bactericidal activity but were less effective against fungi and spores.

Osanai et al. (114) prepared racemic and optically active amphoteric surfactants, *N*-substituted amino acids, and amino sulfate derivatives to determine the physicochemical relations chiral isomers with their chemical structures. The data presented in Tables 4.16 and 4.17 show the antimicrobial action

Table 4.16 Antibacterial Activities of *N*-Alkylated Amino Acid–Type Amphoterics Sodium *N*-Alkyl-*N*-(1-phenylethyl) Aminoacetates and Sodium 3-[*N*-alkyl-*N*-(1-phenylethyl) amino] Propionates

	Minimum inhibitory concentration (μg/ml)											
	(a)			(b)			(c)			(d)		
Tested microorganism	DL-	D-	L-	DL-	D-	L-	DL-	D-	L-	DL-	D-	L-
S. aureus	100	200	100	10	10	25	100	100	100	25	25	25
B. subtilis	200	200	200	25	25	25	100	200	100	25	25	25
M. lutea	25	25	50	10	25	25	50	50	50	25	25	25
E. coli	100	100	100	25	25	50	200	200	200	>200	>200	>200
S. typhi	>200	>200	>200	>200	>200	>200	>200	>200	>200	>200	>200	>200
P. aeruginosa	>200	>200	>200	>200	>200	>200	>200	>200	>200	>200	>200	>200
C. albicans	>200	>200	>200	50	50	50	200	200	200	>200	50	50
S. cerevisiae	>200	>200	>200	50	5	50	100	100	100	25	25	50
T. interdigitale	100	100	100	5	10	10	50	100	100	25	25	25
M. gypseum	100	100	50	5	10	10	50	100	100	25	25	25
P. chrysogenum	>200	>200	>200	10	10	50	200	200	>200	200	50	50
A. niger	>200	>200	>200	50	50	50	>200	200	>200	50	25	50

Source:Ref. 114.

(a): Sodium *N*-decyl-*N*-(1-phenylethyl) aminoacetate; (b): sodium *N*-dodecyl-*N*-(1-phenylethyl) aminoacetate; (c): sodium 3-[*N*-decyl-*N*-(1-phenylethyl) amino] propionate; (d): sodium 3-[*N*-dodecyl-*N*-(1-phenylethyl)amino] propionate.

D-: *dextro*. It is used to show that the groups at the significant asymmetrical carbon atom are placed at the right.

L-: *levo*. It is used to show that the groups at the significant asymmetrical carbon atom are placed at the left.

DL-: compensation as contrasted with *meso*-.

TABLE 4.17 Antimicrobial Activities of *N*-Dodecylated Aminosulfonic Acid–Type Amphoterics Sodium 2-[*N*-alkyl-*N*-(1-phenylethyl)amino]-1-ethanesulfonates and Sodium 3-[*N*-alkyl-*N*-(1-phenylethyl) amino]-1-propanesulfonates

	Minimum inhibitory concentration (μg/ml)					
	(e)			(f)		
Tested Microorganisms	DL-	D-	L-	DL-	D-	L-
S. aureus	>200	>200	>200	25	50	25
B. subtilis	100	200	100	50	50	25
M. lutea	>200	200	>200	25	50	25
E. coli	>200	>200	>200	>200	>200	>200
S. typhi	>200	>200	>200	>200	>200	>200
P. aeruginosa	>200	>200	>200	>200	>200	>200
C. albicans	50	100	100	50	>200	50
S. cerevisiae	25	25	25	200	>200	>200
T. interdigitale	25	25	25	50	50	50
M. gypseum	50	50	50	50	50	50
P. chrysogenum	50	50	50	100	>200	100
A. niger	>200	>200	>200	>200	>200	>200

Source: Ref. 114.
(e): Sodium 2-[N-dodecyl-N-(1-phenylethyl) amino]-1-ethanesulfonate; (f): sodium 3-[N-dodecyl-N-(1-phenylethyl) amino]-1-propanesulfonate.
D-: *dextro*. It is used to show that the groups at the significant asymmetrical carbon atom are placed at the right.
L-: *levo*. It is used to show that the groups at the significant asymmetrical carbon atom are placed at the left.
DL-: compensation as contrasted with *meso*-.

against specific bacteria and fungi. Their work indicated that the hydrophobic long alkyl group significantly influenced antimicrobial activity more than the chiral isomerism, and *N*-dodecylated homologs had more antimicrobial activity than *N*-decylated homologs.

4.5 ACTION OF SURFACTANTS ON MICROBIAL PHYSIOLOGY

Lawrence (47) used bacterial respiration as an index of quaternary action on the metabolism of microorganisms. Several other workers have used respiration to monitor the effect of quats on glycolysis and respiration (115–118). Hotchkiss (119) noted that bacteria treated with a cationic surfactant and certain other surface-active agents would release nitrogen and phosphorus compounds into the surrounding medium. Salton (120) also reported the "leakage" and release of cellular constituents from bacterial cells exposed to quats. These studies indicate that surfactants cause changes in the membrane integrity of bacteria. (See Chapter 2 for more information on the effects of surfactants on microorganisms.)

4.6 SUMMARY

This chapter reviewed much of the published information on the antimicrobial action of surfactants, with an emphasis on drawing correlations between chemical structure and antimicrobial activity. It is important for formulation chemists to have some knowledge of the antibacterial activity of surfactants—both additive effects and interferences—in order to approach formula development in a rational manner.

Generally, antimicrobial preservatives are used for product preservation because they prevent growth of microorganisms that may be harmful or cause defects (spoilage, discoloration, off-odors, etc.). Preservatives may be irritating or sensitizing to some people who use products containing them; consequently, it is desirable to use only the amount necessary for adequate product preservation.

Many surfactants are multifunctional compounds because they have antimicrobial activity in addition to reducing the surface tension of aqueous solutions in which they are used. When viewed in this manner, it is seen that surfactants represent a class of compounds that may contribute to the activity of the preservative system. It may be possible to reduce and/or eliminate preservatives in some formulations by selection of surfactants and physicochemical conditions that prevent/minimize microbial growth. Many aqueous surfactants are supplied to the cosmetic industry as aqueous solutions containing preservatives. It is important that these incidentally added preservatives be considered in formulation of the product and in claims for the finished product. The use of surfactants may allow reduction in the use of preservatives in some formulas and aid in the development of self-preserving cosmetic and drug products.

REFERENCES

1. Y. Abe. Long-chain fatty acids containing ether linkage. I. The antibacterial and fungicidal activities of some new β-alkyloxypropionic acids and their methyl esters. *Lipids 1*(2):141–145 (1966).
2. W. M. Linfield, R. E. Casely, and D. R. Noel. Studies in the development of antibacterial surfactants. II. Performance of germicidal and deodorant soaps, *J. Am. Oil Chem. Soc. 37*:251–254 (1960).
3. N. M. Molnar. Influence of various surfactants on the antimicrobial activity of bromsalan and other ring-halogenated substances. *J. Am. Oil Chem. Soc. 45*:729–737 (1968).
4. J. J. Kabara. Medium-chain fatty acid esters as antimicrobial aents. In: *Cosmetic and Drug Preservation: Principles and Practice*. J. Kabara, ed. Marcel Dekker, New York, 1984.

5. W. Kastner, Pharmacological properties. In: *Anionic Surfactants.* 2nd ed. C. Gloxhuber and K. Kunstler, eds., Marcel Dekker, 1992, pp. 440–444.
6. C. Gloxhuber, ed. *Anionic Surfactants.* Marcel Dekker, New York, 1970.
7. J. G. Voss. Effect of inorganic cations on bactericidal activity of anionic surfactants. *J. Bacteriol. 86*:207–211 (1963).
8. I. Nikodemusz, F. Fodor, and L. Sujbert. *Forum Städte-Hygiene 28*:176–178 (1977).
9. K. Oba and S. Tomiyama. Biochemische Studien über Alkylbenzolsulfonate II. Die Wirkung von Alkylbenzolsulfonaten auf Mikroorganismen. *Fett, Seifen, Anstrichm. 12*:1069 (1966).
10. H. Damm. Grenzflächenaktive Stoffe und Desinfektionsmittel. *3rd Int. Kong. für Grenzflächenaktive Stoffe*, Band IV, Sektion D/1, Nr. 29. Verlag der Universitätsdruckerei, Mainz, 1960, pp. 256–259.
11. F. K. Schaller, G. Shubach, and H. Rieth. Inhibitory action of detergents on pathogenic fungi. *Mykosen 11*(6):419–422 (1968).
12. Cosmetic raw materials. Bulletin of Grillo-Werke Aktiengesellschaft, Duisburg, Germany.
13. Rewocid SBU 185, Bulletin of Rewo—Chemical Group.
14. I. R. Schmolka. The synergistic effects of nonionic surfactants upon cationic germicidal agents. *J. Soc. Cosm. Chem. 24*:577 (1973).
15. J. J. Kabara, D. M. Swieczkowski, A. J. Conley, and J. P. Truant. Fatty acids and derivatives as antimicrobial agents. *Antimicrob. Agents Chemother. 2*:23 (1972).
16. A. J. Conley and J. J. Kabara. Antimicrobial action of esters of polyhydric alcohols. *Antimicrob. Agents Chemother. 4*:501 (1973).
17. J. J. Kabara, R. Vrable, and M. Lie Ken Jie. Antimicrobial lipids: natural and synthetic fatty acids and monoglycerides. *Lipids 9*:753 (1977).
18. J. A. Sands, D. A. Auperin, P. D. Landin, A. Reinhardt, and S. P. Cadden. Antiviral effects of fatty acids and derivatives. In: *Pharmacological Effect of Lipids.* J. J. Kabara, ed. American Oil Chemists' Society, Champaign, IL, 1979, p. 75.
19. L. R. Beuchat. Comparison of antimicrobicidal activities of potassium sorbate, sodium benzoate and glycerol and sucrose, esters of fatty acids. *Appl. Environ. Microbiol. 39*:1178 (1980).
20. I. Shibasaki. Recent trends in the development of food preservatives. *J. Food Safety 4*:35 (1982).
21. A. Kato and K. Arima. Inhibitory effect of sucrose ester of lauric acid on the growth of *E. coli. Biochim. Biophys. Res. Commun. 42*:596 (1971).
22. I. Shibasaki and N. Kato. Combined effects on antibacterial activity of fatty acids and their esters against gram-negative bacteria. In: *Pharmacological Effect of Lipids.* J. J. Kabara, ed. American Oil Chemists' Society, Champaign, IL, 1979, p. 15.
23. A. Kato and L. Shibasaki. Combined effect of different drugs on the antibacterial activity of fatty acids and their esters. *J. Antibact. Antifung. Agents 8*:355 (1975).
24. A. Kato and L. Shibasaki. Combined effect of citric and polyphosphoric acid on the anti-bacterial activity of monoglycerides. *J. Antibact. Antifung. Agents 4*:254 (1976).
25. T. Uchibori, S. Watanabe, and Y. Abe. Study on antimicrobial surface-active agents. IV. Preparation of polyethylene glycol monoesters of some alkyldihydroxybenzoic acids and determination of their antimicrobial activities and surface tension of their aqueous solutions. *J. Jpn. Oil Chem. Soc.* (Yukagaku) *25*(7):399–403 (1976).

26. R. G. Bistline Jr., E. W. Maurer, F. D. Smith, and W. M. Linfield. Fatty acid amides and anilides, syntheses and antimicrobial properties. *J. Am. Oil Chem. Soc. 57*:98–103 (1980).
27. A. F. Novak, J. M. Solar, R. R. Mod, F. C. Magne, and E. L. Skan. Antimicrobial activity and physical characteristics of some N,N-disubstituted decanamides. *J. Am. Oil Chem. Soc. 46*:249–251 (1969).
28. R. R. Mod, F. C. Magne, G. Sumrell, A. F. Novak, and J. M. Solar. 1,3-Diacyl derivatives of imidazoline and hexahydropyrimidine: II. Antimycotic activity. *J. Am. Oil Chem. Soc. 48*:257–258 (1971).
29. R. R. Mod, J. A. Harris, J. C. Arthur, F. C. Magne, G. Sumrell, and A. F. Novak. Free radical addition of dialkyl phosphites to N,N-disubstituted amides of unsaturated fatty acids and screening of the products for antimicrobial activity. *J. Am. Oil Chem. Soc. 49*:634–635 (1972).
30. D. Kenney. Formulating self-preserving cosmetics. *Cosmet. Toiletr. 94*(9):25–26 (1979).
31. A. V. Bailey, G. J. Boudreaux, G. Sumrell, and A. F. Novak. Antimicrobial properties of some fatty ester-amides. *J. Am. Oil Chem. Soc. 55*:835–836 (1978).
32. S. Saito. *J. Med. Sci. 6*:23 (1958).
33. G. W. Kidder, V. C. Dewey, and M. R. Heinrich. The effect of non-ionic detergents on the growth of tetrahymena. *Exp. Cell. Res. 7*:256–264 (1954).
34. H. J. Ferlin and J. V. Karabinos. Bactericidal nonionic detergents. *Trans.* Ill. State Acad. Sci. *47*:86–88 (1955).
35. J. N. Breston. *Oil Gas J. 48*:96, 124, 127 (1949).
35a. J. N. Breston. *Producers Monthly 13*(7):16 (1949).
36. D. S. Orth. Inactivation of preservatives by surfactants. In: *Surfactants in Cosmetics*, 2nd ed. M. M. Rieger and L. D. Rhein, eds. Marcel Dekker, New York, 1996.
37. D. L. Wedderburn. In: *Advances in Pharmaceutical Sciences*, Vol. 1. A. H. Beckett, J. Carless, and H. S. Bean, eds. Academic, New York, 1964, pp. 195–268.
38. W. P. Evans. The solubilisation and inactivation of preservatives by non-ionic detergents. *J. Pharm. Pharmacol. 16*:323–331 (1964).
39. D. H. Johnson. The use of fatty acid derivatives in cosmetics and toiletries. *J. Am. Oil Chem. Soc. 55*:438–443 (1978).
40. Crodesta esters, Information Sheet 11574, Croda, Inc., New York, 1974.
41. Once upon a slime. *Cosmet. Toiletr. 92*(3):94 (1977).
42. P. C. Lerk and H. Sucker. The effect of sucrose laurate and propyl 4-hydroxy-benzoate combinations on microbial growth: II. Evaluation of growth of *Escherichia coli* and *Staphylococcus aureus* in two liquid oral pharmaceutical formulations. *Eur. J. Pharm. Biopharm. 39*(4):158–161 (1993).
43. S. Matsumura, K. Imai, S. Yoshikawa, K. Kawada, and T. Uchibori. Surface activities, biodegradability and antimicrobial properties of n-alkyl glucosides, mannosides and galactosides. *J. Am. Oil Chem. Soc. 67*:996–1001 (1990).
44. T. Uchibori. Preparation of amino group-containing oligosaccharides as biodegradable microbicides. Jap. Patent 4.266.827 to Sanyo Chemical Industries Ltd., 1992.
45. S. Matsumura, Y. Kawamura, S. Yoshikawa, K. Kawada, and T. Uchibori. Surface activities, biodegradability and antimicrobial properties of glucosamine derivatives containing alkyl chains. *J. Am. Oil Chem. Soc. 70*:17–22 (1993).

46. S. Y. Mhaskar and G. Lakshminarayana. Synthesis of diethanolamides of N-lauroyl dipeptides and correlation of their structures with surfactant and antibacterial properties. *J. Am. Oil Chem. Soc. 69*:643–646 (1992).
47. W. M. Linfield. Straight-chain alkylammonium compounds. In: *Cationic Surfactants.* E. Jungermann, ed. Marcel Dekker, New York, 1970, pp. 9–70.
48. R. A. Reck and H. J. Harwood. *Ind. Eng. Chem. 45*:1022 (1953).
49. A. R. Galloppo, N. L. Ibrahim, and S. Mazzanobile. Quaternary ammonium antibacterial dentrifrices with selected calcium abrasives. U.S. Patent 683.149 (1991) through *CA 118*:11539. SmithKline Beckman Corp., 1993.
50. Cationic derivative. Bulletin of CECA S.A., Paris, La Défense, France, 1992.
51. R. G. Harry. *Cosmetic Materials. Their Origin, Characteristics, Uses and Dermatological Action*, Vol. 2. Leonard Hill Ltd., London, 1963, pp. 391–393.
52. F. V. Wells and I. I. Lubowe. *Cosmetics and the Skin.* Reinhold Publishing Corp., New York, 1964, p. 346.
53. P. J. Schaeufele. Advances in quaternary ammonium biocides. *J. Am. Oil Chem. Soc. 61*:387–389 (1984).
54. M. M. Rieger. *Surfactants in Cosmetics.* Marcel Dekker, New York, 1985.
55. S. J. Gutcho. *Surfactants and Sequestrants, Recent Advances.* Noyes Data Corporation, Park Ridge, NJ, 1977.
56. N. S. Bodor. Soft quaternary surface active agents exhibiting antimicrobial activity. U.S. Patent 3.989.711 to Interx Research Corporation, 1976.
57. H. A. Green, J. J. Merianos, and A. N. Petrocci. Microbiocidal polymeric quaternary ammonium compounds. U.S. Patent 3.874.870 to Millmaster Onyx Corporation, 1975.
58. W. J. Suling and W. M. O'Leary. Effect of surfactants on antibiotic resistance. *Antimicrob. Agents Chemother. 8*:334–343 (1975).
59. A. Baydar and R. Johnston. Novel quaternary ammonium salts derived from triglycerides and their application in skin and hair products. *Int. J. Cosmet. Sci. 13*:169–190 (1991).
60. M. L. Schlossman and J. McCarthy. Lanolin and its derivatives. *J. Am. Oil Chem. Soc. 55*:447–450 (1978).
61. Lanoquat 1756 Lanolin Quaternary, Bulletin of Emery Industries, Mauldin, South Carolina, U.S.A., November 1982.
62. J. J. Kabara, W. J. McKillip, and E. A. Sedor. Aminimides: I. Antimicrobial effect of some long chain fatty acid derivatives. *J. Am. Oil Chem. Soc. 52*:316–317 (1975).
63. J. J. Kabara and G. V. Haitsma. Aminimides: II. Antimicrobial effect of short chain fatty acid derivatives. *J. Am. Oil Chem. Soc. 52*:444–447 (1975).
64. J. J. Kabara. Aminimides: III. Antimicrobial effect of various hexadecyl and quaternary derivatives. *J. Am. Oil Chem. Soc. 54*:202–206 (1977).
65. D. Kenney. Antimicrobial compositions for controlling body odor. *Cosmet. Toiletr. 94*(8):13–14 (1979).
66. C. D. Moore. Ampholytic surface active agents. *J. Soc. Cosmet. Chem. 11*:13–25 (1960).
67. E. B. Michaels. Disinfecting compositions containing amphoteric surfactants with high viscosity for personal care products. U.S. Patent 673.631, 1991, through *CA 117*:257977z to E. B. Michaels Research Associates Inc., 1991.

68. C. Ceccarelli and G. Proserpio. Le alchilamidobetaine—Caratteristiche-proprietà-impieghi igienico/cosmetici, Rivista Italiana Essenze, Profumi, Piante officinali, Aromi, Saponi, Cosmetici, Aerosol 193, April 1978.
69. Tego Betaines—Tego Betain L 7, Bulletin of Th. Goldschmidt A. G., Essen, Germany, 1989.
70. D. H. Johnson. The use of fatty acid derivatives in cosmetics and toiletries. *J. Am. Oil Chem. Soc. 55*:438 (1978).
71. A. Schmitz. Disinfecting and cleaning solution for the human skin. U.S. Patent 2.717.850 to Th. Goldschmidt A. G., 1955.
72. A. Schmitz. Germ. Patent 819.402 to Th. Goldschmidt A. G., 1951.
73. A. Schmitz. Verfahren zur Herstellung germicider Ampho-tenside. Germ. Patents 1.070.189, 1957, and 1.190.142, 1965, and Brit. Patent 771.635, 1957, and A. Schmitz and G. Cramer, Germ. Patent 1.020.033 to Th. Goldschmidt A. G., 1957.
74. P. Bert and R. Barbier. Agents tensio-actifs à caractères ampholyte et chèlatant dèrives de l'acide aspartique. *5th Int. Cong. Det. 1*:175–182 (1969).
75. M. Katsumi, F. Tokiwa, and M. Miyazaki. Some observations on the action of a new amphoteric surfactant in washing. *5th Int. Cong. Det. 3*:203–209 (1969).
76. S. Ishikawa, M. Takada, and M. Takai. Cleaning compositions. Japan. Patent 68.609 to Kyarabasu Kenkyusko Co., Ltd., 1976.
77. H. Marumo. Detergent compositions. Germ. Patent 2.222.899, 1972.
78. H. J. Hueck, D. M. M. Adema, and J. R. Wiegmann. Bacteriostatic, fungistatic and algistatic activity of fatty nitrogen compounds. *Appl. Microbiol. 14*:308–319 (1966).
79. Z. W. Dudzinski and R. L. Wakeman. Amphoteric surface active agents. U.S. Patent 3.991.208 to Millmaster Onyx Corporation, 1976.
80. L. Chalmers. Ampholytic surface active agents, their properties and uses. *Specialties 3*(5):3, 5–10 (1967).
81. A. Schmitz and W. S. Harris. Germicidal ampholytic surface-active agents. *Mfg. Chem. 29*:51–54 (1958).
82. G. Prosperpio. I tensidi anfoteri. *Cosmet. News 8*:293–319 (1985).
83. W. M. Linfield. *Anionic Surfactants*, Vol. 7, Part II. Marcel Dekker, New York, 1976, pp. 582–612.
84. M. Fieser, L. F. Fieser, E. Toromanoff,Y. Hirata, H. Heymann, H. Teff, and S. Bhattacharya. Synthetic emulsifying agents. *J. Am. Chem. Soc. 78*:2825–2832 (1956).
85. Y. E. Kameda, E. Toyoura, S. Ohshima, M. Tsuji, and C. Irie. *J. Pharm. Soc. Japan 68*:143 (1948).
86. L. S. Fosdick, J. C. Calandra, R. O. Blackwell, and J. H. Burill. *J. Dental Res. 32*:486 (1953).
87. R. Infante, J. G. Dominguez, P. Erra, R. Julia, and M. Prats. Surface active molecules: preparation and properties of long chain N^{α}-acyl-L-α-amino-ω-guanidine alkyl acid derivatives. *Int. J. Cosmet. Sci. 6*:275–282 (1984).
88. N. Molin. *Physiol. Plant. 17*:456 (1964).
89. R. Yoshida, K. Baba, T. Saitò, and L. Yoshimura. Surfactants derived from amino acids. III. Some surface-active properties and antimicrobial activities of the salts of long-chain N^{α}acyl-L-arginine esters, *J. Jpn. Oil Chem. Soc.* (Yukagaku) *25*:404 (1976).
90. M. R. Infante, J. Garcia Dominguez, P. Erra, and M. R. Julià. *Proceed. XII Congrès IFSSC*, Vol. III, 1982, pp. 99–103.

91. M. R. Infante, P. Erra, M. R. Julià, and J. Garcia Dominguez. Sintesis de tensioactivos derivados de aminoacidos dibasicos con propiedades antimicrobianas. *Proceed. XIV Jornadas C.E.D.*, 1983, pp. 267–281.
92. M. R. Infante, J. Molinero, P. Erra, M. R. Julià, and J. Garcia Dominguez. A comparative study on surface active and antimicrobial properties of some N^{α}- lauroyl-L-α,ω dibasic aminoacids derivatives. *Fette Seifen Anstrichm.* *87*:309–313 (1985).
93. S. Y. Mhaskar, G. Lakshminarayana, and L. Saisree. N-Acyl-L- leucines of biologically active uncommon fatty acids: synthesis and antibacterial activity. *J. Am. Oil Chem. Soc.* *70*:23–27 (1993).
94. S. C. Osanai, S. C. Yukara, and Y. Abe. Preparation and antimicrobial properties of optically active amphoteric surfactants. *J. Jpn. Oil Chem. Soc.* (Yukagaku) *33*(6):372–379 (1984).
95. S. C. Osanai, T. K. Wakisaka, and H. S. Inoue. Synthesis of alkoxyl and phenoxyl substituted glycerides and their antimicrobial properties. *Bokin Bobai* *14*(3):109–116 (1986).
96. W. Zhi Guang, J. Chun Nuan, L. S. Ming. Synthesis of aromatic amino acids for antibacterial surfactants. *J. Am. Oil Chem. Soc.* *70*:109–110 (1993).
97. L. S. Ming and J. Q. Rong. Studies on synthesis and antimicrobial activity of N,N-bis[2-(N′- acylamino)ethyl]glycines and 4-[2-(N-acylamino)ethyl]-2-piperazinones. *J. Jpn. Oil Chem. Soc.* (Yukagaku) *42*:49–51 (1993).
98. K. Narasimhan, P. K. Sehgal, and J. K. Thomas. Condensation of aromatic sulfonic acid with hydrolysates from leather wastes. *J. Am. Leather. Chem. Assoc.* *75*(6):211–215 (1980).
99. Stepan Chemical Co., Protein Derivatives Fact File, Form No. 6178-S, 1978.
100. P. Sokol. Thermal condensation of fatty acids and amines with proteins. *J. Am. Oil Chem. Soc.* *52*:101–102 (1975).
101. B. R. Bluestein and C. L. Hilton. In: *Amphoteric Surfactants.* Marcel Dekker, New York, 1982, p. 265.
102. J. Molinero, M. R. Julia, P. Erra, M. Robert, and M. R. Infante. Synthesis and properties of N^{α}- lauroyl-L-arginine dipeptides from collagen. *J. Am. Oil Chem. Soc.* *65*:975–978 (1988).
103. M. R. Infante, J. Molinero, P. Bosch, M. R. Julia, and P. Erra. Lipopeptidic surfactants: I. Neutral N-lauroyl-l-arginine dipeptides from pure amino acids. *J. Am. Oil Chem. Soc.* *66*:1835–1839 (1989).
104. M. R. Infante, J. Molinero, and P. Erra. Lipopeptidic surfactants: II. Acidic and basic Nα-lauroyl-l-arginine dipeptides from pure amino acids. *J. Am. Oil Chem. Soc.* *69*:647–652 (1992).
105. K. Tsubone, N. Uchida, K. Niwase, and K. Honda. Syntheses of sodium 2-(N-alkyl-N-methylamino)ethanephosphates and their physicochemical properties. *J. Am. Oil. Chem. Soc.* *66*:829–833 (1989).
106. K. Tsubone and N. Uchida. Syntheses of 2-(N-2-hydroxyalkyl- N,N-dimethylammonio)ethyl hydrogen phosphates and their physicochemical and antimicrobial properties. *J. Am. Oil Chem. Soc.* *67*:394–399 (1990).
107. G. E. Davies, J. Francis, A. R. Martin, F. L. Rose, and G. Swain. 1:6-Di-4′-chlorophenyldiguanidohexane ("Hibitane"). Laboratory investigation of a new antibacterial agent of high potency. *Br. J. Pharmacol.* *9*:192–196 (1954).

108. K. Tsubone and N. Uchida. Synthesis of 2-[N-(3-alkoxy-2- hydroxypropyl)-N,N-dimethylammonio]ethyl hydrogen phosphates and their physicochemical and antimicrobial properties. *Jpn. Oil Chem. Soc.* (Yukagaku) *40*:392–399 (1991).
109. Y. Abe and S. Osanai. Long-chain carboxylic acids containing ether linkage. III. The antibacterial and antifungal activities of the amine salts of some β-alkyloxypropionic acids. *J. Am. Oil Chem. Soc. 46*:365–367 (1969).
110. Y. Abe, S. Osanai, and S. Matsummura. Long chain carboxylic acids containing ether linkage. IV. The antibacterial activities of O-(2-alkylaminoethyl)-3-oxypropionic and N-(2-alkyloxyethyl)-3-aminopropionic acids. *J. Am. Oil Chem. Soc. 49*:357–360 (1972).
111. Y. Abe, S. Osanai, and T. Matsushita. Long chain carboxylic acids containing ether linkage. V. N-Methyl substituted glycine- type amphoterics containing alkoxy radical. *J. Am. Oil Chem. Soc. 51*:385–388 (1974).
112. Y. Abe and S. Osanai. Synthesis and antimicrobial properties of N-substituted amino acid-type amphoterics containing a thio ether linkage. *J. Jpn. Oil Chem. Soc.* (Yukagaku) *25*(7):419–423 (1976).
113. L. C. Q. Monteine. Collagen derived amphoteric surfactant. Bulletin of Seppic, Paris, January 1988.
114. S. Osanai, Y. Yoshida, K. Fukushima, and S. Yoshikawa. Preparation of optically active amphoteric surfactants and their surface and antimicrobial properties. *J. Jpn. Oil Chem. Soc.* (Yukagaku) *38*(8):633–641 (1989).
115. B. F. Miller, Z. Baker, and R. W. Harrison. *Proc. Soc. Exp. Biol. Med. 42*:705 (1939).
116. D. Jerchel. Invert soaps. X. Sulfamyltetrazolium salts. Action on the glycolysis by lactic acid bacteria. *Ber. 75*:75–81 (1942).
117. E. J. Ordal and A. F. Borg. Effect of surface-active agents on oxidation of lactate by bacteria. *Proc. Soc. Exp. Biol. Med. 50*:332–336 (1942).
118. M. G. Sevag and O. A. Ross. Studies on the mechanism of the inhibitory action of zephiran on yeast cells. *J. Bacteriol. 48*:677 (1944).
119. R. D. Hotchkiss. The nature of the bactericidal action of surface agents. *Ann. NY Acad. Sci. 46*:479–493 (1946).
120. M. R. J. Salton. The adsorption of cetyltrimethylammonium bromide by bacteria, its action in releasing cellular constituents and its bactericidal effects. *J. Gen. Microbiol. 5*:391–404 (1951).
121. *Quaternary Ammonium Compounds.* Bulletin of Bayer, Germany, 1978.
122. *Registry of Toxic Effects of Chemical Substances.* United States Department of Health and Human Services, Washington, 1985–86.
123. M. H. Angele. Oberflächenaktivität, mikrobielle Wirksamkeit und Anwendung von Ausgewählten Tetraalkyl- stickstoffverbindungen. *Seifen, Oele, Fette, Wachse 101*:273–277 (1975).
124. M. H. Angele. Biozide quaternäre Ammoniumverbindungen: Moderne Aspekte. *Seifen, Oele, Fette, Wachse 104*:433–436 (1978).
125. A. L. L. Hunting. *Encyclopedia of Shampoo Ingredients.* Micelle Press Inc., Cranford, NJ, 1983, pp. 338–339.
126. Cationics. Bulletin of Onyx Chemical Company, Jersey City, NJ, 1976.
127. Agents tensio-actifs dans les preparations cosmetiques. Bulletin of Witco Chemical, Paris.

128. G. Gottardi and A. Crivello, Gli anfoteri disinfettanti. Bulletin of Tego Italiana S.r.l., Milan, 1991.
129. Deriphat amphoteric surfactants. Bulletin of Henkel Corporation, Minneapolis, MN.
130. Genamin CTAC, Bulletin of Hoechst Italia S.p.A., Divisione TA, 1975.
131. Produits chimiquesd, Bulletin of Rewo, Steinau, Germany.

5

Fatty Acids and Esters as Multifunctional Components

Jon J. Kabara
Technology Exchange, Inc.
Galena, Illinois, and
Michigan State University
East Lansing, Michigan

In classical terms, preservation of a cosmetic meant the addition of a chemical that had no other function than to prevent microbial spoilage. It generally is agreed that no single biocide is an "ideal preservative" at customary use levels, and two or more preservatives often are necessary for adequate protection of a formula.

Fatty acids and their corresponding esters have a long history of use as antimicrobial agents (1,2). Soaps (i.e., salts of fatty acids) have been used as cleaning and disinfecting agents for centuries (3). Before 1930, the lack of success in the search for active antibacterial agents other than fatty acids resulted in pessimism as to whether an active agent would ever be found. Two important discoveries changed the course of thinking on the subject. First was a serendipitous finding by Alexander Fleming (4). In 1928, Fleming observed the lysis of *Staphylococcus* colonies in an area surrounding the growth of *Penicillium*, a mold, and thus antibiotics were added to the small arsenal of compounds lethal to microorganisms. Second, the discovery of antimicrobial sulfonilamides, first observed by Trefouel, Nitti, and Boret in Fourneau's laboratory (Pasteur Institute) in 1935, gave hope for finding new and more effective synthetic biocidal agents (5). These two major discoveries in less than a decade gave tremendous impetus to the search for since the biocidal activity of the isolated antibiotics and synthesized germicides was so much greater than that of fatty acids. In proportion to the speed with which new and more powerful drugs were discovered, interest in the action of fatty acid biocides waned. Frantic research activities in the 1950s and 1960s produced hundreds of antibiotics and synthetic products useful to the pharmaceutical field, but these were of little use to the cosmetic and food industries.

Several reasons can be given for the slow or "nonparticipation" of the microbiologist in adopting antibiotics for use in preservation. First, antibiotics and synthetic germicides were not without health risks in terms of toxicity, irritation, and the possible development of resistant strains; second, the cost of these newer agents precluded their use in quantities normally required except in special situations; and third, the use of old standbys was sufficient (6). Interestingly, these industries kept distant for the most part from new discoveries and continued to use old standbys including vinegar (i.e., about 4% acetic acid) from ancient Egypt, sorbic acid isolated from rowanberry oil and reported in 1859 by Hoffman (7), benzoic acid discovered by Fleck (see Ref. 8), and propionic acid, which was recommended for the preservation of bakery goods by Watkins in 1906 (9).

From these few references, it becomes obvious that some fatty acids have played an important historical role in food preservation. Many other fatty acids, even more active than the standbys, have not been fully examined as potential preservatives, nor have detailed studies dealing with their structure-function relations been carried out. The objective of this chapter is to collate and evaluate the literature dealing with antimicrobial fatty acids and their corresponding esters.

5.1 FATTY ACIDS AS ANTIMICROBIAL AGENTS

One of the easiest ways to control the growth of microorganisms is by pH. Microorganisms have specific pH requirements for growth (see Chapter 2). The addition of a lipophilic acid can have synergistic antimicrobial action with lowered pH. At neutral or near-neutral pH values, fatty acids are less effective antimicrobials. In contrast, some esters of fatty acids are nonionic and are inhibitory over a wider pH range than the constituent fatty acids from which they are made.

Fatty acids generally are used in cosmetics as acidulants, co-emulsifiers, or super-fatting agents. Stearic acid is the fatty acid most often used in cosmetics. In view of the 30 or more fatty acids available to the cosmetic formulator, one wonders what special benefit to product integrity and consumer needs could be satisfied by other fatty acids.

Recent attention to essential fatty acids and their effects on skin physiology has increased. Linoleic (ω-6) fatty acid found in vegetable oils and an α- linolenic acid found in fish oils have been hyped as possible beneficial cosmetic agents. This is because dramatic clinical epidermal changes have been noted where there has been a dietary deficiency. Transepidermal water loss (TEWL) has been monitored following topical application; however, the value of correcting skin deficiencies by topical application of lipids per se is doubtful because Bohles et al. (10) and Holman (11) reported that cutaneous application of safflower seed oil to infants and animals, respectively, did not produce a marked reversal of symptoms leading to high TEWL. It is not known whether the absorption of the unesterified fatty acids would have been more effective. As the additional benefits of fatty acids in consumer products are recognized, one becomes aware that the antimicrobial activity of the fatty acid should be considered. The material presented in this chapter examines the structure-function (antimicrobial) relationships of fatty acids.

Much of the early research on the germicidal activity of fatty acids was carried out between 1920 and 1940 (12–14). More recent studies were reviewed by Kabara (15). The antibacterial effects of the fatty acids have been discussed by many authors, and the consensus of opinion is that their action results from the undissociated molecules rather than the anions (16–20). Because of this, the antimicrobial activity of soaps is profoundly affected by pH, since hydrogen ion concentration determines the degree of dissociation of the acid. The effect of dissociation was shown by demonstrating more rapid killing at lower pH, where pH alone was not lethal (21,22). The minimum inhibitory concentrations (MICs) of several fatty acids against *Clostridium welchii* are given in Table 5.1. A rise in pH from 6.5 to 7.5 increases the MICs (i.e., decreases the antimicrobial action) of the short-chain fatty acids (caproic, caprylic, and capric), but decreases the MICs (i.e., increases the antimicrobial action) of the two medium-chain fatty acids (lauric and myristic). The MIC values of $C_{16:0}$–$C_{22:1}$ fatty acids were the same at pH 6.5 and pH 7.5 in these studies. Among the fatty acids tested, those with the lowest MIC values were the polyunsaturated acids; however, the MIC of linoleic acid was somewhat less than that of linolenic acid. Esterification with monohydric alcohols (methanol, ethanol, etc.) raises the MIC of a fatty acid; however, data reported by Tetsumoto (23) indicate that methyl linoleate and methyl linolenate had lower MICs at pH 6.5 than at pH 7.5 (Table 5.1).

One of the problems in earlier studies of fatty acids was the dubious purity of the compounds being tested. For example, the unsaturated fatty acids were not tested for peroxide formation and seldom were chromatographically pure. Support for this comes from evaluating the potency of arachidonic acid. Although Tetsumoto (23) reported that arachidonic acid was active, Kabara et al. (24) showed no antimicrobial activity for this compound (Table 5.2). Also, the use of polysorbate 80 to solubilize fatty acids should be avoided, because polysorbate 80 generally is used by microbiologists to neutralize the antimicrobial action of preservatives.

From our present perspective, a clearer picture is emerging of the structure-function relationships with regard to the antimicrobial properties of fatty acids. The term "antimicrobial" is also intended to cover their antiviral properties because it has been shown that enveloped viruses can be inactivated by treatment with specific fatty acids (25). Chain length, degree and kind of unsaturation and geometric

TABLE 5.1 Minimum Inhibitory Concentration Against *Clostridium welchii*[a]

	Minimal inhibitory concentration (mg/100 ml)	
Fatty acid	pH 6.5	pH 7.5
Caproic (6:0)	1160.00	5810.00
Caprylic (8:0)	721.00	3610.00
Capric (10:0)	172.00	862.00
Lauric (12:0)	1000.00	200.00
Myristic (14:0)	2280.00	457.00
Palmitic (16:0)	2560.00	2560.00
Stearic (18:0)	2850.00	2850.00
Oleic (18:1)	283.00	283.00
Erucic (22:1)	339.00	339.00
Linoleic (18:2)	5.61	5.61
Linoleic (18:3)	27.90	27.90
Arachidonic (20:4)	6.09	6.09
Methyl linoleate	146.00	733.00
Methyl linolenate	293.00	1460.00

[a]Fatty acids and methyl esters in 0.1 M phosphate buffer (pH 6.5 or 7.5) containing 0.1% (wt/vol) Tween 80.
Source: Ref. 23.

isomerism are important determinants of biocidal activity of a fatty acid. Certain generalizations concerning the effect of fatty acids on microorganisms can be made. For instance, all fatty acids generally are ineffective against gram-negative bacteria. Some short-chain (i.e., <8 carbon atoms) fatty acids inhibit both gram-negative and gram-positive bacteria. Lauric acid (C_{12}) is the most active saturated fatty acid against most bacteria, while the C_8 and C_{10} fatty acids generally are more active against yeast and molds.

The effect of unsaturation in a fatty acid is complex, because it is influenced by the number, position, and geometrical isomerism of the double bond(s) (15,24). While monounsaturated fatty acids generally are more active than saturated fatty acids, it depends on the chain length of the fatty acid. Optimum activity is found in palmitoleic acid ($C_{16:1}$), whereas short-chain (i.e., $<C_{14}$) unsaturated fatty acids generally are less active (unpublished data). While the position of the double bond is important for the activity of monounsaturated fatty acids, it is less so for polyunsaturated fatty acids (24). A second double bond increases the antimicrobial activity, but its place in the chain is not significant. With regard to geometrical isomerization, the *cis* isomer always is more inhibitory than the *trans* isomer. Acetylenic derivatives have been found to be more active against fungi than

Table 5.2 Minimal Inhibitory Concentrations (mM) of Saturated and Unsaturated Fatty Acids

Fatty acid	*Staphylococcus aureus*	*Streptococcus* group A	*Candida albicans*
Capric	NI	NI	NI
Caprylic	NI	NI	NI
Capric	2.90	1.45	2.90
Lauric	2.49	0.12	2.49
Myristic	4.37	0.55	4.37
Myristoleic	0.44	0.11	0.55
Palmitic	NI	3.90	NI
Palmitoleic	0.98	0.10	0.49
Stearic	NI	NI	NI
Oleic	NI	1.77	NI
Elaidic	NI	NI	NI
Linoleic	NI	0.09	0.46
Linolenic	1.79	0.35	NI
Linolelaidic	NI	NI	NI
Arachidonic	NI	NI	NI

NI, not inhibitory at the concentrations tested (1.0 mg/ml or 30–60 mM). Note that gram-negative bacteria are not affected.
Source: Ref. 24.

ethylenic derivatives. Also, studies showed that esterification of fatty acids with a monohydric alcohol, such as methanol or ethanol, reduced their activity (24) (see MICs for methyl esters and corresponding free fatty acids in Table 5.1).

Sorbic acid, which was isolated by A. W. Hoffman in 1859 (7), has been omitted from this discussion because it is considered to be a preservative and has been used as a food preservative since 1940.

5.2 MONOESTERS AS ANTIMICROBIAL AGENTS

Esters of monoglycerides are widely used in cosmetics. The most important member of this group has been glycerol monostearate (GMS). Many different grades of this emulsifier are available, and they vary in grade of stearic acid used or in the ratio of mono-, di-, or triester present. The usual cosmetic grade of GMS has a monoester content of 40–45%. When acting as an emulsifier, GMS forms water-in-oil (w/o) emulsions.

The more highly purified monoglyceride (>90% monoester content) behaves differently than the usual grade of GMS, which has higher levels of mono- and diglycerides (26). First, the distilled monoglyceride is more likely to form oil-in-water (o/w) emulsions. Second, the emulsions made with purified GMS have smaller and more even particle size, which results in a more stable system. Uniform, stable emulsions are easier to preserve than unstable systems. Third, specific monoglycerides, such as glyceryl monolaurate (Lauricidin®), are antimicrobial (14,15,24). These monoesters are good examples of multifunctional ingredients which impart body, opacity, and smoothness to the emulsion. Use of monolaurin (Lauricidin®) helps create a "hostile environment" for microorganisms, so this multifunctional chemical becomes part of the preservative system.

While fatty acids are antimicrobial, esterification with a monohydric alcohol (e.g, methanol and ethanol) results in an ester that is inactive. This was not true when an α-hydroxy fatty acid was esterified (24). From these initial results, it was concluded that some hydrophilic group was necessary for biological activity and that the esterification of a polyhydric alcohol would yield active derivatives (27). Additional work indicated the need for a free single or multiple hydroxy group after esterification. One of the more common polyhydric alcohols, glycerol, was esterified and found to be more active than the corresponding fatty acids (Table 5.3). Details of these findings were published by Kabara (14,15,27–29), and in all cases they were confirmed by others (25,30–32).

There appears to be some confusion about which fatty acid ester is the most biocidal. In general, the fatty acid used to esterify the polyol determines the potency of the ester. The structure-function relationship for saturated and unsaturated esters follows the activity of their respective fatty acids, as reviewed earlier. This means that lauric acid (C_{12}) and palmitoleic acid ($C_{16:1}$) form the most active saturated and unsaturated monoesters, respectively (24,25).

The above statement concerning the structure-function relationship is true except where specific organisms and complex products are concerned. Because the monoglycerides can be neutralized by a variety of polymers, including starch and protein, and because binding seems to be related to biological activity, several Japanese researchers have reported that the lower fatty acid esters were more active in foods than the higher homologs (33–35). These reports must be viewed as exceptions found under particular circumstances. Reports by Beuchat (32) and Shibasaki (36) have supported the findings that glyceryl monolaurate is the most active monoglyceride—even in the presence of other foodstuffs. Beuchat (32) compared the effects of glycerol and sucrose esters, benzoate, sorbic acid, and potassium sorbate against *Vibrio parahaemolyticus* and found that the C_{12} monoglyceride was more active than lower- (C_8 or C_{10}) or higher- (C_{14}) chain-length derivatives. Also, the low minimum inhibitory concentration value for monolaurin (5 μg/ml) showed it to be more effective than sodium benzoate (≤300 μg/ml) or sorbic acid (70 μg/ml). Because these esters generally are regarded as inactive against gram-negative bacteria (36), the inhibition of *V. parahaemolyticus* growth was somewhat surprising, even though Kato and Shibasaki (38,39) reported effects on gram-negative bacteria. In these special cases, the Japanese workers demonstrated that chelating acids (citric acid and polyphosphoric acid) were necessary to

TABLE 5.3 Minimum Inhibitory Concentrations (g/ml) for Fatty Acids and Their Corresponding Monoglycerides

	Compound									
Organism[a]	Undecanoic acid	10-Undecenoic acid	10-Undecenoyl monoglyceride	10-Undecynoyl monoglyceride	11-Dodecenoic acid	Dodecanoic acid	Dodecanyol monoglyceride	12-Tridecenoic acid	Tridecanoic acid	Tridecanoyl monoglyceride
Streptococcus faecalis	NI[b]	NI	500	500	NI	500	NI	1000	NI	NI
Streptococcus pyogenes	125	1000	125	125	250	62	8	125	1000	62
Staphylococcus aureus	1000	1000	500	500	NI	500	250	1000	NI	NI
Corynebacterium sp.	31	31	62	62	125	31	16	31	NI	NI
Nocardia asteroides	62	125	125	62	62	62	16	125	1000	125
Candida albicans	1000	1000	250	100	1000	1000	500	1000	NI	NI
Saccharomyces cerevisiae	500	500	250	100	500	1000	250	500	1000	NI

[a] *Escherichia coli* and *P. aeruginosa* were not affected.
[b] NI, minimum inhibitory concentration greater than 1000 g/ml.
Source: Ref. 14.

have an enhancing effect on the antibacterial action of monoglycerides against gram-negative bacteria. Chelating agents are known to interfere with the integrity of the gram-negative bacterial cell envelope and enable preservatives and antibiotics to act on the cells (see Sec. 2.6.4). Apparently, the presence of added chelating agents is not always a prerequisite for demonstrating an inhibitory action of monoglycerides against all gram-negative bacteria, because Beuchat (32) observed that use of the monoester without chelating agents affected *V. parahaemolyticus* in a manner similar to that of gram-positive bacteria.

More complex esters of polyhydric alcohols that have been prepared and used in the food and cosmetic industries include glucose, sucrose, sorbitan, and polyglycerol esters (26). While they have the virtue of exhibiting a range of hydrophobic/lipophilic balance (HLBs) from 3 to 14, they all tend to be mixtures. Most commercial sources of these ester mixtures have surfactant properties, but their biocidal action is low. Reports on their biocidal activity are based on highly purified esters, which generally are not available for incorporation into cosmetic and pharmaceutical products.

The antimicrobial activity for esters of more complex polyhydric alcohols has been reported by several workers (27,31,35). Kato and Arima (35) reported that the sucrose ester of lauric acid was active against a gram-negative organism, whereas Conley and Kabara (27) and Shibasaki and Kato (31) found that these and other esters are active primarily against gram-positive bacteria and fungi. In contrast to data generated using glycerides, the diester possessed the highest activity of the sucrose esters, but was still less active than monolaurin (32,38).

The monoesters of glycerol (Table 5.3) and the diesters of sucrose not only have higher antimicrobial activity than their corresponding free fatty acids, but also compare favorably in activity with commonly used preservatives, such as parabens, sorbic acid, and dehydroacetic acid (Tables 5.4 and 5.5) (35).

Table 5.4 Comparison of Antifungal Activities of Fatty Acid Esters with Some Commonly Used Preservatives

	Minimum inhibitory concentration (μg/ml)[a]		
Food additive	*Aspergillus niger*	*Candida utilis*	*Saccharomyces cerevisiae*
Monocaprin	123	123	123
Monolaurin	137	69	137
Butyl-*p*-hydroxybenzoate	200	200	200
Sodium lauryl sulfate	100	400	100
Sorbic acid	1000	1000	1000
Dehydroacetic acid	100	200	200

[a]By agar dilution method.
Source: Ref. 30.

TABLE 5.5 Comparison of Antibacterial Activities of Fatty Acid Esters with Some Commonly Used Preservatives

Food additive	Minimum inhibitory concentration (μg/ml)[a] *Bacillus subtilis*	*Bacillus cereus*	*Staphylococcus aureus*
Sucrose dicaprylin	74	74	148
Monocaprin	123	123	123
Monolaurin	17	17	17
Butyl-*p*-hydroxybenzoate	400	200	200
Sodium lauryl sulfate	100	100	50
Sorbic acid	4000	4000	4000

[a]By agar dilution method.
Source: Ref. 30.

Another series of food-grade esters was derived from a group of compounds called polyglycerol esters. These linear polymers of glycerol, discovered and advocated for food use by Babayan et al. (2), were supplied to me for biocidal testing. As with sucrose esters, the medium-chain-length fatty acids appeared to be the most active (Table 5.6). It was interesting that, regardless of whether the polyol was tri-, hexa-, or decaglycerol, the fatty acid moiety seemed to determine overall biocidal activity. Generally, as the polyol became bulkier, the spectrum of biocidal activity became narrower and the MICs increased (e.g., the compounds were less effective against test organisms). The laurate esters had smaller MICs (i.e., were more effective against test organisms) than polyglycerol esters with fatty acid moieties with chain lengths up to C_{18}. Other polyol esters have been tested, and except for special applications none have proven to be more useful than the

TABLE 5.6 Minimum Inhibitory Concentrations of Various Polyglycerol Esters (mM) Against *Corynebacterium* sp.

Fatty acid moiety	Polyglycerol: Triglycerol	Hexaglycerol	Decaglycerol
Acetate	>3.55[a]	>1.98[a]	>1.25[a]
Butyrate	>3.55[a]	>1.98[a]	>1.21[a]
Caproate	2.96	>1.79[a]	>1.17[a]
Caprylate	1.37	0.85	1.13
Caprate	0.32	0.39	1.10
Laurate	0.24	0.38	0.53

[a]No inhibitions, maximum concentration tested.
Source: Ref. 27.

monolaurin. Emulsions made with decaglycerol produce stable w/o emulsions with good aesthetic properties.

The term "monolaurin" does not properly define the purity of the monoester; consequently, the use of the trademarked monoester (Lauricidin®) is preferred because it relates to a commercially available emulsifier with 95 ± 1% monoester content. The work of Kabara (37) demonstrated that the purity of the monoglyceride is critical for optimal emulsification and antimicrobial action. Much of the discussion that follows is detailed from sources dealing with food systems because (1) food systems are complex and only safe chemical additives can be used, (2) this information generally is not covered by cosmetic or pharmaceutical literature, and (3) few applications outside of our laboratory can be found in cosmetic literature.

One of the first reports (in 1980) of the commercial use of monolaurin as an emulsifier was to help solve a difficult fungal problem in a diet margarine (40). Since the concentration of salt is lower and the amount of water is higher in diet margarine than in regular margarine, the use of preservative levels of sorbate was limited by taste and off-flavor. The introduction of low levels of monolaurin helped produce an acceptable and marketable product. No product failures have resulted in more than 10 years of using the Lauricidin preservative system in this margarine. The single instance in which fungal growth occurred was when the monoglyceride was inadvertently omitted.

Shibasaki (36) stated "of all lipids evaluated, lauric acid (esters) appear to have the greatest overall antimicrobial activity. Of these derivatives, monolaurin would appear to have the highest potential for use in foods and cosmetics." This may be true for studies carried out in laboratory media, but monolaurin is influenced by the presence of starchy and proteinaceous materials, as is the case for other lipophilic preservatives (30). To avoid this problem, the lower-weight monoglyceride (MC_8) was used in Japanese-style food. It is the observation of this author that the use of MC_8 in foods is not warranted when Lauricidin is used properly, that is, with homogeneous mixing, proper pH range, and effective acidulants (citric and lactic acids). The purity of the monoglyceride as well as the order of addition into a formula is as important as the kind of monoglyceride chosen.

In the possible application of monolaurin to food systems, Kato and Shibasaki (30) demonstrated that lauric acid, monocaprin (MC_{10}), and monolaurin (MC_{12}) significantly enhanced the thermal death of *Escherichia coli* and *Pseudomonas aeruginosa.* The thermodynamic characteristics of bactericidal action of monolaurin and heat were studied by Tsuchido et al. (41). These authors found that the bactericidal effect of increasing concentrations of monolaurin would be more effective in low-temperature/long-term treatment than in high- temperature/short-term treatment. The usefulness of monolaurin in the heat inactivation of spores was substantiated by Kimsey et al. (42). Inactivation of *Bacillus stearothermophilus* 1518 was increased when heated in the presence of 18 mM monolaurin. The

Table 5.7 Effectiveness of Various Saturated Monoglycerides on *B. stearothermophilus* Spores at 120°C

Spores heated in 18 mM	Number of C in chain	Average D_{120} (s)	Range
Control	—	265	236–294
Monocaprylin	8	195	180–210
Monocaprin	10	88	84–91
Monolaurin	12	119	117–120
Monomyristin	14	221	210–228
Monopalmitin	16	244	234–248

Source: Ref. 42.

1518 was increased when heated in the presence of 18 mM monolaurin. The decimal reduction time, or D-value (the time required to reduce the number of viable spores by 1-log or 90%), at 120°C was lowered by 49.8% for this organism, and the D_{120} for *Bacillus subtilis* was reduced by 47.9%. The effect of various saturated monoglycerides on *B. stearothermophilus* spores is given in Tables 5.7 and 5.8. Monocaprin and monolaurin were the most active of the series C_8–C_{16}. Monolaurin was also tested in 10% nonfat dry milk and in the fluid portion of cream-style corn. In these foods, the D_{120} time was lowered by approximately 50%, which means that the rate of death was approximately twice as fast. It was concluded that monolaurin could be used as an additive (emulsifier) to reduce the heat treatment required to achieve commercial sterility of the product. Use of

Table 5.8 Effect of Dodecanoic Acid and Its Monoglyceride on the Loss of Heat Resistance of Spores During Germination[a]

Exposure time (min)	Viability (%) after heat treatment of spores: None	DA[b] (0.2 mM)	GMD[c] (0.05 mM)
0	100	100	100
30	11.5	59	55
60	6.4	35	33

[a]Spores exposed to fatty acid, then heat treated at 85°C for 30 min; viability expressed as percentage of initial spore count (approximately 10^7 CFU/ml).
[b]Dodecanoic acid.
[c]Glycerol monododecanoate (40).
Source: Ref. 42.

for considering monolaurin as part of the preservative system of cosmetic and drug products.

Naruki et al. (43) confirmed earlier studies by demonstrating that fatty acids and monoglycerides inhibited the development of *B. subtilis* var. *niger* spores and that the fatty acid is less active than the corresponding monoglyceride. Their work did not explain why the addition of lipid additives caused a loss of heat resistance of spores during the germination process.

Inactivation of *Clostridium sporogenes* and *Listeria monocytogenes* during cooking was studied by Unda et al. (44). Beef roasts were pumped with brine containing phosphates and acetic acid, sodium lactate, potassium sorbate, or monolaurin. The meat was heated in the bags once or twice to 62.8°C. In roasts recooked after being inoculated with these organisms, glycerol monolaurate and sodium lactate were very effective and significant at ($p < 0.05$) in inhibiting anaerobic bacterial growth during temperature abuse.

In addition to enhancement of thermal death by monolaurin, Takano et al. (45) earlier showed cells of gram-negative bacteria to be significantly injured by freezing. Of the various nonionic derivatives tested, monolaurin was the most active in reducing viability. Freezing with these agents was also found to be lethal for gram-positive bacteria. The carbohydrate esters enhanced damage in the growth phase rather than during the freezing process.

In addition to potentiating the effects of heat or freezing on microorganisms, the addition of Lauricidin in hydrostatic pressure experiments indicate another use for this unusual lipid. In 1992, Butz et al. (46) indicated that Lauricidin destabilized the membrane and made viruses more susceptible to pressure inactivation. Significant effects were obtained with 0.02% Lauricidin. In 1989, Hite (47) first explored the use of high pressure as a method of food processing, but no mention has been found of the use of this technology in the cosmetic or pharmaceutical literature. (For the reader interested in greater detail on the use of hydrostatic pressure in product processing, I recommend reading *Food Technology*, June 1993, pp. 150–172, which covered a symposium on the subject.)

Kato (48) compared the antimicrobial action of fatty acids, monoglycerides, and sugar esters. He found that capric acid and monolaurin were the most active surfactants against an osmophilic yeast. In a series of sugar esters, sucrose monolaurate was more active than sucrose monocaprate. The activity of monolaurin compares with that of lauryl sulfate and was superior to butyl paraben, which is approved as an antimicrobial additive for foods in Japan. Also, monolaurin is approved for use in foods in Japan.

A number of examples from food technology are presented to compare the antimicrobial action of monoglycerides with sorbic acid. Smith and Palumbo (49) investigated the inhibition of aerobic and anaerobic growth of *S. aureus* in a model system. Whereas anaerobic growth of *S. aureus* was inhibited by 2500 ppm sorbate and monolaurin, twice as much was required to inhibit aerobic growth. It was necessary to acidify the system with 20 mg/ml lactic acid. Under the same acidulating conditions, sorbic acid and Lauribic (a mixture of sorbic acid and monolaurin) inhibited both aerobic and anaerobic growth equally at 500 and 750 ppm, respectively. In these experiments, the amount of lactic acid added, not the pH, was critical to growth inhibition. The addition of 10–20 mEq lactic acid resulted in a small change in pH value of 0.4 pH units and no change in microbial concentration. The addition of 25 and 30 mEq lactic acid resulted in slight lowering of pH (i.e., 0.6 pH unit to pH 5.7) but a drop of 4–5 log units in microbial population in the preserved meat emulsion (compared with control levels). It is important to recognize that sorbic acid is both a preservative and an acidulating agent. This factor should be recognized when comparisons between sorbic acid and monlaurin are made.

In a similar model system, Kabara (50) showed that Lauricidin was slightly more active than potassium sorbate when the pH was controlled. At 500 ppm, Lauribic gave greater inhibition than either monolaurin or sorbic acid alone. Inhibition was greater against anaerobic than aerobic growth, and in the latter instance high (>10,000 ppm) concentrations of potassium sorbate and moderate (5000 ppm) amounts of monolaurin were required for a good reduction in the bacterial population. Katagiri et al. (51) reported that a mixture of one part monolaurin, 20 parts of a 1:1 mixture of pyro- and polyphosphate, and 40 parts sorbic acid prevented sausage spoilage for up to 11 days at 25°C when used at a final concentration for sorbic acid of 0.3%. The control sausage spoiled in 3 days. Robach et al. (52) found sorbic acid to be more active than monolaurin in laboratory media, cottage cheese, and a pork homogenate, but they found no advantage of monolaurin–sorbic acid combinations.

Other studies have suggested that combinations of monoglycerides and fatty acids were useful in preserving food systems. Yuda et al. (53) reported that organic acids, such as fumaric acid, helped preserve starch-containing foods when combined with lauric esters of propylene glycol, glycerol, and polysorbate 80. Monoglycerides combined with acetic, citric, tartaric, malic, lactic, and/or fumaric acids have been reported to increase the antimicrobial action of C_8–C_{12} saturated fatty acid monoglycerides (54,55). Other workers have found that glycerol and sucrose fatty acid esters enhance the preservative systems of food samples (8,56–62). A detailed discussion of the mechanism of action can be found in a publication by Kabara (63).

5.3 THE PRESERVATIVE SYSTEMS APPROACH

A systems approach to preservation maximizes the multifunctional properties of chemicals so that fewer chemicals can satisfy more needs of the formulator. For example, monoglycerides have been used as emulsifiers in the food industry for more than 60 years. The technology developed for this class of compounds and their record of safety, in which some compounds are generally recognized as safe (GRAS), allows one to use monoglycerides in foods with great confidence. The same arguments can be raised for the phenolic antimicrobials, methylparaben and propylparaben, and to *t*-butyl hydroxyanisole (BHA) and *t*-butylhydroxytoluene (BHT), which are phenolic antioxidants (63). Food-grade chelators, such as EDTA and citric acid, and acidifying agents, such as lactic and citric acids, can potentiate the effect of preservatives and act as deterrents to the chemical oxidation of foods/cosmetics (63,64).

For many years, Kabara (65,66) has advocated creating a hostile environment with food-grade chemicals. This same system can be applied to cosmetic and pharmaceutical products. As in any systems approach, the number and level of variables used depends upon the problem to be solved. Along with determining which level of material will be used, the amount used in a final formulation is important. A 1:1:1 ratio of monolaurin:antioxidant:chelator has been found to be satisfactory. Although the amount of phenolic antioxidant and EDTA allowed for food use is 1000 and 100 ppm, respectively, the amount of monolaurin that can be added is restricted more by taste than by governmental regulations. For most oral applications, 500 ppm or less of the monolaurin is effective and remains undetected by most taste panelists (J. J. Kabara, unpublished data). Higher use levels need not be discouraged, particularly where special problems exist and the bitter or soapy taste can be masked, or if it is not an issue.

Kato and Shibasaki (39) described the usefulness of acidulants in increasing the antimicrobial spectrum and activity of monoglycerides, particularly monolaurin. Acidulants including citric, lactic, malic, and polyphosphoric acids were selected from the standpoint of safety. These agents enhanced the antibacterial effects of monoglycerides and sucrose esters against *E. coli*, *Serratia marcescens*, *Proteus vulgaris*, and *Salmonella typhimurium*. It is believed that the enhanced bactericidal activity of monlaurin in the presence of acids or chelating agents is due to unfavorable intracellular pH (see Sec. 2.3) and increased uptake of the monoglyceride into the cells (41). In addition, Projan et al. (67) reported that glycerol monolaurate inhibits the synthesis of most staphylococcal toxins and exoproteins. Reducing the production of microbial virulence factors in this manner would be expected to decrease the ability of microorganisms to survive in products containing glyceryl laurate as part of their preservative system.

The preceding GRAS agents are effective to various degrees, but all have limitations when used alone. Fatty acids and their derivatives are less effective against gram-negative organisms; the phenolic antioxidants are active against a wide variety of organisms, but are limited by toxicological considerations; and EDTA has little antimicrobial activity by itself and needs to be considered a potentiator of preservative action rather than a biocide in its own right (68) (see Chapter 9). The systems approach uses the multifunctional aspects of selected chemicals and introduces them into products to create a hostile environment for microorganisms. When BHA and EDTA are used, oxidative deterioration as well as microbial growth can be lowered or even eliminated. Such complete preservative systems require that microbiologists have a good understanding of multifunctional additives and are able to describe the "when, where, and how much" for the optimum usage of chemicals available. Manufacturers must be sensitive to consumer desires for safe and wholesome products without biocides; however, preservatives may not simply be avoided, because manufacturers must ensure that formulas are adequately preserved.

The formulator and microbiologist should think in terms of a preservative system, rather than a chemical per se, to protect a product from spoilage. The advantages of this approach are that 1) chemicals already approved for food are used, 2) the chemicals mentioned are well known to the industry and no new technology for their introduction is needed, 3) the use of preservatives increases the risk of sensitization and skin/eye irritation, and 4) the system becomes "self-preserving." The systems approach to product preservation not only solves a microbiological problem, but may help create a new marketing strategy—use of self-preserving (i.e., preservative-free) products that deemphasize the need for classical preservatives.

5.4 SUMMARY

This chapter presented information on the antimicrobial action of fatty acids and their esters. Although the antimicrobial action of fatty acids is increased with increasing unionized fatty acid (i.e., as the pH is lowered below the pKa of the fatty acid) so that it is available to partition into the cell membrane, the antimicrobial action of fatty acid esters is less pH dependent. Fatty acids generally are more effective against gram-positive bacteria than against gram-negative bacteria. Lauric acid and palmitoleic acid are the most active saturated and monounsaturated acids, respectively. In some formulations, the systems approach enables formulators to make self-preserving products without the use of preservatives.

Both glycerol and sugar esters have antimicrobial properties that have been used to advantage in food and cosmetic products. The structure-function relationship for saturated and unsaturated esters follows the activity of their respective fatty acids, which means that lauric and palmitoleic acids form the most active saturated and unsaturated monoesters, respectively. The purity of the ester is important in determining its product performance and antimicrobial action. Monolaurin, 95+% (i.e., Lauricidin), is reported to have greater antimicrobial action than monoesters with lower levels of purity.

The antimicrobial activity and spectrum of activity of fatty acids and esters are enhanced by decreasing pH and the presence of a chelating agent or an antioxidant. The use of multifunctional components in a systems approach to preservation is the touchstone of self-preserving products. This approach allows the development of products with preservative systems that are both effective and mild.

REFERENCES

1. L. R. Beuchat and D. A. Golden. Antimicrobials occurring naturally in foods, *Food Technol.* (Jan):134–142 (1989).
2. V. K. Babayan, T. G. Kaufman, H. Lehjman, and R. J. Tkaczuk. Some uses and applications of poly-glycerol esters in cosmetics and pharmaceutical preparations. *J. Soc. Cosmet. Chem. 15*:473 (1964).
3. J. J. Kabara. Medium-chain fatty acids and esters as antimicrobial agents. In: *Cosmetic and Drug Preservation. Principles and Practice.* J. J. Kabara, ed. Marcel Dekker, New York, 1984, pp. 275–304.
4. A. Fleming. On the antibacterial action of cultures of a *Penicillium* with special reference to their use in the isolation of *B. influenza. Br. J. Exp. Pathol. 10*:226 (1929).
5. R. R. Mellon, P. Gross, and F. B. Cooper. *Sulfanilamide Therapy of Bacterial Infections.* C. C. Thomas, Baltimore, 1939.
6. E. Lueck. *Antimicrobial Food Additives.* Springer-Verlag, New York, 1980.
7. A. W. Hoffman. Neue fluchige Säure der Vogelbeeren. *Ann. Chem. Pharmacol. 34*:129 (1859).
8. B. Strahlmann. Entdeckungsgeschichte antimikrobieller Konservierungsstaffe für Lebensmittel. *Mitt. Geb. Lebensm. Hyg. 65*:96 (1974).
9. E. J. Watkins. Ropiness in flour and bread and its detection and prevention. *J. Soc. Chem. Ind.* (Lond.) *25*:350 (1906).
10. H. Bohles, M. A. Bieber, and W. C. Heird. Reversal of experimental essential fatty acid deficiency by cutaneous administration of safflower oil. *Am. J. Clin. Nutr. 29*:398–401 (1976).
11. R. T. Holman. Essential fatty acid in human nutrition. In: *Function and Biosynthesis of Lipids.* N. G. Bazan, R. R. Brenner, and N. M. Giusto, eds. Plenum, New York, 1977, pp. 515–534.
12. E. Kodicek. The effect of unsaturated fatty acids on gram- positive bacteria. *Soc. Exp. Biol. Symp. 3*:217 (1949).
13. C. Nieman. Influence of trace amounts of fatty acids on the growth of microorganisms. *Bacteriol. Rev. 18*:147 (1954).
14. J. J. Kabara, R. Vrable, and M. Lie Ken Jie. Antimicrobial lipids: natural and synthetic fatty acids and monoglycerides. *Lipids 9*:753 (1977).
15. J. J. Kabara. In *Antimicrobials in Foods.* P. M. Davidwson and A. L. Branen, ed. Marcel Dekker, New York, 1993, pp. 307–342.
16. C. E. A. Winslow and E. E. Lochridge. The toxic effect of certain acids upon typhoid and colon bacilli in relation to the degree of their dissociation. *J. Infect. Dis. 3*:547 (1906).
17. J. D. Reid. The disinfectant action of certain organic acids. *Am. J. Hyg. 16*:510 (1932).
18. C. Hoffman, T. R. Schweitzer, and G. Dalby. Fungistatic properties of the fatty acids and possible biochemical significance. *Food Res. 4*:539 (1939).
19. A. S. Levine and C. R. Fellers. Action of acetic acid on food-spoilage microorganisms. *J. Bacteriol. 39*:499 (1940).

20. A. Albert. *Selective Toxicity.* Methuen, London, 1960.
21. G. G. Meynell. Some factors affecting the resistance of mice to oral infection by *Salmonella. Proc. R. Soc. Med. 48*:916 (1955).
22. H. N. Prince. Effect of pH on the antifungal activity of undecylenic acid and its calcium salt. *J. Bacteriol. 78*:788 (1959).
23. S. Tetsumoto. Sterilizing action of acids. III. Sterilizing action of saturated monobasic fatty acids. *J. Agric. Chem. Soc. Jpn. 9*:563 (1933)
24. J. J. Kabara, D. M. Swieczkowski, A. J. Conley, and J. P. Truant. Fatty acids and derivatives as antimicrobial agents. *Antimicrob. Agents Chemother. 2*:23 (1972)
25. J. A. Sands, D. A. Auperin, P. D. Landin, A. Reinhardt, and S. P. Cadden. Antiviral effects of fatty acids and derivatives. In: Pharmacological Effect of Lipids. J. J. Kabara, ed. American Oil Chemists Society, Champaign, IL, 1979, p. 75.
26. M. M. Rieger. *Surfactants in Cosmetics.* Marcel Dekker, New York, 1985, pp. 25–26.
27. A. J. Conley and J. J. Kabara. Antimicrobial action of esters of polyhydric alcohols. *Antimicrob. Agents Chemother. 4*:501 (1973).
28. J. J. Kabara. Synergistic microbiocidal composition and methods. U.S. Patent 4,067,997, January 10, 1978.
29. J. J. Kabara. A new preservative for food. *J. Food Safety 4(1):13 (1982).*
30. N. Kato and I. Shibasaki. Comparison of antimicrobial activities of fatty acids and their esters. *J. Ferment. Technol. 53*:793 (1975).
31. I. Shibasaki and N. Kato. Combined effects on antibacterial activity of fatty acids and their esters against gram-negative bacteria. In: *Pharmacological Effect of Lipids.* J. J. Kabara, ed. American Oil Chemists Society, Champaign, IL, 1979, p. 15.
32. L. R. Beuchat. Comparison of anti-*Vibrio* activities of potassium sorbate, sodium benzoate and glycerol and surose esters of fatty acids. *Appl. Environ. Microbiol. 39*:1178 (1980).
33. S. Ueda and H. Tokunaga. Antiseptic effect of capric monoglyceride on pellicle-forming yeast. *Season SC 13*:1 (1966).
34. T. Koga and T. Watanbe. Antiseptic effect of volatile fatty acid monoglycerides. *J. Food Sci. Technol.* (Jpn). *15*:297 (1968).
35. A. Kato and K. Arima. Inhibitory effect of sucrose ester of lauric acid on the growth of *E. coli. Biochem. Biophys. Res. Commun. 42*:596 (1971).
36. I. Shibasaki. Recent trends in the development of food preservatives. *J. Food Safety 4*:35 (1982).
37. J. J. Kabara. Fatty acids and derivatives as antimicrobial agents—a review. In: *Pharmacological Effects of Lipids.* J. J. Kabara, ed. American Oil Chemists Society, Champaign, IL, 1979, p. 1.
38. A. Kato and I. Shibasaki. Combined effect of different drugs on the antibacterial activity of fatty acids and their esters. *J. Antibacteriol. Antifung. Agents* (Jpn.) *8*:355 (1975).
39. N. Kato and I. Shibasaki. Combined effect of citric and polyphosphoric acid on the anti-bacterial activity of monoglycerides. *J. Antibacteriol. Antifung. Agents* (Jpn.) *4*:254 (1976).
40. A. Moustafa and I. R. Agin. Use of monolaurin in a margarine product. Report of ISF/AOCS World Congress, Abstract 42 (1980).
41. T. Tsuchido, T. Saeki, and I. Shibasaki. Death kinetics of *Escherichia coli* in a combined treatment of heat and monolaurin. *J. Food Safety 3*:57 (1981).

42. M. Kimsey, H. R. Adams, and J. J. Kabara. Increased inactivation of bacterial spores in the presence of monolauryl-glyceride. *J. Food Safety 3*:69 (1981).
43. S. Naruki, T. Tsuchido, and I. Shibasaki. Inhibition of germination and out-growth of *Bacillus subtilis* var. niger spores by dodecanoic acid and its monoglyceride. *Ann. Rep. Int. Center Coop. Res. Dev. Micro. Eng.* (Jpn.) 161–171 (1983).
44. J. R. Unda, R. A. Molins, and H. W. Walker. *Clostridium sporogenes* and *Listeria monocytogenes*: survival and inhibition in microwave-ready beef roasts containing selected antimicrobials. *J. Food Sci. 56*:198–205 (1991).
45. M. Takano, A. B. Simbol, M. Yasin, and I. Shibasaki. Bactericidal effect of freezing with chemical agents. *J. Food Sci. 44*:112–115 (1979).
46. P. Butz, G. Habiscon, and H. Ludwig. Influence of high pressure on lipid coated virus. In: *High Pressure and Biotech.* C. Balny, R. Hayashi, K. Heremans, and P. Maason, eds. Colloque INSERNM/John Libby Eurotext Ltd., 1992, pp. 61–64.
47. B. B. Hite. The effect of pressure in the preservation of milk. Bull. 58, W. Va. Agric. Exp. Sta., Morgantown, 1899, pp. 15–35.
48. N. Kato. Antimicrobial activity of fatty acids and their esters against film-forming yeast in soy sauce. *J. Food Safety 3*:121 (1981).
49. J. L. Smith and S. A. Palumbo. Inhibition of aerobic and anaerobic growth of *Staphylococcus aureus* in a model sausage system. *J. Food Safety 2*:221 (1980).
50. J. J. Kabara. Inhibition of *Staphylococcus aureus* in a model agar-meat system by monolaurin. *J. Food Safety 6*:197–201 (1984).
51. K. Katagiri, T. Owaki, and K. Kino. Sausage preservative containing monoglycerides, phosphates, and sorbic acid. Japan Kokai Tokkyo Koho JP 55/15728 (8/15728), February 4, 1980.
52. M. C. Robach, C. Hickey, and E. C. To. Comparison of antimicrobial action of monolaurin and sorbic acid. *J. Food Safety 3*:89–98 (1981).
53. M. Yuda, H. Matsuura, A. Takeuchi, and T. Iwasaki. Preservatives for starch-containing foods. Japan Kokai JP 52/21317 (77/21317), February 17, 1977.
54. K. Katagiri and H. Yamamoto. Food preservatives. Japan Kokai Tokkyo Koho JP 53/148528 (78/148528), December 15, 1978.
55. H. Ogawa. Quality improvement of frozen cooked foods. Japan Tokkyo Koho JP55/2269 (80/2269), January 16, 1980.
56. I. Shibasaki, K. Yuta, H. Matsuura, and T. Iwasaki. Food preservative composition. Japan Tokkyo Koho JP 54/34061 (79/34061), October 24, 1979.
57. R. C. Baker, D. Kline, W. Poon, and D. H. Vadehra. Antimicrobial properties of Lauricidin in mechanically deboned chicken and minced fish. *J. Food Safety 4*:177 (1982).
58. M. A. Hall and A. J. Maurer. Spice extracts, Lauricidin and propylene glycol as inhibitors of *Clostridium botulinum* in turkey frankfurter slurries. *Poultry Sci. 65*:1167–1171 (1986).
59. M. Catsara, J. P. Danjoy, and R. Seynave. Temperature and action of Laurilac on multiplication of *Salmonella* in minced meat. *Bull. Acad. Vet. Fr. 60*:359 (1987).
60. D. D. Duxbury. Combination emulsifier/acidulant extends cheese sauce shelf life. *Food Proc. 46*(9):38–39 (1986).
61. N. Lisker and N. Poster. Antifungal activity of Lauricidin and related compounds. *J. Food Safety 4*:27 (1982).

62. T. Furuta, K. Komori, and A. Ueda. Combined effects of various food additives on the bactericidal activity of ethanol against *Escherichia coli and Staphylococcus aureus. Bokin Bobai* (Jpn.) *12*(2):65–72 (1984).
63. J. J. Kabara. Medium chain fatty acids and esters. In: *Antimicrobials in Foods.* P. M. Davidson and A. L. Branen, eds. Marcel Dekker, New York, 1993, pp. 330–333.
64. A. L. Branen, P. M. Davidson, and B. Katz. Antimicrobial properties of lipids and phenolic antioxidants. *Food Technol. 34*:42 (1980).
65. J. J. Kabara. Multi-functional food-grade preservatives in cosmetics. *Drug Cosmet. Ind. 125*:60 (1979).
66. J. J. Kabara. GRAS antimicrobial agents for cosmetic products. *J. Cosmet. Chem. 31*:1 (1980).
67. S. J. Projan, S. Brown-Skrobot, P. M. Schlievert, F. Vandenesch, and R. P. Novick. Glycerol monolaurate inhibits the production of β-lactamase, toxic shock syndrome toxin-1, and other staphylococcal exoproteins by interfering with signal transduction. *J. Bacteriol. 176*:4204–4209 (1994).
68. M. A. Kenward, M. R. Brown, and J. J. Fryer. The influence of calcium or manganese on the resistance of EDTA, polymyxin B or cold shock, and the composition of *Pseudomonas aeruginosa* grown in glucose or magnesium depletion batch cultures. *J. Appl. Bacteriol. 45*:489 (1979).

6

Biomimetic Phospholipids: Components for Self-Preservation

Dennis L. Fost
Mona Industries, Inc.
Paterson, New Jersey

John I. Yablonski
Bio-Control Consultants, Inc.
Westfield, New Jersey

Phospholipids represent a broad range of compounds that play a number of key roles in mammalian and nonmammalian cellular systems. Natural phospholipids can be subdivided into glycerophospholipids and sphingophospholipids. While both groups are important from a biochemical point of view, the glycerophospholipids have acquired much greater commercial significance, especially in the food industry where they are widely used as emulsifiers. In the personal and health care fields, however, use of commercial glycerophospholipids has been limited by a number of drawbacks such as color instability, formulating difficulties, inhibition of preservative systems, and the high cost of purified phospholipid fractions. Commercial phospholipids can be derived from animal or vegetable sources, but from a practical standpoint, most phospholipids in use today are derived from soybean oil and are most often described generically as lecithins (1). Recently, a new class of phospholipids has been developed. These products have been described as *biomimetic*, in that they mimic structurally and functionally the natural phospholipids (2). This chapter will focus on the structural features of the natural and biomimetic phospholipids, their applications in personal care products, and their use in self-preserved products.

6.1 ROLE OF PHOSPHOLIPIDS IN CELLULAR MEMBRANES

Phospholipids are a key component in cellular membranes, forming a matrix composed of bilayer lamellar aggregates. The bilayer arrangement of the phospholipid molecules is head to head or tail to tail, which gives rise to a stable structure that keeps the cellular membrane intact and also helps control many of the cell's biological functions (3).

Structurally, phospholipids are characterized by multiple hydrophobic tail groups and a highly polar head group, which is conducive to the formation of membrane bilayers and multilamellar aggregates. The chemical structures of the most common phospholipids are as follows:

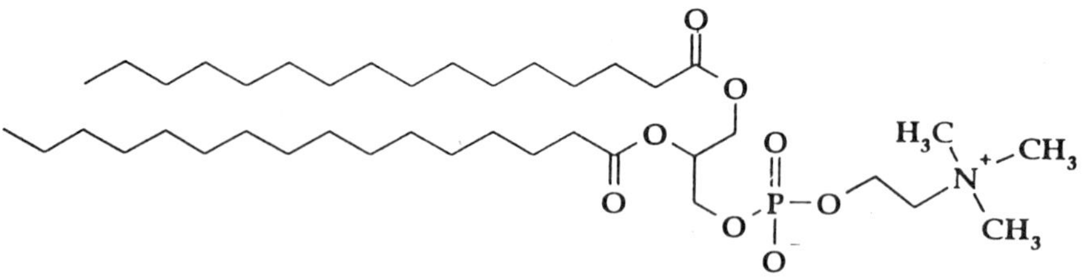

Phosphatidyl Choline

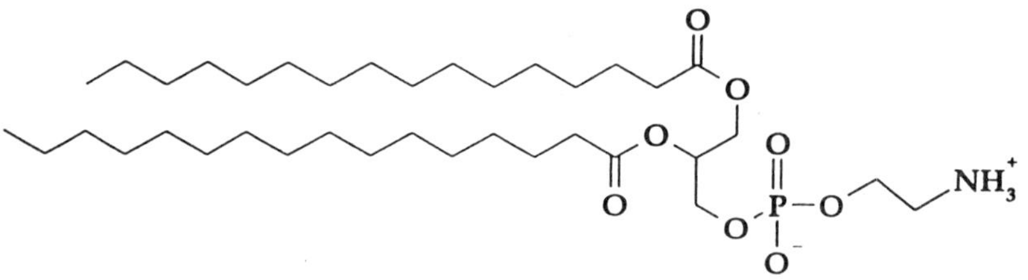

Phosphatidyl Ethanolamine

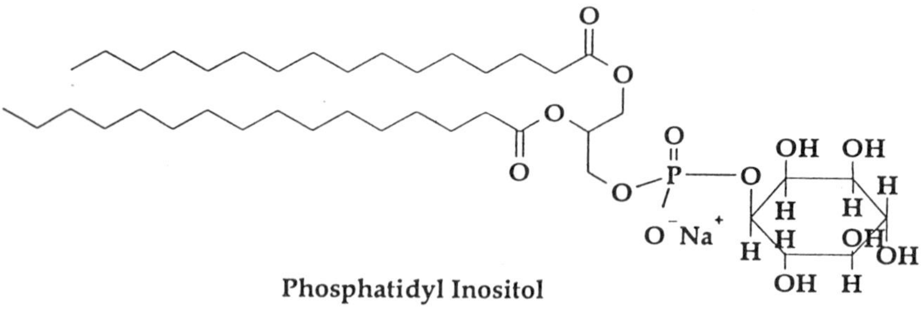

Phosphatidyl Inositol

As can be seen from their chemical structures, the natural glycerophospholipids generally contain two hydrophobic tails composed of long-chain (C_{16}-C_{18}) fatty acid (saturated and/or unsaturated moieties) and a hydrophilic head group that consists of a phosphate ester and a quaternary ammonium, primary ammonium, or sugar alcohol (inositol) group. The common spatial feature among the phospholipids is their overall cylindrical shape, which favors a molecular packing arrangement that provides stable membrane bilayers. In the case of phosphatidyl choline and phosphatidyl ethanolamine, the hydrophilic and zwitterionic head groups provide sites for binding water, ionic materials, and proteins within the cellular membrane matrix (4), thus generating a mechanism for moisture level and osmotic pressure control within the cells (5). Phosphatidyl inositol, on the other hand, plays a role in the phosphoinositide cascade, a series of biochemical reactions that take place in plasma membranes, that controls calcium transport and forms part of the primary action of neurotransmitters (6–8).

In the epidermis, cells go through a process of differentiation which involves the upward movement of cells from the deepest level, the basal cells, to the intermediate level, the granular cells, and finally to the uppermost level, the stratum corneum. In addition, further upward migration to the outer stratum corneum results in desquamation, or the sloughing off of the top layer of cells. During the process of differentiation, the lipid composition in the cells undergoes significant changes. In a study by Lampe et al. (9), human abdominal skin was used as a model to measure the changes in lipid composition during epidermal differentiation (Table 6.1).

In the stratum corneum, the phospholipids have been depleted and replaced by more hydrophobic lipids comprised of free fatty acids, triglycerides, wax esters,

TABLE 6.1 Human Skin Lipid Composition During Epidermal Differentiation

Lipid type	Basal lamina % (w/w)	Stratum granulosum % (w/w)	Stratum corneum % (w/w)
Phospholipids	45	25	5
Sterols	11	12	14
Fatty acids	7	9	19
Triglycerides	12	25	25
Wax esters	5	5	5
Hydrocarbons	9	8	11
Ceramides	4	9	18
Glucosylceramides	4	6	0
Cholesterol sulfate	2	6	2

Source: Ref. 9.

hydrocarbons, and ceramides, which account for 92% of the total lipid composition. These lipids reside in the intercellular spaces where they help hold the protein-enriched, cornified cells together forming what Elias refers to as the "mortar," whereas the corneocytes make up the "bricks" (10). This two-compartment stratum corneum lipid matrix provides essential moisure barrier properties and helps to protect the cells below from environmental damage. The ceramides, which play an extremely important role in controlling the moisture barrier properties of the stratum corneum, are derived via sphingomyelinase-mediated dephosphorylation from another group of phospholipids, namely, the sphingophospholipids. Sphingophospholipids and their dephosphorylated analogs are involved in a variety of cellular functions including cytochemical reactions leading to the formation of lamellar bodies in epidermal granular cells (11). The structural features of this class of phospholipids can best be illustrated with the structure of sphingomyelin:

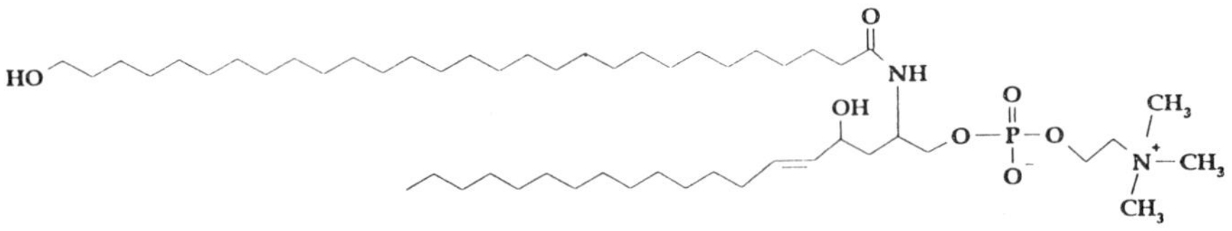

Sphingomyelin

Once again, nature has provided a structure with multiple tail groups, which favor bilayer lamellar aggregation. The ceramides, which contain hydrophobic tail groups with up to 30 carbon atoms, are essential in the formation of a multilamellar matrix in the stratum corneum that fills the intercellular spaces (12).

Under adverse environmental conditions, the stratum corneum moisture barrier properties can be degraded, which, for the most part, is a physicochemical occurrence resulting from the breakdown of the stratum corneum lipid matrix. Even though the ceramides are present in high concentration in the stratum corneum, breakdown of the lipid matrix occurs readily since the ceramides by themselves are not capable of forming bilayers, but need cholesterol sulfate, which is a minor component and quite easily depleted (13). Once microlesions are formed in the skin, damage to the cells below can occur resulting in clinical manifestations such as dermatitis, chapping, etc. For many years, cosmetic scientists have formulated skin treatment products using nonpolar and occlusive materials to replenish the lipids that are lost from the stratum corneum, which can occur, for example, from regular use of detergents to cleanse the skin. Topically effective skin treatment products, however, should ideally utilize a combination of nonpolar occlusive lipids and polar nonocclusive lipids in order to deliver a lipid composition that will interact with the natural lipids in the stratum corneum in such a manner so as to repair the discontinuities in the bilayer lipid structures, restore its moisture barrier properties, and prevent crystallinity and brittleness of the lipid matrix in the intercellular spaces.

6.2 BIOMIMETIC PHOSPHOLIPIDS

While natural phospholipids provide the requisite compatibility with epidermal lipids, their topical efficacy is limited by their minimal substantivity to the skin. In addition, while the natural phospholipids are structurally well suited for forming plasma membranes, they are not adequate for the moisture barrier in the stratum corneum (14). The commercially available phospholipids, such as lecithin, also have the drawback of having an adverse effect on the preservative systems used to protect topical preparations (15,16). Nevertheless, lecithin, which is available commercially in a variety of grades ranging in phosphatidyl choline content from 25 to 90%, has found limited use as an emulsifier in personal and health care applications.

Recently, a new family of phospholipids has been developed in which the arrangement of the phosphate and quaternary groupings has been reversed in order to overcome the disadvantages of the natural phospholipids in topical applications. The term biomimetic has been used to indicate that these materials have been designed to closely simulate, through topical application to the skin, the function of the natural stratum corneum lipids. A broad range of functional properties has been achieved with these phospholipids through the careful selection of specific vegetable oils as raw materials. Rather than interfering with preservative systems as occurs with other commercial phospholipids, the biomimetic phospholipids actually display potent antimicrobial activity without showing significant mammalian toxicity or skin irritation effects (17).

Based on phosphate ester chemistry, the biomimetic phospholipids consist of diester and triester phosphatides, which provide molecules that structurally resemble the natural phospholipids, as illustrated by the following representative diester structure:

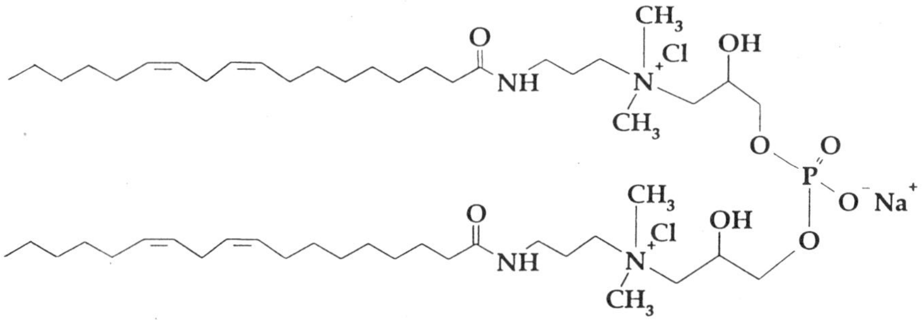

Linoleamidopropyl Phosphatidyl PG-Dimonium Chloride

The most significant structural difference between the natural phospholipids and the biomimetic phospholipids is that the cationic groups are near the middle of the molecule in the biomimetic phospholipids, whereas the cationic group is in a terminal position in the natural phospholipids. As a result, the biomimetic phospholipids not only have multiple hydrophobic groups, but multiple cationic groups that provide them with a high degree of substantivity to proteinaceous substrates. The substantivity of the biomimetic phospholipids has been demonstrated using a rubine dye substantivity test.

An additional feature of several of the biomimetic phospholipids is that they utilize an amide linkage between the fatty acid groups and the polar groups, simulating the ceramide structural characteristics and providing better long-term stability than the natural phospholipids, which contain ester linkages. Due to the ampholytic character of the biomimetic phospholipids, they remain completely compatible with virtually all other types of ingredients including anionic surfactants.

On a commercial basis, five different biomimetic phospholipids have been developed to provide a broad range of performance properties (18). These products have been derived from various vegetable oils as starting materials and are listed in Table 6.2.

TABLE 6.2 Vegetable-Derived Biomimetic Phospholipids[a]

Description	Oil source	Primary R-group
Stearamidopropyl phosphatidyl PG-dimonium chloride	Palm	Stearyl
Linoleamidopropyl phosphatidyl PG-dimonium chloride	Safflower	Linoleyl
γ-Linolenamidopropyl phosphatidyl PG-dimonium chloride	Borage	γ-Linolenyl
Cocamidopropyl phosphatidyl PG-dimonium chloride	Coconut	Cocyl
Coco phosphatidyl PG-dimonium chloride	Coconut	Cocyl

[a]Biomimetic phospholipids commercially available under trade names Phospholipid SV, Phospholipid EFA, Phospholipid GLA, Phospholipid PTC, and Phospholipid CDM from Mona Industries, Inc., 76 E. 24th St., Paterson, NJ, 07544-0425.

In contrast to the natural phospholipids, which are generally made water dispersible by the addition of a surfactant, the biomimetic phospholipids are inherently water soluble or dispersible, making the formulation of these materials into water-based systems quite easy. Typical properties of the biomimetic phospholipids are shown in Table 6.3

6.3 ANTIMICROBIAL ACTIVITY OF THE BIOMIMETIC PHOSPHOLIPIDS

The original research investigation into the creation of the biomimetic phospholipids had as its primary goals the development of a class of uniquely mild and efficacious surfactants that would (1) encompass the highly desirable quality of forming stable emulsions with a wide variety of emollients, oils, and waxes, (2) be compatible with a wide range of natural and synthetic formulation ingredients, (3) perform over a broad range of pH, (4) have excellent skin-cleansing properties, and (5) mimic as closely as possible the various positive attributes and functions of naturally occurring phospholipid compounds. The concepts of broad-spectrum antimicrobial activity and suitability for enhancing formulation resistance to microbial contamination had been originally theorized; however, the magnitude of this application was not readily apparent until further formulation investigation was initiated. It was subsequently observed that this class of materials displayed a relatively broad range of activity against both gram-negative and gram-positive bacteria as well as commonly occurring yeasts and molds (18). The medium chain length biomimetic phospholipids were found to be especially active and they

TABLE 6.3 Typical Properties of Biomimetic Phospholipids

Chemical description	Appearance	Color	Iodine	Nitrogen	Concentration
Stearamidopropyl phosphatidyl PG-dimonium chloride	Paste	3	<1	1.8	41
Linoleamidopropyl phosphatidyl PG-dimonium chloride	Liquid	3–4	22	1.5	30
γ-Linolenamidopropyl phosphatidyl PG-dimonium chloride	Liquid	4–5	12	1.9	36
Cocamidopropyl phosphatidyl PG-dimonium chloride	Liquid	2–3	2	2.7	47
Coco phosphatidyl PG-dimonium chloride	Liquid	1	<1	1.4	37

appeared to possess many of the attributes of the so-called ideal preservative, while avoiding the usual adverse qualities and formulation restrictions normally associated with commonly used antimicrobial compounds. Of particular interest were cocamidopropyl phosphatidyl PG-dimonium chloride and coco phosphatidyl PG-dimonium chloride, which were found to be applicable as functional surfactants, conditioning agents and preservative enhancers (19).

6.4 COCAMIDOPROPYL PHOSPHATIDYL PG-DIMONIUM CHLORIDE AND COCO PHOSPHATIDYL PG-DIMONIUM CHLORIDE

Based on coconut oil as the primary raw material, cocamidopropyl phosphatidyl PG-dimonium chloride and coco phosphatidyl PG-dimonium chloride are readily water soluble and exhibit a surprisingly broad range of functional characteristics including gentle cleansing and foaming properties, anti-irritation effects when combined with anionic surfactants, skin conditioning, and broad-spectrum antimicrobial activity, and they are virtually nonirritating to skin and eyes. To illustrate the range of antimicrobial properties, minimum inhibitory concentration (MIC) data for cocamidopropyl phosphatidyl PG-dimonium chloride against a variety of organisms is presented in Table 6.4.

The effective range of antimicrobial activity against a variety of bacteria, yeast, and molds is in the range of 50–1250 ppm. Both gram-positive and gram-negative bacteria appear to be highly susceptible, which may be related in part to the unique hydrophilic/lipophilic characteristics of the biomimetic phospholipid molecule. Pseudomonads, including *Pseudomonas aeruginosa*, were affected by phospholipid levels of 150–300 ppm.

Somewhat less activity was apparent with the Enterobacteriaceae; however, the activity is well within acceptable and applicable levels of use for most formulations. The antifungal activity demonstrated by the biomimetic phospholipids is considered to be important in their selection as multifunctional materials with preservative activity. *Candida albicans*, *Penicillium spp.*, and *Aspergillus niger* were inhibited at phospholipid levels of ≤800 ppm. The MICs of these compounds are compared with methyl paraben, DMDM hydantoin, and 2-bromo-2-nitropropane-1,3-diol in Table 6.5.

The pH, as measured by a modified bactericidal suspension test (Kill Time method, DGHM, Germany, 1981), does not appear to have significant influence on the antimicrobial activity of the biomimetic phospholipids. As presented in Table 6.6, the biomimetic phospholipids appear to have an effective pH range of 4.0–8.0 with no detectable variation in overall effectiveness (18). The shorter kill times observed at pH 4 and pH 8 may be attributable to the effect of pH on the growth rate and susceptibility of the test organism. The effective pH range would be suitable for most product applications. It should be cautioned, however, that

TABLE 6.4 Antimicrobial Activity of Cocamidopropyl Phosphotidyl PG-Dimonium Chloride (MIC Method)

Test organism	ATCC number	Type	MIC (ppm)
Staphylococcus aureus	6538	Gram +	300
Staphylococcus epidermidis	14990	Gram +	300
Streptococcus faecalis	6569	Gram +	300
Bacillus subtilis	6633	Gram +	150
Micrococcus luteus	4698	Gram +	300
Escherichia coli	8739	Gram –	50
Proteus mirabilis	9921	Gram –	50
Pseudomonas aeruginosa	15442	Gram –	300
Pseudomonas cepacia	25608	Gram –	150
Pseudomonas stutzeri	17591	Gram –	150
Salmonella choleraesuis	10708	Gram –	1250
Enterobacter aerogenes	13048	Gram –	1250
Klebsiella pneumoniae	13883	Gram –	1250
Aeromonas hydrophila	9071	Gram –	50
Candida albicans	10259	Yeast	800
Aspergillus niger	6275	Mold	625
Penicillium spp.	1117	Mold	75

MIC data were developed employing a typical broth dilution protocol (18, 19). Growth conditions (incubation time, temperature, and growth media) were those appropriate for the specific organisms being evaluated.

TABLE 6.5 Antimicrobial Activity of Biomimetic Phospholipids Compared to Conventional Preservatives (MIC Method)

Product	*A. niger* (ppm)	*C. albicans* (ppm)	*Penicillium* spp. (ppm)
Methyl paraben	1000	1000	500
DMDM hydantoin	1400	8700	—
2-Bromo-2-nitropropane-1,3-diol (Bronopol[a])	3200	400	1600
Cocamidopropyl phosphatidyl PG-dimonium chloride	625	800	75
Coco phosphatidyl PG-dimonium chloride	80	160	160

[a]Trademark of Angus Chemical Co.
MIC data were developed employing a typical broth dilution protocol (18, 19).Growth conditions (incubation time, temperature, and growth media) were those appropriate for the specific organisms being evaluated.

Table 6.6 Effects of pH on Bactericidal Activity of Cocamidopropyl Phosphatidyl PG-Dimonium Chloride Against *S. epidermidis* (ATCC 14990)

Time (min)	Colony forming units/ml at: pH4	pH5	pH6	pH7	pH8
0.0	80	52,000	180,000	64,000	60,000
0.5	<10	1,300	3,000	400	<10
1.0	<10	40	600	<10	<10
5.0	<10	<10	<10	<10	<10
10.0	<10	<10	<10	<10	<10

Kill time recovery data was developed employing trypticase soy broth containing inactivators incubated at 32–35°C for 24–48 hours.
Source: Ref. 18. (20–23).

individual formulations may react somewhat differently depending on the interactions of the ingredients (20–23). This may influence the functional pH range for that particular application, and, therefore, the system should be adjusted accordingly.

Extended antimicrobial application studies have been reported (18) indicating the potential skin substantive properties of the biomimetic phospholipids. When evaluated for antimicrobial human skin substantivity effectiveness in a 33% soap slurry, panel studies utilizing a modified finger imprint technique clearly demonstrated the biomimetic phospholipid to be reasonably substantive after multiple finger washes. Data reflecting the residual antimicrobial activity as measured by the finger imprint technique are presented in Table 6.7.

The versatility of cocamidopropyl phosphatidyl PG-dimonium chloride was further confirmed by studies conducted to determine its antimicrobial substantivity characteristics on fiber materials. As demonstrated in Table 6.8, phospholipid-

Table 6.7 Antimicrobial Activity/Substantivity of Cocamidopropyl Phosphatidyl PG-Dimonium Chloride on Skin Tested in Soap Solution (33% by weight)

	Control (soap slurry only)	3% Conc. (as supplied)	5% Conc. (as supplied)
Initial exposure	2.0	4.0	4.0
Rinse 1	1.6	2.6	3.0
Rinse 2	1.0	2.0	2.3
Rinse 3	1.3	1.3	2.0
Untreated	1.3	1.3	1.3

Source: Ref. 18.

TABLE 6.8 Antimicrobial Activity/Substantivity of Cocamidopropyl Phosphatidyl PG-Dimonium Chloride on Fiber (Zone of Inhibition Method)

Swatch (2 × 2 cm)	1% (as supplied)	3% (as supplied)
Untreated control	0	0
Initial treatment	4	4
Wash 1	4	4
Wash 2	3	3
Wash 3	3	3

Procedure: Swatches of fiber material (Fort Howard air-laid web, cellulose/cotton blend) were exposed to the test materials by dipping into the test solution and blotting to remove excess moisture. Imprints were then made on seeded agar plates (inoculated with *S. epidermidis*).The swatches were rinsed and blotted to remove excess moisture. A series of four rinsings and imprints were made, including the initial exposure and imprint. The degree of residual activity/fiber substantivity was qualitatively determined by the size and clarity of the zone of inhibition surrounding the imprints on the seeded agar plates as compared to the untreated controls.
Grading system: 0, No Activity; 1, Slight; 2, Moderate; 3, Good; 4, Excellent

treated fiber swatches continued to exhibit antimicrobial residual effects after several rinses. The degree of fiber substantivity was determined by means of the *zone of inhibition method*, whereby the effects were measured by the clarity of the zones surrounding the imprints on seeded agar plates. It is theorized that increasing the percentage of natural fiber versus synthetic fiber in the blend would further enhance the degree of substantivity, thus allowing for significantly greater and more pronounced long-term antimicrobial activity.

The second coconut oil–based biomimetic phospholipid, coco phosphatidyl PG-dimonium chloride, is similar in chemical structure to cocamidopropyl phosphatidyl PG-dimonium chloride, but does not contain the amidopropyl linkage. Removal of the amidopropyl moiety reduces the size and molecular weight of the molecule, thus theoretically increasing its ability to penetrate the cell wall. In various tests, it has been found to be even more antimicrobially active than the cocamidopropyl-based phospholipid, and yet it retains similar mild cleansing and conditioning properties. Comparative antimicrobial properties (MICs) are shown in Table 6.9.

Due to its relatively higher degree of activity against microorganisms, coco phosphatidyl PG-dimonium chloride can be used to advantage in substantially reducing the preservative requirements in many personal care formulations. These effects are illustrated by the data in Table 6.10, in which this compound was shown

Table 6.9 Comparative Antimicrobial Data (MIC Method)

Test organism	ATCC #	Type	Cocamidopropyl phosphatidyl PG-dimonium chloride (ppm)	Coco phosphatidyl PG-dimonium chloride (ppm)
S. aureus	6538	Gram-positive	300	20
E. coli	8739	Gram-negative	50	160
P. aeruginosa	15442	Gram-negative	300	160
S. choleraesuis	10708	Gram-negative	1250	40
C. albicans	10259	Yeast	800	160
A. niger	6275	Mold	625	80

MIC data were developed employing a typical broth dilution protocol (18, 19). Growth conditions (incubation time, temperature, and growth media) were those appropriate for the specific organisms being evaluated.

to enhance the overall preservative activity in a creme formulation challenged with mixed bacteria and mixed fungi inocula.

The data in Table 6.10 show that the combination of 0.5% coco phosphatidyl PG-dimonium chloride and 0.05% imidazolidinyl urea gave adequate preservation, with decimal reduction times (D-values) of <4 hours for bacteria and <4.8 hours for the fungi. The preservative efficacy of this formulation was superior to the creme base that contained 0.2% imidazolidinyl urea. Preservative efficacy testing of the same creme base containing 0.5% biomimetic phospholipid and 0.05% methyl paraben (and no imidazolidinyl urea) gave D-values of 7.1 hours for the mixed bacteria inoculum and <4.8 hours for the fungi.

One of the most interesting properties of the biomimetic phospholipids is their low order of mammalian toxicity. Table 6.11 shows the toxicological profile of cocamidopropyl phosphatidyl PG-dimonium chloride and coco phosphatidyl PG-dimonium chloride. In addition, these compounds show anti-irritant activity in anionic surfactant systems using the Eytex Rapid Membrane Assay (trademark of In Vitro International, Inc.). In this method, the Draize Equivalency Score was reduced by approximately 18% using an 80:20 mixture of sodium lauryl sulfate and cocamidopropyl phosphatidyl PG-dimonium chloride, as compared with sodium lauryl sulfate alone.

Studies have been conducted to determine whether the biomimetic phospholipids are inactivated by substances used to inhibit antimicrobial activity of preservatives. The data in Table 6.12 indicate that, for the most part, these materials have no apparent adverse effect on the antimicrobial activity of the phospholipid in MIC assays using *Staphylococcus aureus* and *Escherichia coli*. Although a slight increase in MIC was observed in the presence of 5% horse serum with *S. aureus*,

Table 6.10 Accelerated Preservation Study of Coco Phosphatidyl PG-Dimonium Chloride with Germall® 115

	Number of days				
Test sample	0	1	2	5[a]	7
Creme base⁺ 0.1% Germall® 115					
Mixed bacteria	1.0×10^6	1300	1100	720	240
Mixed fungi	1.0×10^5	700	520	510	300
Creme base⁺ 0.2% Germall® 115					
Mixed bacteria	1.0×10^6	1100	1500	2700	2300
Mixed fungi	1.0×10^5	200	<10	<10	<10
Creme base⁺ 0.05% Germall® 115 + 0.5% coco phosphatidyl PG-dimonium chloride					
Mixed bacteria	1.0×10^6	<10	<10	<10	<10
Mixed fungi	1.0×10^5	<10	<10	<10	<10

Creme base formula:	% by wt.
Steareth-20	1.4
Glycerin	5.0
Steareth-2	4.6
Isopropyl myristate	8.0
Cetyl alcohol	2.0
Water	79.0
	100.0

Test organisms
Mixed bacteria:
P. aeruginosa ATCC 15442
E. coli ATCC 8739
S. aureus ATCC 6538
Mixed fungi:
A. niger ATCC 9642
P. luteum ATCC 9644
C. albicans ATCC 10231

Germall® 115 (imidazolidinyl urea) is a trademark of Sutton Laboratories, Div.of ISP, Inc.
[a]Reinoculation.

it was not considered to be significant. It is believed that similar effects would be observed with cocamidopropyl phosphatidyl PG-dimonium chloride due to its closely related chemical structure. Additional investigation is needed to determine potential interactions of the biomimetic phospholipids with other ingredients, such as anionic surfactants, since one would expect an anionic/cationic interaction would interfere with antimicrobial action.

Continued research into the multifunctional characteristics of the biomimetic phospholipids led to the evaluation of these substances for potential antiviral properties. Both cocamidopropyl phosphatidyl PG-dimonium chloride and coco phosphatidyl PG-dimonium chloride exhibited significant antiviral properties against human immunodeficiency virus, herpes simplex type 2, and human influenza A virus using U.S. EPA Pesticide Assessment Guidelines, Subdivision G: Product Performance, 1982, Sec. 91–30, Pp. 72–76 (25).

Determination of antimicrobial action by incorporation of test materials into growth medium revealed a qualitative improvement in antimicrobial action against

TABLE 6.11 Toxicological Profile of Coconut-Based Biomimetic Phospholipids

Property	Cocamidopropyl phosphatidyl PG-dimonium chloride	Coco phosphatidyl PG-dimonium chloride
Oral LD_{50} (rats)	>5 ml/kg	No data available
Human occluded patch test, 48 hour, 50 test subjects, 5% by wt	48 out of 50—no erythema; 2 out of 50—barely perceptible erythema	48 out of 50—no erythema; 2 out of 50—barely perceptible erythema
Eytex® (a trademark of In Vitro International, Irvine, California, USA), rapid membrane assay, 3% by wt	Draize equivalency score: 12.2 (classification: minimal/mild)	Draize equivalency score: 18.6 (classification: mild)
Skin sensitization [Magnusson and Kligman Guinea Pig maximization (24)], 5% by wt	Not a sensitizer	No data available
Skin sensitization (repeated insult occluded patch test), 50 subjects, 3% by wt	No data available	Not a sensitizer
Cytotoxicity (agar diffusion test using mouse fibroblast L929 cell line)	Noncytotoxic (2.0% by wt): meets requirements of USP XXII	Noncytotoxic (0.25% by wt): meets requirements of USP XXII

S. aureus, *E. coli*, *P. aeruginosa*, and *C. albicans* when 0.05% stearamidopropyl phosphatidyl PG-dimonium chloride was added to 0.05% methyl paraben or 0.05% imidazolidinyl urea (Table 6.13). Studies such as this provide a rationale for using stearamidopropyl phosphatidyl PG-dimonium chloride as a component of product preservative systems.

Numerous scientific articles have been published showing the potential benefits of essential fatty acids (EFA) in the skin (26–28). Linoleamidopropyl phosphatidyl PG-dimonium chloride is derived from safflower oil and contains the highest level of linoleic acid obtainable in a phospholipid derived from a natural oil. This compound has MICs of 40, 150, 150, and 300 ppm for *S. aureus*, *C. albicans*, *A. niger*, and *Penicillium spp.*, respectively, which suggests that it also has potential application as a component of product preservative systems.

Table 6.12 Antimicrobial Inactivation Studies with Coco Phosphatidyl PG-Dimonium Chloride (MIC Method)

Inactivator	Test organism	MIC (ppm)
Distilled water	*S. aureus* (ATCC 6538)	20
5% Horse serum	*S. aureus* (ATCC 6538)	80
Human saliva	*S. aureus* (ATCC 6538)	20
Sodium sulfite	*S. aureus* (ATCC 6538)	20
Sodium thiosulfate	*S. aureus* (ATCC 6538)	20
Triton X-100	*S. aureus* (ATCC 6538)	20
Distilled water	*E. coli* (ATCC 8739)	160
Carbopol 940 (carbomer)	*E. coli* (ATCC 8739)	160
Xanthan gum	*E. coli* (ATCC 8739)	160

Table 6.13 Microbial Growth Inhibition of Stearamidopropyl Phosphatidyl PG-Dimonium Chloride (Agar Incorporation Method)

Sample	*S. aureus* (ATCC 6538)	*E. coli* (ATCC 8739)	*P. aeruginosa* (ATCC 15442)	*C. albicans* (ATCC 10259)
Methyl paraben, 0.05%	1	0	0	1
Germall® 115, 0.05%	1	0	0	3
Stearamidopropyl phosphatidyl PG-dimonium chloride, 0.05% plus methyl paraben, 0.05%	3	3	3	3
Stearamidopropyl phosphatidyl PG-dimonium chloride, 0.05% plus Germall® 115*, 0.05%	2	3	3	3
Control	0	0	0	0

Procedure: Stearamidopropyl phosphatidyl PG-dimonium chloride, methyl paraben, and Germall 115 were individually incorporated into trypticase soy agar at a concentration of 0.05%. Combinations of stearamidopropyl phosphatidyl PG-dimonium chloride/methyl paraben and stearamidopropyl phosphatidyl PG-dimonium chloride/Germall 115* were also incorporated into trypticase soy agar at a 1:1 ratio (i.e., 0.05% each). Test organisms were diluted in phosphate buffer to a level of approximately 5.0×10^3.The treated plates, along with appropriate controls, were streaked with 0.1 ml of each test organism. Bacterial plates were incubated at 30–35°C for 24–48 hours. Yeast plates were incubated at 28–30°C for 48–72 hours. Plates were read after incubation.

Grading system: 3, Excellent Activity; 2, Moderate Activity; 1, Slight Activity; 0, No Activity.

Germall 115 (imidazolidinyl urea) is a trademark of Sutton Laboratories, Div. of ISP, Inc.

6.5 POTENTIATION OF ANTIMICROBIAL ACTIVITY OF FATTY ALCOHOLS

Various fatty alcohols—and especially myristyl (C_{14}) and cetyl (C_{16}) alcohols—have been widely used as viscosity modifiers or emulsion stabilizers. Recent investigations have demonstrated that these materials exhibit a slight to moderate degree of antimicrobial activity on their own. When combined with cocamidopropyl phosphatidyl PG-dimonium chloride or linoleamidopropoyl phosphatidyl PG-dimonium chloride at certain concentrations in a 1:1 ratio, an increased antimicrobial effect on a variety of microorganisms is observed. Gram-positive skin microflora, such as *Staphylococcus epidermidis* and *Corynebacterium pseudodiptheriticum*, were found to be especially sensitive to this combination. These fatty alcohols appear to have only slight to moderate activity against gram-negative bacteria and fungi. It was demonstrated that the antimicrobial action of the phospholipid/fatty alcohol blend was not inactivated or inhibited by nonionic ethoxylated surfactants, hydrocolloids such as polyvinylpyrrolidine (PVP), or cellulose derivatives It was believed that this potentiation of antimicrobial activity has application in self-preserving products.

It is known that perfume oils have some degree of antimicrobial activity (29). Enhanced antimicrobial effectiveness was achieved using combinations of the biomimetic phospholipids, myristyl alcohol, and perfume oils by combining selected oils and formulating mixtures that were reasonably active against bacteria, yeasts, and molds (30). These studies demonstrated that combining the phospholipid with specially formulated fragrances and the C_{14}–C_{16} fatty alcohols resulted in a dramatic increase in antimicrobial activity against several different types of microorganisms, including *P. aeruginosa*. Use of this combination appears to have potential for application in self-preserving products.

6.6 THEORETICAL MECHANISM FOR ANTIMICROBIAL ACTION

It has been theorized that the antimicrobial characteristics demonstrated by the biomimetic phospholipids may be based upon antimicrobial activity mechanisms similar to, yet inherently different from, those of the quaternary ammonium compounds. Although they are not "quats" in the true sense, the phospholipids appear to possess some of the same abilities in their theoretical mechanism of action. The activity of quats and similar types of antimicrobial materials is due, at least in part, to their ability to penetrate the cell wall/envelope of the microorganism. It is known that quat mixtures containing various chain-length compounds generally function against a much broader spectrum of microorganisms that single chain-length compounds (31). This is primarily a function of the relative efficiency of mixed chain-length compounds in penetrating the many lipid barriers composing the cell wall/envelope and membrane. In addition, the cationic polar groups are believed to be more likely to penetrate the cell membrane/porins (see Sec. 2.2) than are nonionic or anionic polar groups (32,33). Once present within the cytoplasmic membrane of the cell, they are thought to interfere with cellular metabolism and/or membrane transport.

Considering the fact that different types of microorganisms have significantly different cell wall/envelope structures, the effectiveness of an antimicrobial agent against a broad spectrum of microorganisms may be largely a product of its ability to pass easily into a variety of dissimilar barrier layers. For example, gram-negative bacteria have a cell envelope that is primarily lipophilic in nature, which should lead to greater susceptibility to antimicrobial compounds that are lipophilic. In general this is true; however, the outer components of the gram-negative bacterial envelope contains a series of both hydrophilic regions (porins) and lipid layers. The composition of the aqueous/lipid layers in the outer membrane often determines the organism's degree of resistance to antimicrobials. Substances of low water solubility may be unable to penetrate through the water barriers and would be unavailable to interact with the cytoplasmic membrane. On the other hand, substances with low oil solubility may not be able to cross the lipophilic barriers of the outer membrane and move into the membrane via the aqueous layers and/or porins. It would appear that the optimum inhibitory action would be achieved by a molecule that possess the balance between hydrophilicity and lipophilicity to maximize its ability to negatively impact the cell. Another desirable feature of the biomimetic phospholipids is that they have a high degree of substantivity to proteinaceous material, such as are present in the cell wall of bacteria, which should result in a high affinity for the cell wall of bacteria.

6.7 SUMMARY

Phospholipids are essential components in cell membranes because they are a key component in the bilayers comprising cell membranes and their breakdown products are involved with calcium transport and intracellular signaling. This chapter presented information on biomimetic phospholipids, a new family of phospholipids in which the arrangement of the phosphate and quaternary groupings have been reversed.

Biomimetic phospholipids exert antimicrobial action against gram-positive and gram-negative bacteria, yeasts, and molds and have MICs in the range of 50–1250 ppm. One percent cocamidopropyl phosphatidyl PG-dimonium chloride was shown to be rapidly effective against *S. epidermidis* in culture media over the pH range 4 to 8. The addition of 0.5% coco phosphatidyl PG-dimonium chloride to 0.05% methyl paraben or 0.05% imidazolidinyl urea increased the efficacy of the preservative system of a creme over that obtained using either 0.2% methyl paraben or 0.2% imidazolidinyl urea alone. Also, combining a biomimetic phospholipid with C_{14}–C_{16} fatty alcohols increased the antimicrobial action of the preservative system. The phospholipids are inherently mild and possess anti-irritant potential so that their use may deliver an overall benefit of improved performance and mildness along with antimicrobial action.

Salient features of the biomimetic phospholipids include their water and oil solubility characteristics, which provide significant functional benefits in formulation, and their antimicrobial activity. The effects of phospholipids on cells are believed to result in interference with cell membrane integrity, transport, and metabolic processes, which allows other materials (including fatty alcohols and preservatives) to more effectively penetrate the cell. The data presented suggest application of this new family of phospholipids for enhancing the preservative system. This may enable the formulator to reduce or eliminate conventional antimicrobials in the development of mild, self-preserving products.

REFERENCES

1. Kirk-Othmer, Lecithin, *Encyclopedia of Chemical Technology*, Vol. 14 3rd ed. 1981, pp. 250–269.
2. Technical Bulletins #1016b, January 1993; #1019b, May 1994; #1018b, June 1994. Mona Industries, Inc., Paterson, NJ.
3. J. F. Kuo, ed. *Phospholipids and Cellular Regulation*, Vol. 1. CRC Press, Inc., Boca Raton, FL, 1985, pp. 3–59.
4. L. Stryer. Introduction to biological membranes. In: *Biochemistry*, 3rd ed. W. H. Freeman and Co., New York, 1988, pp. 283–292.
5. A. Finkelstein. Water movement through lipid bilayers, pores, and plasma membranes: theory and reality. In: *Distinguished Lecture Series of the Society of General Physiologists*, Vol 4. John Wiley & Sons, New York, 1987, pp. 93–106.
6. M. J. Berridge, Inositol triphosphate and calcium mobilization. In: *Calcium and the Cell.* Ciba Foundation Symposium 122. John Wiley & Sons, New York, 1986, pp. 39–57.
7. R. H. Michell. Inositol phospholipids and cell surface receptor function. *Biochim. Biophys. Acta 215*:81–147 (1975).
8. L. Stryer. *Biochemistry*, 3rd ed. W. H. Freeman and Co., New York, 1988, pp. 985–988.
9. M. A. Lampe, M. L. Williams, and P. M. Elias. Human epidermal lipids: characterization and modulations during differentiation. *J. Lipid Res. 24*:131 (1983).
10. P. M. Elias. Structure and function of the stratum corneum permeability barrier. *Drug Develop. Res. 13*:97 (1988).
11. G. K. Menon, S. Grayson, and P. M. Elias. Cytochemical and biochemical localization of lipase and sphingomyelinase activity in mammalian epidermis. *J. Invest. Dermatol. 86*:591 (1986).
12. P. W. Wertz, D. C. Swartzendruber, W. Abraham, K. C. Madison, and D. T. Downing. Essential fatty acids and epidermal integrity. *Arch. Dermatol. 123*:1381 (1987).
13. P. W. Wertz, W. Abraham, J. Jandmann, and D. T. Downing. Preparation of liposomes from stratum corneum lipids. *J. Invest. Dermatol. 87*:582–584 (1986).
14. G. Brooks and B. Idson. Skin lipids. *Int. J. Cosmetic Sci. 13*:103–113 (1991).
15. A. D. Russel, I. Ahonkai, and D. T. Rogers. Microbiological applications of the inactivation of antibiotics and other antimicrobial agents. *J. Appl. Bacteriol. 46*:207–245 (1979).
16. A. D. Russel. Neutralization procedures in the evaluation of bactericidal activity. In: C. H. Collins, ed. *Disinfectants: Their Use and Evaluation of Effectiveness*. Academic Press, Inc., London, 1981, pp. 45–59.

17. D. L. Fost. Multifunctional biomimetic phospholipids: natural base ingredients for personal care products. *Asian COSFA '95 Convention Proceedings*, Bangkok, Thailand 1995.
18. J. I. Yablonski and D. L. Fost. Synthetic phospholipids: a new class of multifunctional preservative/surfactants. In: *Cosmetics Proceedings*. Reed Exhibitions, Richmond, United Kingdom, 1991.
19. J. I. Yablonski and D. L. Fost. Synthetic phospholipid-II. Multifunctional preservative/surfactants—an update. Presented at the IVth International Meeting on Cosmetic Dermatology, Rome, November 1991.
20. J. J. O'Neill, P. L. Peelor, A. F. Peterson, and C. H. Strube. Application of the Ferguson principle to the selection of sparingly soluble preservatives. *Dev. Indust. Microbiol. 19*:335–345 (1978).
21. J. J. O'Neill and C. A. Mead. The parabens: bacterial adaptation and preservative capacity. *J. Soc. Cosmet. Chem 33*:75–84 (1982).
22. N. A. Allawala and S. Riegelman. Phenol coefficients and the Ferguson principle. *J. Am. Pharm. Assoc. XLIII*: 93–97 (1954).
23. W. P. Evans. Applicability of the Ferguson principle to systems of mixed preservatives. *J. Pharm. Pharmacol. 17*:217–221 (1965).
24. B. Magnusson and A. M. Kligman. The identification of contact allergens by animal assay, the guinea pig maximization test method. J. Invest. Dermatol. *52*:268–276 (1970).
25. D. L. Fost and J. A. Komor. Phospholipid virucidal compositions. U.S. Patent 5,286,719 (1994).
26. C. Prottey, P. J. Hartop, J. G. Black, and J. I. McCormack. The repair of impaired epidermal barrier function in rats by the cutaneous application of linoleic acid. *Br. J. Dermatol. 94*:13 (1976).
27. P. Morganti and S. D. Randazzo. Essential fatty acids and skin aging. *J. Appl. Cosmetol. 3*:211–222 (1985).
28. N. J. Lowe. Essential fatty acid deficient hairless mouse: the effects of topical agents on the epidermis. *Br. J. Dermatol. 97*:39–47 (1977).
29. W. Sturm. Deosafe fragrances: fragrances with deodorizing properties. *Cosmet. Toiletr. 94*:35–48 (1979).
30. G. W. Holzner. New ideas for body deodorants—Part 1. Technical research report from Firmenich S. A., Geneva, 1992.
31. J. J. Merianos. Quaternary ammonium antimicrobial compounds. In: S. S. Block, ed. *Disinfection, Sterilization, and Preservation*, 4th ed. Lea & Febiger, Philadelphia, 1991; pp. 225–255.
32. H. Nikaido and M. Vaara. Molecular basis of bacterial outer membrane permeability. *Microbiol. Rev. 49*:1–32 (1985).
33. R. Benz, K. Janko, and P. Jauger. Ionic selectivity of pores formed by the matrix protein (porin) of *Escherichia coli. Biochim. Biophys. Acta 551*:238–247 (1979).

7

Use of Antioxidants in Self-Preserving Cosmetic and Drug Formulations

A. Larry Branen and P. Michael Davidson
University of Idaho
Moscow, Idaho

The term antioxidants generally refers to compounds that are used to stop the free radical lipid oxidation process. Antioxidants can be classified as primary antioxidants and synergists (or potentiators). The primary antioxidants delay onset or slow the rate of autoxidation by inhibiting the formation of free radicals (initiation) or interrupting propagation. Initiation is inhibited through decomposition of peroxides, metal chelation, or singlet oxygen inhibition (free radical acceptors). The antioxidant is able to terminate free radical propagation by reacting with peroxy radicals, which prevents them from reacting with another unsaturated fatty acid. The production of a new free radical is thereby stopped and the autoxidation reaction is delayed.

Antioxidant systems containing multiple antioxidants or antioxidants used in combination with metal chelators often demonstrate synergism. This occurs when two compounds together show greater activity than the sum of the activities of the compounds alone. Mixed free radical acceptors show synergism when one is more effective than the other. This allows the more active free radical acceptor to donate a hydrogen to the fatty acid free radical. The less active free radical acceptor then donates a hydrogen to the more active of the pair, causing regeneration of the more active species. Metal chelating agents act as synergists and are the primary synergist-type antioxidants. Metals such as copper and iron are pro-oxidants, i.e., they increase the rate of oxidation through hydroperoxide decomposition, formation of free radicals, and activation of singlet oxygen. Metal chelators reduce pro-oxidation (see Chapter 9).

The effectiveness of antioxidants in lipid systems is dependent upon their oil solubility, volatility, and stability during processing. They must be oil soluble to have access to the free radicals. They must not be highly volatile or they may be lost in processing, and they must be stable to the heat used in processing. The choice of antioxidant depends upon its ease of incorporation, carry-through, pH sensitivity, production of discoloration or off-flavors or off-odors, availability, multifunctionality, and cost.

Multifunctionality is a major consideration in developing self-preserving cosmetic and drug products. In addition to preventing the oxidation of lipids, certain antioxidants can serve other functions in cosmetic and food products. For example, antioxidants play an important role in human nutrition and physiology. Both ascorbic acid and tocopherols are important vitamins and play an essential role in preventing oxidation in human cells. These compounds, especially the tocopherols, are considered the first line of defense against lipid peroxidation in the body and, through their ability to scavenge for free radicals, can prevent a number of problems associated with aging, arthritis, circulation, cataracts, air pollution, strenuous exercise, inflammation, and cancer (1–3).

Bacteria, yeasts, and molds are capable of growing in many aqueous cosmetic and drug products, and viruses and protozoa can be carried by them. In addition to preventing physiological and nutritional problems, several naturally occurring and synthetic phenolic antioxidants and chelating agents provide appreciable antimicrobial activity (4,5). The objective of this chapter is to summarize information on synthetic and natural antioxidants, to discuss information on antimicrobial action, which is taken primarily from model systems or food products, and to relate this to the use of antioxidants in self-preserving cosmetic and drug products.

7.1 SYNTHETIC ANTIOXIDANTS

7.1.1 Antioxidant Properties

The primary synthetic antioxidants are phenolic compounds that have the ability to donate an electron and stop the propagation of free radicals. Butylated hydroxyanisole (BHA), butylated hydroxytoluene (BHT), tertiary butylhydroquinone (TBHQ), and propyl gallate (PG) are the most common phenolic antioxidants used in foods and cosmetics. BHA, with a molecular weight of 180.0, exists as the 2- and 3-isomers of tertiary-butyl-4-hydroxyanisole, with the 3-isomer predominating (5). BHT, also known as 2,6-ditertiary-butyl-*p*-cresol, has a molecular weight of 220.36, while TBHQ and PG (*n*-propyl ester of 3,4,5-tri-hydroxybenzoic acid) have molecular weights of 166.22 and 212.0, respectively. All of these compounds except PG are fairly soluble in lipids but insoluble in water. Phenolic antioxidants are marketed as powders or tablets and are reported to be very stable upon storage (6). In the United States, BHA (21 CFR 182.3169) and BHT (21 CFR 182.3173) are generally recognized as safe (GRAS) for use in food at a maximum concentration of 0.02% (200 μg/ml), based on the weight of oil or fat content (7). In addition to approval in the United States, phenolic antioxidants are approved for use in foods in at least 60 other countries (6).

Phenolic compounds are good electron donors and therefore good antioxidants. Monohydric (BHA, BHT) and polyhydric (TBHQ, PG) phenolic antioxidants are relatively stable due to resonance delocalization and lack of positions suitable for attack by molecular oxygen. Hydroquinones (e.g., TBHQ) form stable semiquinone resonance hybrids. Monohydric phenolics are sterically hindered, which further reduces initiation following hydrogen donation, and have radical intermediates with moderate resonance delocalization.

7.1.2 Antimicrobial Activity

Although the primary function of these compounds is to delay autoxidation of unsaturated lipids in products, they also show significant antimicrobial activity (Table 7.1). BHA is the most effective antimicrobial of the phenolic antioxidants. Chang and Branen (8) found that enteropathogenic *Escherichia coli*, *Salmonella typhimurium*, and *Staphylococcus aureus* were inhibited by concentrations of 400, 400, and 150 μg/ml, respectively. This and other studies generally demonstrated that gram-positive bacteria were more susceptible to BHA than gram-negative bacteria. Some exceptions are *Vibrio parahaemolyticus* (9), *Pseudomonas fluorescens* (10), *Aeromonas hydrophila* (11), *Pseudomonas fragi*, and *Edwardsiella tarda* (12).

BHA has also been shown to be more effective than the other phenolic antioxidants against both molds and yeast. Chang and Branen (8) first showed that 250 μg/ml BHA totally inhibited growth and toxin production of *Aspergillus parasiticus*. BHA has also been found to be effective against several other species of *Aspergillus*, *Byssochlamys*, *Penicillium*, and *Geotrichum* (13–18). Rico-Munoz and Davidson (17), Knox et al. (19), and Eubanks and Beuchat (20–22) found that BHA was effective against growth of *Saccharomyces cerevisiae* and increased heat inactivation and decreased recovery of heat-treated *S. cerevisiae* cells. BHA has also been found to be active against viruses (23–26), mycoplasmas (27), and protozoa (28–31). The use of BHA and BHT in preservative systems is discussed in Chapter 5.

TBHQ has antimicrobial activity similar to that of BHA (Table 7.1). Studies on the antibacterial activity of BHT have shown that it is generally less effective than either BHA or TBHQ (12,18,32–36). PG generally has been demonstrated to have little effectiveness, however, Chung et al. (37) reported that propyl gallate was an effective inhibitor of a number of bacterial pathogens. McCarthy (38) also reported that both BHT and PG potentiated the antimicrobial effects of the zinc ion against several microorganisms important in cosmetics. Gailani and Fung (12) determined the effectiveness of 1:1 combinations of TBHQ, BHA, BHT, and PG against 24 strains of

Table 7.1 Concentrations of the Phenolic Antioxidants BHA and TBHQ Necessary for Complete Inhibition or Delayed Growth of Bacteria in Laboratory Media[a]

	Concentration (μg/ml)	
Microorganism	BHA	TBHQ
Gram-positive		
Arthrobacter spp.	100	100
Bacillus cereus	100–200	—
Bacillus megaterium	100–200	—
Bacillus pumilis	—	50–100
Bacillus subtilis	100–2500	50–100
Clostridium botulinum	25–50	—
Clostridium perfringens	150	—
Lactobacillus brevis	100	300
Listeria monocytogenes	128–300	64–105
Pediococcus pentosaceus	—	15–20
Pediococcus spp.	100	100
Sarcina lutea	150	100
Staphylococcus aureus	50–400	25–100
Streptococcus faecalis	400	>500
Gram-negative		
Aeromonas hydrophila	100–150	—
Arizona spp.	200	300
Citrobacter freundii	350	>500
Edwardsiella tarda	150	150
Enterobacter aerogenes	350	500
Escherichia coli	200–400	450
Klebsiella pneumoniae	400	500
Proteus vulgaris	250	250
Pseudomonas aeruginosa	400	500
Pseudomonas fluorescens	100–300	500
Pseudomonas fragi	100–>400	100
Salmonella senftenberg	>10,000	—
Salmonella typhimurium	150–400	—
Serratia marcescens	>500	400
Vibrio angularum	250	100
Vibrio parahaemolyticus	50	—
Yersinia enterocolitica	250	250

[a]pH, incubation temperature, and time varies.
Source: Ref. 5.

bacteria. The most efficacious combination at 400 μg/ml was TBHQ/PG, which inhibited all 24 strains, followed by BHA/PG (23 of 24 strains), BHT/TBHQ (14 of 24), BHA/TBHQ (13 of 24), BHT/PG (12 of 24), and BHA/BHT (10 of 24).

Many environmental factors have been found to affect the antimicrobial activity of the phenolic antioxidants. The presence of lipid in products or test media has been shown to dramatically decrease the effectiveness of phenolic antioxidants. Rico-Munoz and Davidson (17) measured the decrease in antimicrobial activity of BHA and TBHQ against five microorganisms in the presence of corn oil. The antimicrobial activity of BHA was virtually eliminated by as little as 1.5–3.0% corn oil in microbiological media. In contrast, TBHQ demonstrated some inhibition at levels of 4.5% corn oil. Since TBHQ is more polar than BHA, these results suggested that the decrease in activity was simply a result of solubilization of the BHA into the lipid phase.

Protein can also decrease the antimicrobial activity of BHA. Eklund et al. (39) demonstrated that 40 mg/ml bovine serum albumin essentially eliminated the lethal effect of approximately 600 μg/ml BHA against *S. typhimurium* at 10°C, and Rico-Munoz and Davidson (17) found that, in the presence of casein, the antimicrobial activity of both BHA and TBHQ was decreased. Minerals also affect the antimicrobial activity of some phenolic antioxidants. Stern et al. (40) demonstrated that NaCl potentiated the effect of BHA against *S. aureus*, while Raccach et al. (41) showed that NaCl increased the effect of TBHQ against a fermentation by *Pediococcus pentosaceus* or *Lactobacillus plantarum*. Rico-Munoz and Davidson (42) found that Ca^{2+} increased the inhibitory activity of BHA against *S. aureus* A100. No effect of Mg^{2+} on BHA was detected, nor was there an effect of either Ca^{2+} or Mg^{2+} on the inhibitory activity of TBHQ.

Another major factor affecting antimicrobial activity of phenolic antioxidants is temperature. Wanda et al. (26), Davidson and Branen (10), and Eklund et al. (39) found decreased activities for BHT and BHA at lower temperatures. It was theorized that lower temperatures decreased the solubility of the phenolic antioxidants into the lipid of the cell membrane or envelope, which in turn decreased the compound's effectiveness (26). Lower temperatures also decrease microbial metabolism and the speed of chemical reactions, which contribute to the decrease in effectiveness of most antimicrobials at lower temperatures.

The effect of pH on the antimicrobial activity of BHA has been shown to be highly dependent upon the microorganism tested. *Clostridium perfringens* (43) was found to be more susceptible at limiting pH values (5.5 and 8.5), whereas *S. aureus* (44) was most susceptible at a neutral pH. *Aspergillus flavus* conidia (14) were prevented from outgrowth by BHA to a greater extent at pH 3.5 than at pH 5.5. In contrast, *A. flavus* spores were found to be equally susceptible at pH values ranging from 4 to 9 in a study by Ahmad and Branen (13). Reddy and Pierson (45) found little difference in inhibition of *C. botulinum* growth and toxin production by BHA, TBHQ, or BHT between pH 6.0 and 7.0.

The phenolic antioxidants have generally demonstrated additive or synergistic effects when used with other antimicrobial preservatives. Davidson et al. (46), Robach and Stateler (47), and Lahellec et al. (33) found synergistic activity of both BHA and TBHQ with potassium sorbate against *S. aureus* and *S. typhimurium*. With nitrite, sorbate, and parabens, BHA has been shown to exhibit at least additive inhibition against *C. perfringens* (43), *S. aureus*, *S. typhimurium* (48), and *S. cerevisiae* (19). In contrast to previous studies, Knox et al. (19) found slight antagonism between BHA and sodium metabisulfite against *S. cerevisiae*.

Overall it would appear that the antimicrobial activity of phenolic antioxidants can be significantly limited by the presence of other components in a formulation. The method of application to cosmetic or drug products, however, may increase the activity in products. These compounds are generally applied by dissolving them in ethanol, propylene glycol, or dimethyl sulfoxide prior to addition to products. The greatest potential for use of phenolic compounds would most likely be in combination with other antimicrobials as a part of a total preservative system.

7.1.3 Anticancer Activity of Phenolic Antioxidants

Both BHA and BHT have been found to exhibit potent anticarcinogenic activity and are capable of inhibiting tumor development in various organs of rodents induced by more than three dozen different carcinogens (49–52). Milner (52) stated that BHA and BHT appear to accelerate catabolism and excretion of carcinogens by enzyme modification, which leads to protection from chemically induced tumors. BHA and BHT were found to inhibit intestinal carcinogenesis by azoxymethane in rats (53). The phenolic antioxidants PG and TBHQ were effective to a limited extent in inhibiting hepatotoxicity by *N*-dimethylnitrosamine (54). Partial protection against the hepatotoxicity of pyrrolizidine alkaloids was also demonstrated with BHA in the diet (55). Black and Mathews-Roth (56) reported that BHT and certain carotenoids may prevent photocarcinogenesis by UV irradiation and could thus be important in preventing skin cancer.

Although BHA is reported to prevent cancer, it has also been reported to induce carcinoma in the forestomach of rats, mice, and hamsters but not in the stomach of dogs, guinea pigs, or monkeys, which do not have a forestomach (49). Huang and Ferraro (49) concluded that it is unlikely that BHA is carcinogenic in the gastric epithelium of humans at the level of BHA consumed in the diet.

7.2 NATURALLY OCCURRING ANTIOXIDANTS

Natural antioxidants occur in all higher plants and in all parts of the plant and are typically phenolic or polyphenolic compounds (57). According to Pratt (57), natural antioxidants are endogenous compounds found in one or more components of foods or food additives isolated from natural sources and include ascorbic acid, amino acids, organic acids, melanoidins, peptides, tocopherols, carotenoids, and flavonoids and other phenolic compounds. The use of natural antioxidants is not new, and studies on their use were initiated more than 100 years ago (57).

7.2.1 Tocopherols

Tocopherols, which exhibit vitamin E activity, are the primary naturally occurring antioxidants in vegetable oils. They are fat soluble due to their long alkyl chains. The antioxidant effectiveness of the various forms of tocopherols in decreasing order of activity is delta-tocopherol>gamma-tocopherol>beta-tocopherol>alpha-tocopherol. The vitamin E activity of tocopherols is inverse to their antioxidant effectiveness, with alpha-tocopherol having the greatest activity. The addition of an acetate group also increases the activity for nutritive purposes but decreases usefulness for preservation.

Tocopherols are free radical acceptors and become oxidized to quinones. They are somewhat heat sensitive and may be lost during processing. Although tocopherols are used in a variety of products, they generally are less effective as food antioxidants than other synthetic compounds (58). Studies have also shown that high levels of tocopherols (>1000 ppm based on oil content) can have pro-oxidant effects, and their usefulness in oils is sometimes limited since unsaturated vegetable oils already contain significant quantities of tocopherols naturally (59).

Vitamin E is considered the major lipid-soluble antioxidant in biological systems and protects cell membranes from free radical damage. The antioxidant defense mechanisms attributed to vitamin E have recently been reviewed by Packer and Landvik (2). Results of animal and human studies have demonstrated that vitamin E has a significant role in preventing or minimizing peroxidative damage in biological systems. The peroxidative damage from various conditions including aging, arthritis, circulatory conditions, cataracts, strenuous exercise, and air pollution have all been reported to be limited or prevented by vitamin E (2). Tocopherols have also been reported to prevent the oxidative impact of sun damage to skin and have been used to augment the efficacy of sunscreens (60). The acetate form is often recommended since it is nonirritating and will penetrate the epidermis and dermis and be converted to the usable unesterified form (60). Like other antioxidants, tocopherols have also been reported to decrease carcinogenicity based on tissue culture, animal, and epidemiological studies (49). Alpha-tocopherol is reported to inhibit both chemical and radiation-induced cell transformation (49).

7.2.2 Ascorbic Acid

As indicated above, the term "antioxidant" is normally used to refer to a compound that is used to control free-radical mediated lipid oxidation. Although ascorbic acid does inhibit oxidation reactions, it is normally not referred to as a primary antioxidant to prevent lipid oxidation but rather as a synergist (61). Ascorbic acid and its sodium and calcium salts can act in a number of ways to prevent oxidation reactions, including: 1) serving as oxygen scavengers by removing oxygen from solution, 2) serving as reducing agents by transferring hydrogen to oxygen or other compounds, making them unavailable for further reactions, and 3) by acting as metal chelators to remove active metal catalysts from the reaction. Because of these properties, ascorbic acid has been used as an antibrowning additive to prevent enzymatic browning of cut surfaces of fruits and vegetables. In this application, it acts as a reducing agent by transferring hydrogen atoms back to quinones that are formed in enzymatic oxidation of phenolic compounds (62). This action, along with its ability to chelate metals, makes it useful as a synergist to prevent lipid oxidation. Ascorbic acid apparently can regenerate phenolic antioxidants by supplying hydrogen atoms to phenoxy radicals that form when the phenolic antioxidants yield hydrogen atoms to the lipid oxidation chain reaction (62). The ability to chelate metals such as copper and iron can also enhance the action of the phenolic antioxidants. The ability of ascorbic acid to act as a lipid antioxidant, however, is limited since it is water soluble and thus not available in the lipid phase where oxidation is taking place. To make ascorbic acid less polar, it is often esterified to fatty acids. Ascorbyl palmitate (6-O-palmitoyl-L-ascorbic acid) is a fat-soluble ester of ascorbic acid, which is often used as a synergist with tocopherols. The ascorbates and ascorbyl palmitate are GRAS for use in foods in the United States.

The nutritional need for vitamin C, or foods containing it, has been recognized for centuries. Ascorbic acid or dehydroascorbic acid show vitamin C activity and are necessary in the human diet to prevent scurvy. It is generally recognized that smoking, alcoholism, and general stress conditions increase the body's demand for vitamin C (63). Just as it acts as a reducing agent in food, it serves as a nonspecific reducing agent in products and in the body. By acting on metal ions to keep them in the reduced and active state, it enhances the absorption and use of iron and other metals. It also plays a role in the synthesis of collagen and other vital cell compounds such as thyroxin and in the functioning of the immune system (64). Vitamin C also helps to regenerate vitamin E in the same way it acts with phenolic antioxidants in foods (64). This function has also been proposed for use in sunscreens where a combination of vitamin C and vitamin E can be used to augment the efficacy of sunscreens (60). Although evidence is less conclusive about the role of vitamin C in preventing cancer in comparison to the lipid-soluble phenolic antioxidants, it has also been reported to be effective in preventing certain cancers (64). Although vitamin C could act via several mechanisms, it has been confirmed that ascorbate and ascorbyl palmitate reduce the formation of nitrosamines in the curing process of both frankfurters and bacon (65) and likely can prevent formation of these same compounds in cosmetic and drug products.

As indicated above, ascorbic acid is used extensively to prevent the enzymatic browning that occurs in fruits and vegetables. The ability of ascorbic acid to lower the pH, combined with its chelating and reducing capabilities, also make it useful as an antimicrobial. For example, ascorbic acid inhibits *C. botulinum* in cured meats in conjunction with nitrite (65). Ascorbic acid generally has not been used as part of cosmetic and drug preservative systems due to its instability in water; however, ascorbic acid may have application in systems with low oxidation/reduction potential. Isoascorbate (erythorbate) is the D isomer of ascorbic acid and has a similar but more effective impact than ascorbate. It has been used to significantly reduce the amount of nitrite needed for an antibotulinal effect (65). It should also be pointed out that ascorbate and isoascorbate accelerate color development, uniformity, and stability of the nitrosomyoglobin pigment formed in meat curing. Isoascorbate is also often added with ascorbic acid when ascorbic acid is used to fortify foods. Although isoascorbate has no vitamin C activity, it acts to protect the loss of ascorbic acid from oxidation since it is more readily oxidized than ascorbic acid.

7.2.3 Citric Acid

Citric acid and its sodium, potassium, and calcium salts are metal chelating agents that act as synergists with the phenolic antioxidants. As with ascorbic acid, citric acid can bind the copper and iron that catalyze oxidation. Citric acid can be used in lipid emulsions but its use is limited because of its water solubility. It can be modified to form esters with isopropanol and glycerol that can be used in fats and oils (65).

Citrate and ascorbic acid are effective antimicrobials because of their ability to lower the pH and chelate metals. When used as an acidulant, citric acid has shown activity against species of *Salmonella*, *Staphylococcus*, *Lactobacillus* and against various *Aspergillus* fungi (65). Acidic pH is an effective means of controlling microbial growth as long as the organisms cannot utilize the acidulant for growth and/or have enzyme systems for raising the pH (see Sec. 2.3).

7.2.4 Other Phenolic Compounds (Excluding Parabens)

Phenolic compounds are widespread in nature (66) and may be found in a great variety of plant and animal systems. The phenolic compounds are complex materials with thousands of compounds possible based on the chemical location of several functional groups. They generally are classified into four or five groups. Some of the major groups that are useful in cosmetic and drug formulations are listed in Table 7.2 along with selected examples of each group. The simple phenols include the monophenols (e.g., *p*-cresol, 3-ethylphenol, 3,4-dimethylphenol), the diphenols (e.g., hydroquinone) and the triphenols (e.g., gallic acid). Hydroxycinnamic acids are almost exclusively derived from *p*-coumaric, caffeic, and ferulic acids and usually occur in various conjugated forms such as esters and glycosides. The flavonoids are probably the most prominent phenolics consisting of the catechins,

Table 7.2 Major Classes of Phenolic Compounds and Selected Examples

Class	Examples
Simple phenols	Guaiacol, hydroquinone, vanillin
Phenolic acids	Gallic acid, quinic acid, protocatechuic acid
Hydroxycinnamic acids	Coumaric acid, caffeic acid, ferulic acid, chlorogenic acid
Flavonoids	
Anthocyanidins/Anthocyanins	Malvidin/malvidin-3-glucoside
Chalcones	Butein, okanin
Flavanols	Catechin, epicatechin
Flavanonols	Taxifolin
Flavones	Apigenin
Flavonols	Quercetin, quercitrin
Flavonones	Narinigin
Isoflavone	Daidzein, genistein
Tannins	
Hydrolyzable	Gallic/ellagic acid polymers
Condensed	Epicatechin/catechin polymers

proanthocyanins, anthocyanidins and flavones, flavonols, and their glycosides (58). The tannins are polymers of either gallic acid, which are referred to as hydrolyzable tannins, or catechin and epicatechin, which are referred to as condensed tannins. The hydrolyzable tannins are sometimes referred to as tannic acids.

Many of the phenolic compounds have antioxidant activity in lipid-aqueous systems. In particular, the flavones, flavonols, flavonones, flavanonals, the tannins, and cinnamic acid appear to be some of the most effective antioxidants (57,67,68). The antioxidant action of flavonoids and cinnamic esters is primarily related to their ability to serve as free radical acceptors and chain-breakers, although the flavonols can chelate metals and act as synergists as well. The position and degree of hydroxylation of the phenols are of primary importance in antioxidant activity with the 3′,4′-dihydroxy configuration possessing the most activity (57). One of the most effective compounds is reported to be 7-*n*-butoxy-3,2′,5′-trihydroxyflavone (57). It should be pointed out that the 3-glycosides, which are the major forms found in plants, possess the same antioxidant activity as the corresponding aglycone when the glycosyl substitution is a monosaccharide, but have reduced activity if the substitution is a disaccharide.

The major limitation to use of the naturally occurring phenolic compounds is their low solubility in lipids, especially where an aqueous phase that attracts the compounds is present. However, suspending the compounds in a lipid-aqueous system or increasing their solubility in lipids through alkylation or esterification can overcome this problem (57). Some lipophilic esters of hydroxycinnamic acids occur naturally in plant tissue and contribute to the antioxidant properties of such products as oats and other cereal grains (57).

Since phenolics are often found in plants, plant extracts are often used as additives rather than the pure compounds. Although such extracts are often more costly and not as effective as pure compounds, their natural occurrence is an advantage. Most extracts have been used for centuries in foods or even for antiseptic or medicinal purposes and many are already approved by regulatory agencies for use. Spice extracts are currently the most popular because of the high antioxidant activity shown for spices such as clove, cinnamon, sage, rosemary, mace, oregano, allspice, and nutmeg (59). Many of these extracts also show strong synergism with BHA. The only spice extract commercially available in the United States is rosemary. The activity of rosemary extract apparently is due to at least six different compounds, with carnosic acid and rosmaric acid (derivatives of caffeic acid) being primarily responsible for the activity (57, 69).

Several other plants contain antioxidants, including soybeans, tea and tea extracts, oat flour, and the desert plant chia. Chia seeds have high quantities of several phenolic acids, with caffeic acid, gentisic acid, and ferulic acid contributing the greatest to the antioxidant activity (57). Oat flour was one of the first natural plant products used for its antioxidant properties (57). Recently a new company, Nurture, Inc., has been formed to utilize the antioxidant properties of oat flour and oat flour extracts for preservation of both foods and cosmetics (70). The antioxidants of oats include tocopherols and various hydroxycinnamate esters of long-chain alcohols, ω-hydroxy fatty acids, or glycerol (71). The lipophilic nature of these esters increases their contact with the lipid phases of a product and thus increases the antioxidant effectiveness.

Because these compounds are derivatives of phenol, many have been evaluated for their antimicrobial activity. Some of the major compounds that display both antioxidant and antimicrobial activity are listed in Table 7.3. Simple phenols have been evaluated primarily to compare their activity with phenol. A detailed discussion of the relative activity of many of these compounds may be found in Prindle (72). Several generalizations can be made about the structure-activity relations of simple phenols (72–74):

1. *n*-Alkyl chains increase activity of the compounds in proportion to their length (gram-negative bacteria are less susceptible to alkyl chains of more than five carbons).
2. The position of the alkyl chain may or may not influence activity.
3. *n*-Alkyl chains are more effective than branched chains.
4. Polyhydroxy derivatives are generally less effective than monohydroxy compounds.
5. Separation of the alkyl group from the phenol nucleus by an oxygen (e.g., methoxy-) decreases activity.

TABLE 7.3 Phenolic Compounds with Reported Antioxidant and Antimicrobial Activity

Class of compound	Examples	Ref. Antioxidant	Ref. Antimicrobial
Simple phenols	Guaiacol	58	72
	Hydroquinone	58	72
	Vanillin	58	72
Phenolic acids	Gallic acids and esters	57, 67	37, 78, 82–84
	Quinic acid	57, 67	82–84
Hydroxycinnamic acids	Coumaric acid and esters	57, 67	79–81, 85
	Caffeic acid and esters	57, 67	81, 85, 86
	Ferulic	57, 67	80, 87
	Chlorogenic	57, 67	77
Flavonoids	Chalcones	57, 67	97
	Flavones	57	97
	Flavonols	57	92, 97
	Flavonones	57	97
Tannins		68	35, 82, 92, 99–106

It must be remembered that all of these effects are limited to microbiological media and do not necessarily relate directly to cosmetics or drug formulations.

Phenolic acids, including derivatives of *p*-hydroxybenzoic acid (protocatechuic, vanillic, gallic, syringic, ellagic), *o*-hydroxybenzoic acid (salicylic), and hydroxycinnamic acids (*p*-coumaric, caffeic, ferulic, sinapic), are found in plants (66,75,76) and have been shown to have antimicrobial activity. Salicylic acid may be used as an active ingredient in over-the-counter drugs for treatment of acne, dandruff, seborrheic dermatitis, and psoriasis. Grodzinska-Zachwieja and Kahl (77) discovered that an extract isolated from chicory root and containing chlorogenic and isochlorogenic acids was relatively effective against both gram-negative and gram-positive bacteria. They concluded that the active portion of the molecule was caffeic acid. Reddy and Pierson (78) tested gallic acid and its esters for their effectiveness against *C. botulinum* growth and showed greatest inhibition with the isobutyl, isoamyl, *p*-octyl, and *p*-dodecyl esters. Chung et al. (37) showed that propyl gallate inhibited *Alcaligenes faecalis*, *Enterobacter aerogenes*, *E. coli*, *Klebsiella pneumoniae*, *Listeria monocytogenes*, *Proteus vulgaris*, *Salmonella enteritidis*, *Salmonella paratyphi*, and *Shigella flexneri*. Lyon and McGill (79) tested the antimicrobial activity of caffeic, cinnamic, ferulic, salicylic, sinapic, and vanillic acids on *Erwinia carotovora*, a microorganism responsible for soft rot of potatoes. Total inhibition of the microorganism was achieved in nutrient broth with 0.5 mg/ml cinnamic, ferulic, salicylic, or vanillic acids. Herald and Davidson (80) found that ferulic acid and *p*-coumaric acid were inhibitory to the gram-positive bacteria *Bacillus cereus* and *S. aureus* but were much less effective against the gram-negative *P. fluorescens* and *E. coli*. In comparison with their corresponding acids, alkyl esters of hydroxycinnamic acids were found to be good inhibitors of *P. fluorescens* (81). The most effective were methyl caffeoate, ethyl caffeoate, propyl caffeoate, methyl *p*-coumarate, and methyl cinnamate. In contrast to previous studies with other gram-positive bacteria, Payne et al. (35) showed that quercetin, coumaric acid, ferulic acid, caffeic acid, and gallic acid had no effect on the growth of eight strains of *L. monocytogenes*. Although many antioxidants may have little effect when used alone, it is quite possible that potentiation of preservative action would be observed when used as part of a preservative system.

Many studies with hydroxycinnamic acids have evaluated their antifungal properties. In tests with chlorogenic, isochlorogenic, caffeic, gallic, quinic, and ellagic acids, Schanderl (82) and Sikovec (83,84) established that the compounds, alone and in combination, inhibited yeast growth. Baranowski et al. (85) studied the effect of caffeic, chlorogenic, *p*-coumaric, and ferulic acids at pH 3.5 on the growth of *S. cerevisiae*. Caffeic and chlorogenic acid had little effect, however, *p*-coumaric and ferulic acids completely inhibited the microorganism. The degree of inhibition was inversely related to the polarity of the compounds. Caffeic and chlorogenic acids have been shown to inhibit certain *Fusarium* species (86), and ferulic acid inhibited aflatoxin B_1 and G_1 production by *A. flavus* and *A. parasiticus* (87). In contrast to other studies, Leifertova et al. (88) found little inhibition of nine species of yeast and molds by the hydroxycinnamic acid derivatives using a disk assay system. Differences in the results obtained in this study may have been due to the antimicrobial assay technique. Due to the reduced solubility of the hydroxycinnamic acid derivatives in polar solvents, it is possible there was little diffusion of the test compounds from the disk into the microbiological medium (85).

Anthocyanins are a subgroup of the flavonoids that occur naturally in fruits. Masquelier (89) found that *E. coli* and *S. typhi* were inhibited primarily by malvidin, an anthocyanidin found in red wine. Pratt et al. (90) and Powers et al. (91) demonstrated the antimicrobial effects of anthocyanins and anthocyanidins from grapes and strawberries against several bacterial strains. Marwan and Nagel (92) identified the antimicrobial compounds associated with cranberries. Benzoic acid, proanthocyanidins, and flavonols were found to account for 66% of cranberry microbial inhibition against *Saccharomyces bayanus*, with the latter two being the most important. A compound isolated from green olives and identified as the phenolic glycoside oleuropein, or its aglycone, was shown to inhibit a number of microorganisms including *Lactobacillus plantarum*, *Leuconostoc mesenteroides*, *P. fluorescens*, *Bacillus subtilis*, *Rhizopus* spp., and *Geotrichum candidum* (93–96). Schraufstatter (97) investigated the antimicrobial effect of chalcone, flavonone, flavone, and flavonol. All compounds except flavonol had a bacteriostatic effect on *S. aureus* in their tests. Natural sources of these compounds were also found to have slight bacteriostatic activity against the microorganism.

Polyphenolics are present in rinds and barks of many plants. Tannic acid was the first polyphenolic to be recognized for its fungistatic properties (98). The tannins have been tested for their antimicrobial effectiveness against wine yeasts and viruses. Turbovsky et al. (99) found that wine yeasts, wild yeasts, and vinegar bacteria were unaffected or affected only slightly by tannins at concentrations as high as 20 mg/ml. Schanderl (82), however, reported that 2 mg/ml of tannin halted wine yeast fermentations earlier than a control. Singleton and Esau (100) stated that the antibacterial effect of red and white wines was proportional to the amount of flavonoid tannin present. Payne et al. (35) found that tannic acid completely inhibited growth of *L. monocytogenes* at 256 μg/ml. Chung and Murdock (101) also showed tannic acid to have antimicrobial activity against *A. hydrophila, E. coli, L. monocytogenes, S. enteritidis, S. aureus*, and *Streptococcus faecalis*. They further demonstrated that gallic and ellagic acids, components of tannic acid, were not inhibitory, suggesting that tannic acid served as a natural defense mechanism for plants. Konowalchuk and Speirs (102–105) published a series of reports on the antiviral activity of extracts of blueberries, crabapples, strawberries, red wines, grape juice, apple juice, and tea. Some extracts inactivated poliovirus, coxsackievirus, echovirus, reovirus, and herpes simplex virus. They concluded that the primary viral inhibitors in these products were tannins. Tannic acid was found to have the greatest antiviral activity against echovirus, poliovirus, and herpes simplex virus. Cliver and Kostenbader (106) confirmed the work of Konowalchuk and Speirs (103) on the antiviral activity of grape juice. They concluded that the inhibitory factor of grape juice did not degrade viruses and that the effect was reversible.

Many studies involving naturally occurring phenolic compounds have been carried out with plant extracts. These studies supply more direct application information since some of the potential interfering factors are present in the test system. However, most research has involved use of purified compounds in microbiological media. While these studies provide some information on antimicrobial activity, it is possible that compounds added to cosmetic and drug formulations will have no activity due to interfering factors or adverse environmental conditions. Therefore, evaluation of single or multiple naturally occurring compounds in actual use situations must be performed. In addition, information should be collected on the antimicrobial activity of these compounds under various environmental conditions (temperature, pH, water activity).

Due to the tremendous variation in the type of phenolic compounds found naturally, there is also wide variation in the toxicological effects of these compounds. A discussion of the toxicology of each of these phenolic compounds would be impractical; an excellent review on this subject was published by Singleton

(107). Due to the ubiquity of these compounds, people in the United States are estimated to have an average daily intake of up to 1 g of flavonoids (49). Although several in vitro tests have shown the flavonoids to be mutagens, no evidence has been found to indicate that these compounds are carcinogenic in in vivo studies. On the contrary, flavonoids are reported to inhibit tumor development in several experimental models through several different mechanisms (49). The hydroxylated flavonoids have been found to inhibit metabolic activation, inactivate ultimate carcinogens, inhibit generation of active oxygen species, inhibit arachidonic acid metabolism, inhibit protein kinase C, and reduce the bioavailability of carcinogens. Although no direct epidemiological studies have correlated the content of flavonoids with low cancer incidence, the flavonoids could contribute, along with vitamins A and C and the carotenes, to the low cancer incidence accompanying increased consumption of fruits and vegetables.

In addition to flavonoids, the hydroxycinnamic acids, the phenolic catechins in tea, and the rosemary phenols have all been reported to be anticarcinogenic in animal studies (49). The major studies on hydroxycinnamic acids have involved caffeic acid, ferulic acid, and chlorogenic acid, all of which block the formation of nitrosamines. There is little definitive epidemiological information to support the anticarcinogenicity of these compounds; however, studies have shown some reduction of cancer in heavy coffee and tea drinkers (49).

7.3 SUMMARY

Many synthetic and naturally occurring antioxidants are available for use in product preservation. While most of the information available is based on model test systems or food products, the same principles apply for the use of antioxidant materials in cosmetic and drug formulations. The choice of antioxidants depends upon the ease of incorporation, the chemical and physical properties of the antioxidant materials and the product, and the packaging. Safety, compliance with governmental regulations, and cost must also be considered.

Multifunctionality of components is a major consideration in developing self-preserving products, and many antioxidants have been shown to have antimicrobial action. The successful use of BHT with a chelating agent and monolaurin demonstrates the use of an antioxidant in a systems approach for self-preserving products. After preservative efficacy testing in the laboratory, the final selection of the components to use in the preservative system should involve testing that demonstrates maintenance of preservation during manufacturing, shipment, marketing, and use by the consumer.

It is hoped that the information presented in this chapter will give formulators and microbiologists some insights into the use of antioxidants in self-preserving cosmetic and drug products.

REFERENCES

1. M. Gerber, S. Richardson, F. Favier, and A. C. De Paulet. In: *Antioxidants in Therapy and Preventive Medicine.* I. Emerit, L. Packer, and C. Auclair, eds. Plenum Press, New York, 1990, p. 129.
2. L. Packer and S. Landvik. Vitamin E in biological systems. In: I. Emerit, L. Packer, and C. Auclair, eds. *Antioxidants in Therapy and Preventive Medicine.* Plenum Press, New York, 1990, p. 93.
3. K. H. Schmidt and W. Bayer. Efficacy of vitamin E as a drug in inflammatory joint diseases. In: I. Emerit, L. Packer, and C. Auclair, eds. *Antioxidants in Therapy and Preventive Medicine.* Plenum Press, New York, 1990, p. 147.
4. P. M. Davidson and V. K. Juneja. Antimicrobial Agents. In: A. L. Branen, P. M. Davidson, and S. Salminen, eds. *Food Additives.* Marcel Dekker, Inc., New York, 1990, pp. 83–137.
5. P. M. Davidson. Parabens and phenolic compounds. In: P. M. Davidson and A. L. Branen, eds. *Antimicrobials in Foods*, 2nd ed. Marcel Dekker, Inc., New York, 1993, pp. 263–306.
6. E. R. Sherwin. Antioxidants. In: A. L. Branen, P. M. Davidson, and S. Salminen, eds. *Food Additives.* Marcel Dekker, Inc. New York, 1990, p. 139.
7. Code of Federal Regulations. Title 21. Food and Drugs, Parts 170–199. Office of Federal Regulations, National Archives Records Services, General Service Administration, Washington, DC, 1991.
8. H. C. Chang and A. L. Branen. Antimicrobial effects of butylated hydroxyanisole (BHA). *J. Food Sci. 40*:349 (1975).
9. M. C. Robach, L. A. Smoot, and M. D. Pierson. Inhibition of *Vibrio parahaemolyticus* 04:K11 by butylated hydroxyanisole. *J. Food Prot. 40*:549 (1977).
10. P. M. Davidson and A. L. Branen. Inhibition of two psychrotrophic *Pseudomonas* species by butylated hydroxyanisole. *J. Food Sci. 45*:1603 (1980).
11. V. Venugopal, A. C. Pansare, and N. F. Lewis. Inhibitory effect of food preservatives on protease secretion by *Aeromonas hydrophila. J. Food Sci. 49*:1078 (1984).
12. M. B. Gailani and D. Y. C. Fung. Antimicrobial effects of selected antioxidants in laboratory media and in ground pork. *J. Food Prot. 47*:428 (1984).
13. S. Ahmad and A. L. Branen. Inhibition of mold growth by butylated hydroxyanisole. *J. Food Sci. 46*:1059 (1981).
14. L. R. Beuchat and W. K. Jones. Effects of food preservatives and antioxidants on colony formation by heated conidia of *Aspergillus flavus. Acta Aliment. Acad. Sci., Hung. 7*:373 (1978).
15. J. C. Davis, F. A. Draughon, and P. M. Davidson. Inhibition of toxin production in patulin-producing fungi by phenolic antioxidants and insecticides. Annual Meeting of Institute of Food Technologists, Las Vegas, NV, 1982.
16. D. Y. C. Fung, S. Taylor, and J. Kahan. Effect of butylated hydroxyanisole (BHA) and butylated hydroxytoluene (BHT) on growth and aflatoxin production of *Aspergillus flavus. J. Food Safety 1*:39 (1977).
17. E. Rico-Munoz and P. M. Davidson. The effect of corn oil and casein on the antimicrobial activity of phenolic antioxidants. *J. Food Sci. 48*:1284–1288 (1983).
18. R. Sankaran. Comparative antimicrobial action of certain antioxidants and preservatives. *J. Food Sci. Technol. 13*:203 (1976).

19. T. L. Knox, P. M. Davidson, and J. R. Mount. Evaluation of selected antimicrobials in fruit juices as sodium metabisulfite replacements or adjuncts. Annual Meeting of Institute of Food Technologists, Anaheim, CA, 1984.
20. V. L. Eubanks and L. R. Beuchat. Effects of antioxidants on growth, sporulation and pseudomycelium production by *Saccharomyces cerevisiae. J. Food Sci. 47*:1717 (1982).
21. V. L. Eubanks and L. R. Beuchat. Increased sensitivity of heat-stressed *Saccharomyces cerevisiae* cells to food-grade antioxidants. *Appl. Env. Microbiol. 44*:604 (1982).
22. V. L. Eubanks and L. R. Beuchat. Combined effect of antioxidants and temperature on survival of *Saccharomyces cerevisiae. J. Food Prot. 46*:29 (1983).
23. J. Cupp, P. Wanda, A. Keith, and W. Snipes. Inactivation of the lipid-containing bacteriophage PM2 by butylated hydroxytoluene. *Antimicrob. Agents Chemother. 8*:698 (1975).
24. K. S. Kim, H. M. Moon, V. Sapienza, R. I. Carp, and R. Pullarkat. Inactivation of cytomegalovirus and Semliki forest virus by butylated hydroxytoluene. *J. Infect. Dis. 138*:91 (1978).
25. W. Snipes, S. Person, A. Keith, and J. Cupp. Butylated hydroxytoluene inactivates lipid-containing viruses. *Science 188*:64 (1975).
26. P. Wanda, J. Cupp, W. Snipes, A. Keith, T. Rucinsky, L. Polish, and J. Sands. Inactivation of the enveloped bacteriophage Φ by butylated hydroxytoluene and butylated hydroxyanisole. *Antimicrob. Agents Chemother. 10*:96 (1976).
27. T. H. Vardaman, J. D. May, and J. H. Drott. Studies on the effect of butylated hydroxytoluene on *Mycoplasma synoviae. Poult. Sci. 57*:1526 (1978).
28. J. G. Surak, R. L. Bradley, A. L. Branen, and E. Shrago. Effects of butylated hydroxyanisole on *Tetrahymena pyriformis. Food Cosmet. Toxicol. 14*:277 (1976).
29. J. G. Surak, R. L. Bradley, A. L. Branen, E. Shrago, and W. E. Ribelin. Effect of butylated hydroxytoluene on *Tetrahymena pyriformis. Food Cosmet. Toxicol. 14*:541 (1976).
30. J. G. Surak. Monotertiary butyl hydroquinone effects on *Tetrahymena pyriformis. Life Sci. 20*:1735 (1977).
31. J. G. Surak and R. G. Singh. Butylated hydroxyanisole (BHA) induced changes in the synthesis of polar lipids and in the molar ratio of tetrahymenol to polar lipids in *Tetrahymena pyriformis. J. Food Sci. 45*:1251 (1980).
32. M. C. Robach and M. D. Pierson. Inhibition of *Clostridium botulinum* types A and B by phenolic antioxidants. *J. Food Prot. 42*:858 (1979).
33. C. Lahellec, D. Y. C. Fung, and F. E. Cunningham. Growth effect of sorbate and selected antioxidants on toxigenic strains of *Staphylococcus aureus. J. Food Prot. 44*:531 (1981).
34. M. A. Al-Khayat, G. Blank, and C. Biliaderis. Germination and outgrowth of *Bacillus subtilis* spores in the presence of selected antioxidants. *J. Food Prot. 50*:206 (1987).
35. K. D. Payne, E. Rico-Munoz, and P. M. Davidson. The antimicrobial activity of phenolic compounds against *Listeria monocytogenes* and their effectiveness in a model milk system. *J. Food Prot. 52*:151–153 (1989).
36. A. E. Yousef, R. J. Gajewski, and E. H. Marth. Kinetics of growth and inhibition of *Listeria monocytogenes* in the presence of antioxidant food additives. *J. Food Sci. 56*:10–13 (1991).
37. K.-T. Chung, S. E. Stevens, W.-F. Lin, and C. I. Wei. Antimicrobial properties of

tannic acid, propyl gallate and related compounds. ASM General Meeting, New Orleans, 1992.

38. T. J. McCarthy, J. J. Zeelie, and D. J. Krause. The antimicrobial action of zinc ion/antioxidant combinations. *J. Clin. Pharm. Ther. 17*(1):51–54 (1992).
39. T. Eklund, I. F. Nes, and R. Skjelkvale. Control of *Salmonella* at different temperatures by propyl paraben and butylated hydroxyanisole. In: T. A. Roberts, G. Hobbs, J. H. B. Christian, and N. Skovgaard, eds. *Psychrotrophic Microorganisms in Spoilage and Pathogenicity*. Academic Press, London, 1981, p. 377.
40. N. J. Stern, L. A. Smoot, and M. D. Pierson. Inhibition of *Staphylococcus aureus* growth by combinations of butylated hydroxyanisole, sodium chloride, and pH. *J. Food Sci. 44*:710 (1979).
41. M. Raccach, M. E. Schilz, and S. L. Kovac. Combined effect of monotertiary butylhydroquinone and sodium chloride on lactic acid bacteria. *J. Food Prot. 47*:591 (1984).
42. E. Rico-Munoz and P. M. Davidson. The effect of calcium and magnesium on the antibacterial activity of phenolic antioxidants against *Staphylococcus aureus* A100. *J. Food Sci. 49*:282 (1984).
43. K. J. Klindworth, P. M. Davidson, C. J. Brekke, and A. L. Branen. Inhibition of *Clostridium perfringens* by butylated hydroxyanisole. *J. Food Sci. 44*:564 (1979).
44. M. Ayaz, L. O. Luedecke, and A. L. Branen. Antimicrobial effect of butylated hydroxyanisole and butylated hydroxytoluene. *J. Food Prot. 43*:4 (1980).
45. N. R. Reddy, M. D. Pierson, and R. V. Lechowich. Inhibition of *Clostridium botulinum* by antioxidants, phenols and related compounds. *Appl. Env. Microbiol. 43*:835 (1982).
46. P. M. Davidson, C. J. Brekke, and A. L. Branen. Antimicrobial activity of butylated hydroxyanisole, tertiary butylhydroquinone and potassium sorbate in combination. *J. Food Sci. 46*:314 (1981).
47. M. C. Robach and C. L. Stateler. Inhibition of *Staphylococcus aureus* by potassium sorbate in combination with sodium chloride, tertiary butylhydroquinone, butylated hydroxyanisole or ethylenediamine tetracetic acid. *J. Food Prot. 43*:208 (1980).
48. M. D. Pierson, L. A. Smoot, and K. R. Vantassell. Inhibition of *Salmonella typhimurium* and *Staphylococcus aureus* by butylated hydroxyanisole and the propyl ester of p-hydroxybenzoic acid. *J. Food Prot. 43*:191 (1980).
49. M. T. Huang and T. Ferraro. Phenolic compounds in food and cancer prevention In: M. T. Huang, C. T. Ho, and C. Y. Lee, eds. *Phenolic Compounds in Food and Their Effects on Health II, Antioxidants and Cancer Prevention*. American Chemical Society, Washington, DC, 1992, p. 8.
50. L. W. Wattenberg. Effect of dietary constituents on the metabolism of chemical carcinogens. *Cancer Res. 35*:3326 (1975).
51. L. W. Wattenberg. Inhibition of chemical carcinogenesis. *J. Natl. Cancer Inst. 60*:11 (1978).
52. J. A. Milner. Dietary antioxidants and cancer. *Contemp. Nutr. 10*(10):1 (1985).
53. E. K. Weisburger, R. P. Evarts, and M. L. Wenk. Inhibitory effect of butylated hydroxytoluene (BHT) on intestinal carcinogenesis in rats by azoxymethane. *Food Cosmet. Toxicol. 15*:139 (1977).
54. B. D. Astill and L. T. Mulligan. Phenolic antioxidants and the inhibition of hepatotoxicity from N-dimethylnitrosamine formed in situ in the rat stomach. *Food Cosmet. Toxicol. 15*:167 (1977).

55. C. L. Miranda, D. R. Buhler, H. S. Ramsdell, P. R. Cheeke, and J. A. Schmitz. Protective effects of butylated hydroxyanisole against lethality and hepatotoxicity of pyrrolizidine (Senecio) alkaloids. *Toxicologist 1*:133 (1981).
56. H. S. Black and M. M. Mathews-Roth. Protective role of butylated hydroxytoluene and certain carotenoids in photocarcinogenesis. *Photochem. Photobiol. 53*:707 (1991).
57. D. E. Pratt. Natural antioxidants from plant material. In: M. T. Huang, C. T. Ho, and C. Y. Lee, eds. *Phenolic Compounds in Food and Their Effects on Health II, Antioxidants and Cancer Prevention.* American Chemical Society, Washington, DC, 1992, p. 54.
58. C. T. Ho. Phenolic compounds in food. In: M. T. Huang, C. T. Ho, and C. Y. Lee, eds. *Phenolic Compounds in Food and Their Effects on Health II, Antioxidants and Cancer Prevention.* American Chemical Society, Washington, DC, 1992, p. 2.
59. B. F. Haumann. Antioxidants: firms seeking products they can label as "natural". *Inform. 1*(12):1002 (1990).
60. R. Wilson. Antioxidants to augment the efficacy of sunscreens. *DCI* (August):32 (1992).
61. W. W. Nawar. Lipids. In: O. R. Fennema, ed. *Food Chemistry.* Marcel Dekker, Inc., New York, 1985, p. 139.
62. R. C. Lindsay. Food additives. In: O. R. Fennema, ed. *Food Chemistry.* Marcel Dekker, Inc., New York, 1985, p. 629.
63. J. Augustine and F. E. Scarbrough. Nutritional additives. In: A. L. Branen, P. M. Davidson, and S. Salminen, eds. *Food Additives.* Marcel Dekker, Inc., New York, 1990, p. 33.
64. G. M. Wardlaw and P. M. Insel. *Perspectives in Nutrition.* Mosby-Year Book, Inc., St. Louis, 1993.
65. S. Doores. pH control agents and acidulants. In: A. L. Branen, P. M. Davidson, and S. Salminen, eds. *Food Additives.* Marcel Dekker, Inc., New York, 1990, pp. 477–510.
66. J. B. Harborne and N. W. Simmonds. The natural distribution of the phenolic aglycones. In: J. B. Harborne, ed. *Biochemistry of Phenolic Compounds.* Academic, London, 1964, pp. 77–127.
67. F. Shahidi and P. K. J. P. D. Wanasundara. Phenolic antioxidants. *Crit. Rev. Food Sci. Nutr. 32*(1):67–103 (1992).
68. T. Okuda, T. Yoshida, and T. Hatano. Antioxidant effects of tannins and related polyphenols. In: M. T. Huang, C. T. Ho, and C. Y. Lee, eds. *Phenolic Compounds in Food and Their Effects on Health II, Antioxidants and Cancer Prevention.* American Chemical Society, Washington, DC, 1992. p. 87.
69. N. Nakatani. Natural antioxidants from spices. In: M. T. Huang, C. T. Ho, and C. Y. Lee, eds. *Phenolic Compounds in Food and Their Effects on Health II, Antioxidants and Cancer Prevention.* American Chemical Society, Washington, DC, 1992, p. 72.
70. R. Baum. Nurture Inc. stakes future on oats, other grains. *Chem. Eng. News* (October 11):18 (1993).
71. L. H. Dimberg, O. Theander, and H. Lingnert. Avenanthramides—a group of phenolic antioxidants in oats. *Cereal Chem. 70*(6):637 (1993).
72. R. F. Prindle. Phenolic compounds. In: S. S. Block, ed. *Disinfection, Sterilization, and Preservation*, 3rd ed. Lea and Febiger, Philadelphia, 1983, p. 197.
73. E. G. Klarmann and V. A. Shternov. Bactericidal value of coal-tar disinfectants.

Limitation of the *B. typhosus* phenol coefficient as a measure. *Ind. Eng. Chem. 8*:369 (1936).
74. C. M. Suter. Relationships between the structures and bactericidal properties of phenols. *Chem. Rev. 28*:269 (1941).
75. K. Herrmann. Occurrence and content of hydroxycinnamic and hydroxybenzoic compounds in foods. *CRC Crit. Rev. Food Sci. Nutr. 28*:315 (1989).
76. J. A. Maga. Simple phenol and phenolic compounds in food flavor *CRC. Crit. Rev. Food Sci. Nutr. 10*:323 (1978).
77. A. Grodzinska-Zachwieja and W. Kahl. Bacteriostatic action of chicory (*Chicorium intybus* L.). II. The action of chlorogenic and isochlorogenic acid. *Acta Biol. Crac. Ser. Bot. 9*:87 (1966).
78. N. R. Reddy and M. D. Pierson. Influence of pH and phosphate buffer on inhibition of *Clostridium botulinum* by antioxidants and related phenolic compounds. *J. Food Prot. 45*:925 (1982).
79. G. D. Lyon and F. M. McGill. Inhibition of growth of *Erwinia carotovora* in vitro by phenolics. *Potato Res. 31*:461 (1988).
80. P. J. Herald and P. M. Davidson. The antibacterial activity of selected hydroxycinnamic acids. *J. Food Sci. 48*:1378 (1983).
81. J. D. Baranowski and C. W. Nagel. Properties of alkyl hydroxycinnamates and effects on *Pseudomonas fluorescens*. *Appl. Env. Microbiol. 45*:218 (1983).
82. H. Schanderl. Der Einfluss von Polyphenolen und Gerbstoffen auf die Physiologie der Weinhefe und der Wert des pH-7 Testes für die Auswahl von Sektgrundweinen. *Mitt. (Klosterneuburg) 12A*:265 (1962).
83. S. Sikovec. Der Einfluss einiger Polyphenole aud die Physiologie von Weinhefen. I. Der Einfluss von Polyphenolen auf den Verlauf der alkoholischen Grung insbesondere von Emg rungen. *Mitt. (Klosterneuberg) 16*:127 (1966).
84. S. Sikovec. Der Einfluss von Polyphenolen auf die Physiologie von Weinhefen. II. Der Einfluss von Polyphenolen auf die Vermehrung and Atmung von Hefen. *Mitt. (Klosterneuberg) 16*:272 (1966).
85. J. D. Baranowski, P. M. Davidson, C. W. Nagel, and A. L. Branen. Inhibition of *Saccharomyces cerevisiae* by naturally occurring hydroxycinnamates. *J. Food Sci. 45*:592 (1980).
86. E. Valle. On anti-fungal factors in potato leaves. *Acta Chem. Scand. 11*:395 (1957).
87. J. R. Chipley and N. Uraih. Inhibition of *Aspergillus* growth and aflatoxin release by derivatives of benzoic acid. *Appl. Environ. Microbiol. 40*:352 (1980).
88. I. Liefertova, N. Hejtmankova, H. Hlava, J. Kudrnacova, and F. Santavy. Antifungal and antibacterial effects of phenolic substances: a study of the relationship between the biological activity and the constitution of the investigated compounds. *Acta Univ. Palacki. Olomuc. Fac. Med. 74*:83 (1975).
89. J. Masquelier. The bactericidal action of certain phenolics of grapes and wine. In: J. W. Fairbairn, ed. *The Pharmacology of Plant Phenolics*. Academic, New York, 1959, pp. 123–131.
90. D. E. Pratt, J. J. Powers, and D. Somaatmadja. Anthocyanins. I. The influence of strawberry and grape anthocyanins on the growth of certain bacteria. *Food Res. 25*:26 (1960).
91. J. J. Powers, D. Somaatmadja, D. E. Pratt, and M. K. Hamdy. Anthocyanins. II. Action of anthocyanin pigments and related compounds on the growth of certain microorganisms. *Food Technol. 14*:626 (1960).

92. A. G. Marwan and C. W. Nagel. Microbial inhibitors of cranberries. *J. Food Sci. 51*:1009 (1986).
93. H. P. Fleming, W. M. Walter, and J. L. Etchells. Isolation of a bacterial inhibitor from green olives. *Appl. Microbiol. 18*:856 (1969).
94. H. P. Fleming, and J. L. Etchells. Occurrence of an inhibitor of lactic acid bacteria in green olives. Appl. Microbiol. 15:1178 (1967).
95. B. Juven, Z. Samish, and Y. Henis. Identification of oleuropein as a natural inhibitor of lactic fermentation of green olives. *Isr. J. Agric. Res. 18*:137 (1968).
96. B. Juven and Y. Henis. Studies on the antimicrobial activity of olive phenolic compounds. *J. Appl. Bacteriol. 33*:721 (1970).
97. E. Schraufstatter. Die bakteriostatische Wirking von Chalkon, Flavanon, Flavon und Flavonol. *Experientia 4*:484 (1948).
98. A. M. Cruickshank and D. R. Perrin. Pathological function of phenolic compounds in plants. In: J. B. Harborne, ed. *Biochemistry of Phenolic Compounds.* Academic Press, London, 1964, p. 511.
99. M. W. Turbovsky, F. Filipello, W. V. Cruess, and P. Esau. Observations on the use of tannin in wine making. *Fruit Prod. J. 14*:106 (1934).
100. V. L. Singleton and P. Esau. *Phenolic Substances in Grapes and Wine, and Their Significance.* Academic, New York, 1969.
101. K.-T. Chung and C. A. Murdock. Natural systems for preventing contamination and growth of microoganisms in foods. *Food Microstructure 10*:361 (1991).
102. J. Konowalchuk and J. I. Speirs. Antiviral activity of fruit extracts. *J. Food Sci. 41*:1013 (1976).
103. J. Konowalchuk and J. I. Speirs. Virus inactivation by grapes and wines. *Appl. Environ. Microbiol. 32*:757 (1976).
104. J. Konowalchuk and J. I. Speirs. Antiviral effect of commercial juices and beverages. *Appl. Environ. Microbiol. 35*:1219 (1978).
105. J. Konowalchuk and J. I. Speirs. Antiviral effect of apple beverages. *Appl. Environ. Microbiol. 36*:798 (1978).
106. D. O. Cliver and K. D. Kostenbader. Antiviral effectiveness of grape juice. *J. Food Prot. 42*:100 (1979).
107. V. L. Singleton. Naturally occurring food toxicants: phenolic substances of plant origin common in foods. *Adv. Food Res. 27*:149 (1981).

8

Aroma Chemicals as Preservatives

Jon J. Kabara
Technology Exchange, Inc.
Galena, Illinois, and
Michigan State University
East Lansing, Michigan

8.1 HISTORY

The idea of using aroma chemicals to create self-preserving formulas is not a great stretch for the imagination. Spices and essential oils have long been used to prevent the deterioration of products. The chemistry of these fragrances is complex, but generally the active extracts are comprised of one or more of the following: alcohols, phenols, esters, acids, aldehydes, and/or terpenes. It is believed that the evaporation of essences from plant surfaces serves as a defense mechanism against infection by bacteria, fungi and pests. Although it is impossible to establish a date when plants and plant products were first used for medicinal reasons, papyri exist recording the medicinal use of plants dating back to around 2800 B.C. Even before this time, Shen Nung, a Chinese emperor nearly 5000 years ago, was responsible for the first known herbals. (For a full history of herbals in medicine, see Refs. 1 and 2. Also, an excellent review of antimicrobials (3) provides a broad review of natural antimicrobial agents from plant, animal, and microbial systems.)

As a result of early alchemy, essences and essential oils of plants were incorporated into medicine, food, and cosmetics for their flavoring and therapeutic qualities. Urdang (4) states that reference to essential oils is scarce and vague in early publications. The Greek historian Herodotus (484–425 B.C.) and the Roman naturalist Pliny (23–79 A.D.) mention oil of turpentine. Plants that contain essential oils are mentioned in the Bible. There is historical reference to the trade in spices, odoriferous oils, and fatty extracts, which was carried on in the Orient and in Greek and Roman commerce. A Catalan physician, Arnold de Villanova (1235–1311), is believed to be the first to have described, in print, the actual distillation of essential oils. His praise of the remedial qualities of distilled aromatic waters resulted in an awareness of the value of essential oils.

Aromatic substances have long been used in one form or another to combat disease. During the great plague of the Middle Ages, many kinds of fumigants were used to destroy the "aura" or poison of the disease. Perfumed candles were used in sickrooms and hospitals. These contained roses, cloves, storax, frankincense, lemon peels, juniper berries, musk, and ambergris. Such mixtures dissolved in rose water were formed into candles with gum tragacanth. Aromatics were the best antiseptics available at the time. Up until the nineteenth century, medical practitioners carried a little container filled with aromatics on the top of their walking sticks. This acted as a personal antiseptic and would be held up to the nose when visiting any contagious cases. Herbalists applied their aromatic materials to certain ailments by intuition and empirical experience.

Further reviewing the history of essential oils, Urdang (4) writes: "In the 17th and 18th centuries it was chiefly the pharmacists who improved methods of distillation and made valuable investigations into the nature of essential oils." Of special importance was the work of the French apothecaries (Charas, Lemery, Geoffroy, Rouelle, Demachy, and Baume) their German colleagues (Newmann, Wiegleb, and Green), and the German-Russian pharmacist Bindheim.

In the nineteenth century, there was considerable research on the nature of essential oils and aromatic compounds. Physical and chemical procedures were established for the analysis of aromatics. Standards were prepared for widespread use and reference. Of considerable importance in the further development of the chemistry of volatile oils were the investigations of the French chemist Berthelot (1827–1907), who helped describe the hydrocarbons in these oils. Dumas (1800–1884) made a scientific investigation of stearoptens. Around 1866, the word *terpene* was mentioned in a textbook written by Kekule (1829–1896).

In the United States, the production of essential oils was started using three indigenous American plants: sassafras, American wormseed, and wintergreen. These oils, along with oil of turpentine, were introduced into the first *U.S. Pharmacopoeia*, published in 1820. The current *U.S. Pharmacopoeia* lists no essential oils, while the *National Formulary* (13th ed.) shows only eight. Between 1920 and 1940, the *U.S. Pharmacopoeia XVIII* listed 24 essential oils. The great successes achieved by organic chemists contributed to the deemphasis of natural mixtures, which were difficult to both use and reproduce.

Until the nineteenth century, an exact herbal prescription had never been possible because of the widely varying strengths of the active principles possessed by each individual plant. That the general climate, local weather conditions, habitat, soil, time of gathering, and methods of preserving all have an effect upon the properties of the plant was appreciated by most practitioners. The customs and superstitions surrounding the growing and harvesting of herbs, although on the face of it bizarre, in fact often concealed a sensible method for preserving these properties as well as possible. Indeed, Dioscorides, writing about herbals nearly 2000 years ago, noted that "it is proper to use care . . . in the gathering of herbs each at its due season, for it is according to this that medicines either do their work or become quite ineffectual." Chemistry made it possible to extract a measured quantity of the active principle from a plant and to make it up into exact doses. Chemistry also provided the means for synthetically reproducing the isolated principles, which has led to the modern development of thousands of potent synthetic drugs. Although many of the synthetic drugs are of great value, there has been little time to discover the long-term action of such complex drugs, and only too often the benefits are countered by unpleasant side effects. During the twentieth century these problems led to a revival of interest in medical botany and our quest for safer, "greener" (i.e., natural) agents. The literature dealing with the use and isolation of natural active substances is reviewed in the following sections. It is of special note that in this age of high technology, many consumers have "returned to nature" in the belief that nature, and not the organic chemist, has a better product for our needs.

8.2 ANTIMICROBIAL ACTIVITY OF ESSENTIAL OILS AND FRAGRANCES

Early attention was drawn to the curative powers of herbal antiseptics used in folk medicine. These remedies suffered in that many of the plant parts (roots, bark, fruit, leaves, etc.) were called for in recipes, but not necessarily described in detail. Adding to the confusion were the source and strength of the herbal extract. Historically, to obviate this problem, apothecaries and physicians grew herbs in their own gardens.

Attempts to apply some science to discovering the principle in remedies were fraught with problems. It could be that certain components were included in a remedy either for ritualistic or cosmetic purposes. Another explanation for negative findings may be due to the method of screening for antimicrobial effects. Most studies used the conventional agar streak or disk methods, which have serious limitations. The strength of the extract, the migration of the active ingredient in the agar, and/or the use of mixed microbial cultures influenced the results. Only those active principles that migrated from the paper or agar well onto the agar and formed an area of no growth could be labeled active. Thus, hydrophilic agents readily measure as being active, while similarly active lipophilic agents that do not migrate or migrate poorly would be classified as inactive. A 1969 study of five plants showed antibacterial activity (5).

An 1887 report on the biological activity of spice extracts was made by Chamberland, who tested 100 essential oils against the spores of *Bacillus anthracis* (6). Many early studies dealing with the antimimicrobial activity of aromatic substances claimed comparisons to phenol. Martindale (7) studied the phenol coefficient of a number of essential oils. The following are a few of his results:

Oil	Phenol coefficient
Oil of clove	9
Oil of thyme	15
Oil of sandalwood	1.5
Oil of cinnamon	9
Oil of rose	6
Oil of lemon	4
Oil of rosemary	6
Oil of eucalyptus	4
Oil of lavender	5

Hoffman and Evans (8) reported on the effect of spices as food preservatives. Macht and Kunkel (9) in 1920 and Dyche-Teague (10) in 1924 described the antimicrobial effects of volatile oils or their vapors. The latter reported on the phenol coefficient of alcoholic perfumes using a mixed culture of nasal bacteria, *Micrococcus catarrhalis.* In the same year Bryant (11) also reported on the activity of alcoholic perfumes. His test organism was a pure culture of *Escherichia coli communis.*

Rideal et al. (12) published an extensive list of the Rideal-Walker coefficients of essential oils about the time that Dyson (13) reviewed the physiological aspects of a long list of essential oils. One must not overlook the important contributions of Gattefosse on the antimicrobial activity of essential oils, especially lavender, which continued over a period of many years, part of which were summarized in his books *Aromatherapie* (14) and *Essential Oils as Vegetable Hormones* (15), which cover the years 1680–1933. Naves (16) reported that the U.S. Food and Drug Administration phenol coefficient test method was better than that of the French Hygienic Laboratory when applied to essential oils. Potter (17) found spice oils often to be more effective antiseptics than some well-known aromatic compounds. McLachlen (18) made reference to essential oils as antiseptics. Rolle and Mayr (19) reported that conifer oils were useful as disinfectants against *Mycobacterium tuberculosis*, *E. coli*, and *Staphylococcus aureus.*

The most extensive studies were reported by Maruzzella et al. (20–26). This group of workers published some 18 papers over the period of several years on the broad subject of essential oils and fragrances as antimicrobial agents. Maruzzella and Henry (21) presented data on 35 volatile oils, 5 fixed oils, 5 infused oils, and

95 combinations. The filter paper disk method was used. The five test organisms used were *Salmonella typhosa*, *Micrococcus citreus*, *Proteus morganii*, *Bacillus brevis*, and *Micrococcus pyogenes Var. albus*. Of the 35 oils tested, 33 were found to have some antibacterial activity with at least one of the organisms. Details on the inhibitory activity of oils singly and in combination are given in Tables 8.1 and 8.2. Of the 136 essential oil vapors tested, 114 were found to have some antimicrobial activity against at least one of the 5 test organisms. Of the active 114 oil vapors, 84 produced zones of inhibition against all of the test organisms (20).

In 1960 Maruzzella also studied the action of essential oil vapors on bacterial and fungal growth (23). He reported the results as follows:

> Early studies and more recent observations have clearly demonstrated that certain essential oils possess markedly antibacterial vapors. When one considers the wide use of essential oils in perfumes, cosmetics and pharmaceutical preparations, generally added to enhance the ornamental value of the preparation or to mask unpleasant odors, little attention is given to their germicidal properties. In many instances, the selection of an essential oil with the desired scent and demonstrated antimicrobial activity would certainly give the product some germicidal properties. Since essential oils are generally associated with the vaporous state, studies of the antibacterial-antifungal properties of the vapors assist in the selection of oils with germicidal properties.

As with many prophetic suggestions, Maruzella was ahead of his time in seeing the usefulness of aroma chemicals in helping preserve cosmetic and pharmaceutical formulas. In 1972 Munzing and Schels (27) furnished test results for the potential replacement of preservatives in cosmetics by essential oils. They pointed out that the inhibitory activity of the oils was on the order of magnitude of grams per liter rather than milligrams per liter (see Table 8.3). It is obvious from their statement that they did not see the virtue of a systems approach to product preservation.

Other authors, however, did find enhancement of biocidal activity by aroma chemicals. In 1973, Richards and McBride (28) investigated the inhibitory action of 3-phenylpropan-l-ol, 2-phenylethanol, and benzyl alcohol against *Pseudomonas aeruginosa*. They found that 3-phenylpropan-l-ol was the most effective and benzyl alcohol was the least effective, as shown by: 1) growth rate studies using subinhibitory concentrations of the alcohols, 2) determination of minimum inhibitory concentrations (MICs), and 3) determination of sterilization times. The three compounds enhance the bactericidal action of benzalkonium chloride against *P. aeruginosa* in the same ranking order. It was suggested that 3-phenylpropan-1-ol may be a suitable preservative for oral suspensions and mixtures (27). Aliphatic alcohols also were shown by several authors to be effective antimicrobial agents (29–31).

Table 8.1 Inhibitory Activity of Oils

	Zone of inhibition (mm)[a]					
Volatile oils	*S. typhosa*	*M. citreus*	*P. morganii*	*B. brevis*	*M. albus*	Sum[b]
Amber, rectified	0[c]	2	0	2	2	6
Anise	0	3	0	4	2	9
Bay	0	7	3	3	3	16
Bergarmot	4	7	0	6	3	20
Cajuput	0	10	0	11	5	26
Caraway	0	10	0	5	3	18
Cedar wood	0	0	0	0	0	0
Chenopodium	0	5	3	3	3	14
Cinnamon	0	11	8	12	10	41
Clove	0	3	0	3	2	8
Cubeb	0	0	0	2	0	2
Dwarf pine needle	3	3	5	3	3	17
Eucalyptus	5	10	10	12	8	45
Geranium Algerian	2	12	2	5	2	23
Hemlock	1	0	2	0	0	3
Juniper berries, rectified Dutch	5	2	9	2	3	21
Juniper tar	0	10	0	2	3	15
Lemon	0	0	0	13	0	13
Myrrh	0	0	0	0	0	0
Neroli	5	3	4	10	3	25
Niaouli	0	10	6	10	3	29
Origanum red	0	15	0	15	10	40
Peppermint	5	0	0	5	2	12
Petigrain, Paraguay	3	0	3	4	2	12
Pimento	0	10	0	4	3	17
Pinus sylvestris	8	5	0	6	5	24
Rectified tar	0	10	0	0	6	16
Rosemary	5	7	0	5	6	23
Sandalwood	0	3	0	2	2	7
Sassafras	0	3	0	3	2	8
Spike lavender	0	5	0	5	3	13
Sweet birch	0	0	0	2	0	2
Sweet orange	0	0	0	2	0	2
Turpentine	2	3	8	4	3	20
Wintergreen	0	5	0	3	0	8

[a]Measurement from disk edge to zone edge.
[b]Sum of zones of inhibition for each oil.
[c]Zone of inhibition absent.
Source: Ref. 21.

TABLE 8.2 Inhibitory Activity of Combinations of Oils

Oil combinations	Zone of inhibition (mm)[a]					
	S. typhosa	*M. citreus*	*P. morganic*	*B. brevis*	*M. albus*	Sum[b]
Volatile (1:1)						
Anise–clove	0[c]	3	0	4	2	9
Anise–hemlock	0	3	0	4	2	9
Chenopodium–cedar wood	0	10	0	14	7	31
Cinnamon–clove	0	10	0	10	10	30
Cinnamon–cubeb	0	10	0	13	10	33
Cinnamon–dwarf pine needle	0	15	0	15	5	35
Cinnamon–juniper berries	0	10	0	15	5	30
Cinnamon–niaouli	0	10	0	11	10	31
Cinnamon–*Pinus sylvestris*	0	10	0	12	3	25
Cinnamon–sandalwood	0	10	0	12	5	27
Cinnamon–sweet birch	0	7	0	10	3	20
Cinnamon–turpentine	0	10	0	15	14	39
Cubeb–cedar wood	0	0	0	1	0	1
Cubeb–dwarf pine needle	0	2	4	3	2	14
Cubeb–juniper berries	3	2	5	2	3	15
Cubeb–*Pinus sylvestris*	2	3	0	2	1	8
Dwarf pine needle–cedar wood	3	2	4	2	2	13
Eucalyptus–anise	0	5	0	5	5	15
Eucalyptus–cajuput	0	4	0	12	5	21
Eucalyptus–cedar wood	3	5	0	5	5	18
Eucalyptus–chenopodium	0	9	0	8	6	23
Eucalyptus–cinnamon	0	12	0	15	15	42
Eucalyptus–clove	0	5	0	5	5	15
Eucalyptus–cubeb	5	5	4	5	6	25
Eucalyptus dwarf pine needle	8	10	0	8	6	32
Eucalyptus–juniper berries	8	9	0	10	5	32
Eucalyptus-myrrh	3	5	0	5	3	16
Eucalyptus–niaouli	0	8	0	12	5	25
Eucalyptus–*Pinus sylvestris*	7	10	0	15	3	35
Eucalyptus–rosemary	9	8	0	10	9	36
Eucalyptus–sandal wood	3	5	5	4	5	22
Eucalyptus–sweet orange	7	5	0	6	4	22
Eucalyptus–wintergreen	0	5	0	5	5	15
Juniper berries–cedar wood	2	0	3	2	2	9
Juniper berries–dwarf pine needle	2	2	7	2	2	15
Niaouli–clove	0	5	0	4	3	12
Niaouli–cubeb	0	2	3	3	1	9
Niaouli–*Pinus sylvestris*	0	8	0	12	7	27
Niaouli–rectified tar	0	6	0	6	4	16
Niaouli–sweetbirch	0	4	0	4	3	11
Niaouli–wintergreen	0	3	0	4	3	10
Peppermint–pimento	0	4	0	3	3	10
Turpentine–clove	3	5	0	4	4	16

Table 8.2 Continued

Oil combinations	Zone of inhibition (mm)[a] *S. typhosa*	*M. citreus*	*P. morganic*	*B. brevis*	*M. albus*	Sum[b]
Turpentine–hemlock	0	0	4	2	3	9
Turpentine–juniper berries	3	2	9	2	3	19
Turpentine–origanum red	0	10	0	0	8	18
Turpentine–wintergreen	0	5	0	4	3	12
Volatile (1:1:1)						
Cedar wood–cubeb–bay	0	3	2	3	2	10
Eucalyptus–chenopodium–anise	0	5	0	3	4	12
Eucalyptus–chenopodium–cedar wood	4	6	0	4	5	19
Eucalyptus–chenopodium–hemlock	5	6	0	5	4	20
Eucalyptus–chenopodium–peppermint	3	5	0	4	3	15
Eucalyptus–cinnamon–clove	0	4	3	4	3	14
Eucalyptus–cinnamon–dwarf pine needle	8	11	0	15	18	51
Eucalyptus–cinnamon–juniper berries	10	12	0	17	19	58
Eucalyptus–cinnamon–niaouli	7	14	0	15	15	51
Eucalyptus–cinnamon–rosemary	0	5	3	6	4	18
Sassafras–origanum red-chenopodium	0	5	3	3	3	14
Turpentine–Algerian geranium–rosemary	0	5	3	6	5	19
Turpentine–neroli–Algerian geranium	0	2	2	3	5	12
Wintergreen–sweet birch–lemon	0	2	0	2	0	4
Wintergreen–sweet orange–sassafras	0	4	0	3	0	7
Volatile and fixed (1:1)						
Castor–cedar wood	0	0	0	0	0	0
Castor-eucalyptus	0	5	0	4	5	14
Castor–origanum red	0	2	0	3	0	5
Castor–turpentine	0	0	0	1	0	1
Cod liver–cedar wood	0	0	0	0	0	0
Cod liver–eucalyptus	0	5	0	4	7	16
Cod liver–origanum red	0	3	0	4	3	10
Cod liver–turpentine	0	0	0	2	2	4
Olive–rectified amber	0	0	0	2	0	2
Olive–bay	0	2	0	3	3	8
Olive–bergamot	0	1	0	2	2	5

TABLE 8.2 Continued

Oil combinations	Zone of inhibition (mm)[a] *S. typhosa*	*M. citreus*	*P. morganic*	*B. brevis*	*M. albus*	Sum[b]
Olive–cajuput	0	5	0	4	5	14
Olive–caraway	0	0	0	2	0	2
Olive–cedar wood	0	0	0	0	0	0
Olive-chenopodium	0	2	0	3	0	5
Olive–clove	0	3	0	2	2	7
Olive–eucalyptus	0	5	0	4	8	17
Olive–juniper tar	0	2	0	2	2	6
Olive–origanum red	0	4	0	5	2	11
Olive–peppermint	0	0	2	2	0	4
Olive–sassafras	0	0	0	1	0	1
Olive–turpentine	0	2	4	1	0	7
Olive–wintergreen	0	0	0	0	0	0
Vitamin K_1–eucalyptus	0	7	10	4	6	27
White mineral–cedar wood	0	0	0	0	0	0
White mineral–eucalyptus	0	7	0	5	7	19
White mineral–origanum red	0	7	0	8	0	15
White mineral–turpentine	0	2	6	2	3	13
Volatile and infused (1:1)						
Asafetida–eucalyptus	0	3	3	3	5	14
Burdock–eucalyptus	0	3	5	2	5	15
Henbane–eucalyptus	0	6	5	5	5	21
Lobelia–eucalyptus	0	5	7	5	5	22
Mullein-eucalyptus	0	6	3	3	5	17

[a]Measurement from disk edge to zone edge.
[b]Sum of zones of inhibition for each oil.
[c]Zone of inhibition absent.
Source: Ref. 21.

One of the more ambitious programs to screen aroma chemicals for biocidal activity was carried out in 1979 by Morris et al. (32). The study was designed to provide quantiative data using a three-stage study. The most active fragrance material was incorporated into a soap and then tested in handwashing panel studies. In addition, several fragrance ingredients were incorporated into soap and tested. The ingredients added to the soap fragrance and their MIC values are presented in Table 8.4. The test as carried out had some technical limitations, and the reader is encouraged to refer to the original paper (32). Because of these problems, no skin-degerming effect was measured. The authors suggested that perhaps a better test for the usefulness of aroma chemicals and essential oils as active ingredients would be in cosmetic products that stay on the skin or where the active antimicrobial materials are used as preservatives rather than "active material." Table 8.5 gives the full results of their zealous study, while the microbiological activity of the 521 fragrance raw materials is summaried in Table 8.6. The screen identified 309 materials with activity against at least one of the test organisms, and 212 were

TABLE 8.3 Effectiveness of Aroma Chemicals in a Cosmetic Oil-Water Lotion[a]

	Staphylococcus aureus		*Escherichia coli*		*Proteus vulgaris*		*Pseudomonas aeruginosa*	
	1:500	1:1000	1:500	1:1000	1:500	1:1000	1:500	1:1000
Citral	0	<30	0	0	0	0	R	R
Orange oil	<30	>300	R	R	>300	R	R	R
Lemon oil	0	>300	R	R	R	R	R	R
Bergamot oil	>300	>300	R	R	>300	R	R	R
Linalool	0	204	0	0	R	R	R	R
Spiköl	0	R	0	R	R	R	R	R
Geraniol	0	0	0	R	R	R	R	R
Nerol	0	0	0	0	R	R	R	R
Phenylethyl alcohol	R	R	R	R	R	R	>300	R
Eugenol	0	<30	0	0	0	0	R	R
Zimtaldehyde	0	0	0	0	0	0	0	0
Vetiveröl Bourbon	<30	<30	R	R	R	R	R	R
Sandelholzöl ostind.	<30	<30	R	R	R	R	R	R
Cedrol	0	0	R	R	R	R	R	R
Abs. M. d. Chêne Daksa	0	0	R	R	0	R	R	R
Benzyl acetate	194	R	0	R	0	R	>300	R
α-Amylzimtaldehyde	220	>300	R	R	>300	R	R	R
Hydroxycitronellal	>300	R	R	R	R	R	R	R
Linalylformiat	<30	>300	R	R	R	R	R	R
Linalyl acetate	R	R	R	R	R	R	R	R
Aldehyde C 12 MNA	0	0	R	R	R	R	R	R
Decylalcohol	0	0	R	R	R	R	R	R
Dimethylphthalate	R	R	R	R	R	R	R	R
Tonalide	R	R	R	R	R	R	R	R
Cumarin	178	R	50	R	R	R	R	R
Heliotropin	0	<30	>300	R	R	R	R	R
Benzyl benzoate	R	R	R	R	R	R	R	R
Lilial	<30	<30	R	R	R	R	R	R
Terpineol	0	<30	0	0	0	R	R	R
Thymol	0	0	0	0	0	0	30–300	R
Eucalyptus oil	36	R	R	R	R	R	R	R
Peppermint oil	0	<30	0	R	0	0	R	R
Rosmarinöl	>300	R	R	R	R	R	R	R
Salbeiöl dalmat.	<30	>300	R	R	R	R	R	R
Galbanumöl	52	R	R	R	R	R	R	R
Methylsalicylate	>300	>300	0	0	0	0	R	R
Phenol	200	—	>300	—	>300	—	200	—

[a]0 = no growth, <30 = <30 cal, >300 = >300 cal, R = complete overgrowth as in control.
Source: Ref. 27.

TABLE 8.4 Composition of Soap Fragrance for Hand-Degerming Test

Ingredient	Percentage in fragrance	MIC (ppm)
Aldehyde, C_{11}, undecylenic	0.4	50 (*S. aureus*)[a]
Amyl cinnamyl alcohol	18.0	50 (*S. aureus*)
Anethol	10.0	100 (*C. albicans*)
p-*tert*-Butyl-*m*-cresol	0.5	50 (*S. aureus*)
Citronellol coeur	18.0	100 (*C. albicans*)
Cuminyl acetate	5.0	100 (*C. albicans*)
Isobutyl quinoline	0.1	50 (*C. albicans*)
Isocitral	0.5	100 (diphtheroid)
Lemma (Schiff base)	5.0	100 (*S. aureus*)
Methyl isoeugenol	5.0	100 (diphtheroid)
Methyl lavender ketone	7.0	100 (*S. aureus*)
Mousse, absolute ess.	4.0	100 (*S. aureus*)
Musk ketone	5.0	100 (diphtheroid)
Musk xylol	5.0	50 (diphtheroid)
Ocmea (Schiff base)	8.0	50 (*S. aureus*)
Patchouli oil, dark	3.0	100 (*S. aureus*)
Rosalva	2.0	100 (C. *albicans*)
Sandalwood	0.5	50 (*S. aureus*)
Veramoss	3.0	100 (diphtheroid)

[a]Indicates organism for which the MIC was determined. In some cases this MIC applies to more than one organism.
Source: Ref. 32.

found to be active against a diphtheroid organism isolated from human skin. Again, it must be emphasized that the size of the zone of inhibition on the Petri plate is not necessarily a good indicator of the relative antimicrobial activity of the test material. The size of the cleared zone by this method is dependent on the solubility and rate of diffusion of the sample in the aqueous medium. The above data held some surprises, since previous work had indicated, contrary to these findings, that gram-negative organisms were less sensitive than gram-positive ones (21,33). Also, the activity against the diphtheroid organisms was probably affected by the inclusion of Tween 80 in the growth medium. Tween 80 is known to inhibit the effect of lipophilic germicides.

One aspect of aroma chemicals that has received much attention is the use of fragrances that not only have deodorizing properties but that are also antimicrobial. An example of this application is given in a provocative article by Sturm (34). Deosafe fragrances were formulated by the deliberate combination of aromatic chemicals and antimicrobial essential oils. Sturm used as examples the inhibition of *Staphylococcus* by decyl alcohol, citral, eugenol, and cinnamaldehyde as well as by thyme, clove, and spice oil. Strum found that the method developed by Eleiss for the determination of the contact growth index (CGI) provided exact and reproducible results. Using this CGI method, perfumery raw materials that gave a

Table 8.5 Antimicrobial Activity of Fragrance Materials in Petri Plate Cultures

Organism	Number tested	Number positive	Percentage positive
S. aureus (gram-positive)	521	162	31.1
E. coli (gram-negative)	521	136	26.1
C. albicans (yeast)	521	149	28.6
Diphtheroid	212	101	47.6
All four organisms	212[a]	64	30.2
Three organisms	212	89	42.0
Two organisms	212	118	55.7
One organism	212	144	67.9
Negative	212	68	32.1

[a]Only those materials tested against all four organisms are included in this total.
Source: Ref. 32.

CGI value of zero at a concentration of <3 mg/16 cm^2 were citronellol, geraniol, para-cresol, para-toluyl aldehyde, undecyl alcohol, undecylenic alcohol, cinnamic aldehyde, iso-eugenol, clove oil, clove bud oil, pimento oil, rose oil, cossia oil, and sandalwood oil.

In this approach, results were supplied by a sniff test. The deodorizing properties, with respect to intensity and duration of effect, of a "normal" fragrance, a deosafe fragrance, and Irgasan DP 300 were compared (31). These deosafe fragrance compositions can be formulated to yield safe and effective deodorizing agents. A recent publication gave evidence that a mixture of three natural compounds (farnesol, glyceryl monolaurate, and phenoxyethanol) kills corynebacteria associated with axillary malodor. The mixture was recommended as an effective deodorant by its successful use in practice (35).

In 1975, Isacoff reported that two fragrance mixtures were formulated based on the antimicrobial quality of the fragrance ingredient (36). In this instance, the aroma chemicals were added to aid preservation and not incorporated as "active" ingredients. The products examined had no need for the usual cosmetic preservatives. After addition of the perfumes, the cosmetic products were rendered free of microbial growth within 96 hours of the challenge. This work illustrates how fragrances can be used as the preservative system component in a self-preserving product.

8.3 AROMA CHEMICALS AS NATURAL INGREDIENTS

While the average consumer equates a "natural" substance with a lack of or low toxicity, it is obvious that natural products are not inherently safe. Digitalis (from the foxglove plant) and cobra venom are extreme examples of natural products that

TABLE 8.6 Fragrance Materials with Antimicrobial Activity

Chemicals	*S. aureus*		*E. coli*		*Candida albicans*		*Diphtheroid*	
	Zone diameter (mm)	MIC (ppm)	Zone diameter (mm)	MIC (ppm)	Zone diameter (mm)	MIC (ppm)	Zone diameter (mm)	MIC (ppm)
Acetanilide	0[b]	NT[c]	13	NT	11	NT	NT	NT
N-Acetyl methyl anthranilate	0	>1000	0	>1000	0	>1000	0	>1000
Aldehyde, C_8	0	500	12	> 500	12	500	12[pd]	1000
Aldehyde, C_9	0	NT	0	NT	13	NT	NT	NT
Aldehyde, C_{11} undecylenic	0	50	0	>1000	22	50	17	500
Aldehyde, C_{16}	0	NT	15	NT	0	NT	NT	NT
Aldehyde, C_{18}	0	>1000	15	1000	20	500	15	1000
Allyl amyl glycolate	0	NT	0	NT	11P	NT	NT	NT
Amaryllide	14	>1000	13	1000	19	1000	17	1000
Ambrarome, absolute	13	NT	0	NT	0	NT	NT	NT
Amyl cinnamic aldehyde coeur	0	>1000	0	>1000	0	>1000	0	500
Amyl cinnamyl alcohol	0	> 50	0	>1000	0	> 500	12	500
Amyl salicylate	0	>1000	0	>1000	0	>1000	0	500
Amyris oil	13	NT	0	NT	0	NT	NT	NT
Anethol, USP	0	500	0	500	0	100	0	500
Anisyl acetate	0	>1000	13	>1000	12P	1000	0	1000
Armoise essence	0	1000	0	>1000	0	1000	0	>1000
Arras aldehyde, 50%	20	500	14	1000	17	500	26	500
Aubepine	0	>1000	18	>1000	11	1000	0	1000
Auralva (Schiff base)	11	NT	12	NT	12	NT	NT	NT
Balsam copaiba, USP	0	>1000	0	>1000	0	>1000	0	>1000
Balsam Peru oil	18	>1000	15	>1000	11	>1000	14	500
Basil oil	0	500	0	500	0	500	0	1000

(continued)

TABLE 8.6 Continued

Chemicals	*S. aureus* Zone diameter (mm)	*S. aureus* MIC (ppm)	*E. coli* Zone diameter (mm)	*E. coli* MIC (ppm)	*Candida albicans* Zone diameter (mm)	*Candida albicans* MIC (ppm)	*Diphtheroid* Zone diameter (mm)	*Diphtheroid* MIC (ppm)
Bay oil	18	500	13	1000	20	500	14	1000
Benzaldehyde	0	1000	0	1000	0	1000	0	>1000
Benzoic acid	21	1000	28	1000	21	1000	28	1000
Benzoin coeur	18	>1000	16	1000	10	>1000	12	1000
Benzophenone	0	NT	0	NT	12	NP	NT	NT
Benzyl acetate	0	>1000	0	>1000	0	1000	0	1000
Benzyl alcohol	0	>1000	12P	>1000	0	>1000	0	500
Benzyl benzoate	0	>1000	0	>1000	0	>1000	0	500
Benzyl propionate	0	>1000	0	1000	0	500	0	1000
Benzyl salicylate	0	>1000	0	>1000	0	>1000	0	1000
Bergamot MPF	0	1000	0	>1000	0	500	0	1000
Beta gamma hexenyl formate	13	NT	17	NT	0	NT	NT	NT
Beta naphthyl anthranilate	16	1000	20	1000	22	500	15	1000
Beta pinene coeur	0	500	0	>1000	0	1000	0	>1000
Birch tar, rectified	14	500	14	>1000	17	1000	14	500
Boise de rose filtered	0	>1000	17	1000	0	500	0	1000
Borneol	0	1000	0	1000	0	500	12	500
Camphene 46	0	1000	0	>1000	0	1000	0	>1000
Camphor oil white	12	500	*[e]	>1000	0	500	12	>1000
Caraway oil	0	500	0	>1000	0	500	0	500
Cardamom oil, Guatemala	0	>1000	0	>1000	0	500	0	500
Carvone, dextro	10	1000	11	1000	0	500	0	1000
Carvone, laevo	0	>1000	11	1000	11	500	0	1000
Cashmeran	0	500	0	>1000	0	>1000	14	500

Castoreum, absolute, 50%	11	NT	0	NT	0	NT	NT	NT
Cedarleaf oil	0	1000	0	1000	0	500	0	>1000
Cedarwood, white	0	1000	0	>1000	0	>1000	0	500
Cedrone S.	12	NT	0	NT	0	NT	NT	NT
Cedrus Atlantica coeur	11	500	0	>1000	0	>1000	11	1000
Celery seed oil	0	NT	0	NT	12	NT	NT	NT
Chamomile oil	0	1000	0	>1000	0	500	0	>1000
Cinnamalva	0	>1000	19	1000	13	500	0	1000
Cinnamic alcohol	14	>1000	19	>1000	27	500	16	1000
Cinnamon leaf oil, Ceylon	18	500	17	1000	14	500	12	500
Ciste, colorless	12	NT	0	NT	0	NT	NT	NT
Citral, dimethyl acetal	0	500	0	>1000	33	500	14	500
Citral, refined	15	500	14	500	46	500	16	500
Citronama (Schiff base)	0	NT	0	NT	11	NT	NT	NT
Citronella, Formosa, Java	11	NT	0	NT	17	NT	NT	NT
Citronellal	0	500	0	>1000	12	500	0	500
Citronellol coeur	12	>1000	10	1000	48	100	18	500
Citronellyl acetate	0	>1000	0	>1000	0	>1000	0	500
Citronellyl ethyl ether	11	NT	0	NT	0	NT	NT	NT
Citronellyl isobutyrate	0	>1000	0	>1000	0	>1000	0	>1000
Citrus oil, distilled	0	1000	0	>1000	0	500	0	>1000
Clove leaf oil	14	500	19	1000	19	500	15	500
Clove bud oil	16	500	16	1000	18	500	14	500
Cocal	15	500	14	>1000	11	1000	15	500
Coniferan	0	1000	0	>1000	0	>1000	0	500
Coriander oil	0	1000	11	1000	0	500	0	1000
Corn mint oil	10	NT	0	NT	0	NT	NT	NT
Coronal, beta	12	NT	0	NT	0	NT	Nt	NT
Cortex aldehyde, 50%	17	1000	21	500	16	>1000	16	500
Coumarin	13	>1000	15	>1000	16	1000	12P	1000

(continued)

TABLE 8.6 Continued

Chemicals	*S. aureus*		*E. coli*		*Candida albicans*		*Diphtheroid*	
	Zone diameter (mm)	MIC (ppm)	Zone diameter (mm)	MIC (ppm)	Zone diameter (mm)	MIC (ppm)	Zone diameter (mm)	MIC (ppm)
Cumin oil	0	500	0	1000	0	500	0	1000
Cuminyl alcohol	15P	1000	14	500	23	500	16	1000
Cuminyl acetate	0	>1000	*	>1000	0	100	0	500
Cyclamal extra	0	NT	0	NT	23P	NT	NT	NT
2-Cyclohexyl cyclohexanone	0	NT	0	NT	10	NT	NT	NT
Cyclosia base	14	>1000	16	>1000	14	>1000	13	1000
Cymene coeur	0	1000	0	>1000	0	500	0	>1000
Cypress oil, French	0	1000	0	>1000	0	1000	0	>1000
Decalactone	15	>1000	0	>1000	12	1000	14	1000
n-Decanol	20	NT	0	NT	13	NT	NT	NT
Dibenzyl ether	0	>1000	0	>1000	0	>1000	0	500
Dibutyl sulfide, 10%	0	NT	16P	NT	0	NT	NT	NT
Diethyl phthalate	0	>1000	0	>1000	14P	500	12P	1000
Dihydro cuminyl alcohol	14	1000	16	500	23	500	19	500
Dimethyl anthranilate	0	1000	0	1000	14P	500	0	500
Dimethyl benzyl carbinol	0	>1000	16P	>1000	10	1000	0	>1000
Dimethyl octanol	0	NT	0	NT	12	NT	NT	NT
Dimethyl phenyl acetaldehyde	14	NT	0	NT	0	NT	NT	NT
Dimethyl phenyl ethyl carbinol	0	>1000	14	1000	21	1000	15	1000
Dimethyl phthalate	0	NT	11	NT	12	NT	NT	NT
Dimethyl sulfide	0	1000	0	>1000	0	500	0	1000
Diphenyl oxide	0	>1000	0	>1000	0	500	0	500
Dipropylene glycol	0	>1000	0	>1000	0	>1000	0	>1000
p-Ethyl acetophenone	0	NT	0	NT	13P	NT	0	NT

Ethyl benzaldehyde	0	500	11	500	11	500	0	500
Ethyl benzoate	0	1000	0	1000	0	500	0	1000
Ethyl-3-hydroxy-3-phenyl propionate	0	NT	13	NT	12	NT	NT	NT
Ethyl linalool	18	500	14P	1000	11	500	20	500
Ethyl methacrylate	0	>1000	0	>1000	0	>1000	0	>1000
Ethyl phenyl glycidate	0	NT	0	NT	11	NT	NT	NT
Ethyl vanillin	14	>1000	18	1000	19	1000	15	1000
Eucalyptus oil, 70–75%	0	>1000	0	>1000	0	1000	0	>1000
Eugenol, USP	16	500	21	500	22	500	15	500
Fennel oil, sweet	0	500	0	500	0	500	0	500
Fir balsam, absolute	12	500	15P	1000	0	>1000	0	1000
Fraistone	0	NT	0	NT	26P	NT	NT	NT
Furfural	12	>1000	11	>1000	0	1000	0	>1000
Galaxolide	0	1000	0	>1000	0	>1000	0	500
Galbanum coeur	0	1000	0	>1000	0	>1000	0	500
Geraniol cocur	14	1000	16	500	38	500	20	500
Geraniolene, light	13	NT	0	NT	11P	NT	NT	NT
Geranium, African	12	NT	0	NT	19	NT	NT	NT
Geranoxy acetaldehyde, 50%	0	NT	0	NT	13	NT	NT	NT
Geranyl benzoate	0	>1000	0	>1000	0	>1000	0	>1000
Geranyl methyl tiglate	0	NT	0	NT	14	NT	NT	NT
Geranyl propionate	0	>1000	0	>1000	0	>1000	0	500
Grapefruit oil	0	500	0	>1000	0	500	0	>1000
Guaiene	0	500	0	>1000	0	>1000	12	500
Guaicwood oil	13	500	0	>1000	0	>1000	0	500
Hay, absolute	12	NT	0	NT	0	NT	NT	NT
Hedione	0	>1000	0	>1000	0	1000	12	500
Helional	0	500	14	1000	14	500	16	500
Heliotropyl acetate	0	NT	13P	NT	12P	NT	NT	NT

(continied)

Table 8.6 Continued

Chemicals	*S. aureus*		*E. coli*		*Candida albicans*		*Diphtheroid*	
	Zone diameter (mm)	MIC (ppm)	Zone diameter (mm)	MIC (ppm)	Zone diameter (mm)	MIC (ppm)	Zone diameter (mm)	MIC (ppm)
n-Hexanol	0	NT	11	NT	0	NT	NT	NT
Hexyl cinnamic aldehyde	0	>1000	0	>1000	0	>1000	0	500
Hydratropal acetone	18	NT	0	NT	21	NT	NT	NT
Hydratropic alcohol, white	12	>1000	14	1000	13	1000	11	>1000
Hydroxy citronellal dimethyl acetal	0	NT	0	NT	11	NT	NT	NT
Hydroxy citronellal	20	>1000	16	>1000	13	1000	14	>1000
Hyssop oil	11	NT	0	NT	0	NT	NT	NT
Indisan	12	500	0	>1000	0	>1000	10	500
Indole	18	1000	23	500	17	500	22	500
Iralia	11	NT	0	NT	0	NT	NT	NT
Iritone	10	NT	0	NT	0	NT	NT	NT
Isoamyl pentenoate	12	NT	0	NT	0	NT	NT	NT
Iso beta gamma hexenyl acetate	18	>1000	18	1000	12	1000	12	>1000
Isoborneol	0	>1000	0	500	0	500	0	1000
Isobutyl benzyl carbinol	11	500	12	1000	15	500	14	500
Isobutyl cinnamate	11	NT	0	NT	0	NT	NT	NT
Isobutyl furyl propionate	0	NT	0	NT	10	NT	NT	NT
Isobutyl quinoline	29	100	13	>1000	26P	50	17	100
Isocitral	19	500	11	1000	20	500	18	100
Isoeugenol	23	500	18	500	14	500	14	500
Isoeugenyl benzoate	0	>1000	0	>1000	0	>1000	0	>1000
Isojasmone	12	NT	0	NT	12	NT	NT	NT
Isomuguet aldehyde, 50%	25	NT	0	NT	11	NT	NT	NT

Isopropyl cyclohexyl propanol	15	NT	0	NT	18	NT	NT	NT
Isopropyl quinoline	29	500	13	500	16	100	20	500
Isopulegol M Extra	10	1000	12	1000	0	1000	0	1000
Jasmonate	0	NT	0	NT	11	NT	NT	NT
Jasmone, *cis*	19	1000	13	1000	21	500	15	500
Jasmutone	11	500	12	>1000	13	500	14	500
Labdanax	0	NT	13	NT	12	NT	NT	NT
Labdanol	11	NT	0	NT	0	NT	NT	NT
Labdanum resin, absolute	14	500	0	>1000	0	>1000	12	1000
Lactone HB	17	1000	12	500	16	500	14	500
Lauryl alcohol	0	1000	0	>1000	0	>1000	10	500
Laurine extra	17	>1000	12	>1000	13	1000	0	1000
Lavandin abrialis	0	NT	12	NT	0	NT	NT	NT
Lavandulol	11	1000	12	1000	13	500	0	500
Lavender, absolute, Camilli	0	>1000	0	>1000	0	1000	0	1000
Lemma (Schiff base)	18	100	11	>1000	12	500	12	500
Lemon oil, Calif.	0	500	0	>1000	0	500	0	>1000
Lemongrass	13	500	14	500	15	500	17	500
Lime oil, washed	0	1000	0	>1000	0	500	0	1000
Limonene	0	>1000	0	>1000	0	500	0	>1000
Linalool oxide	0	NT	10	NT	0	NT	NT	NT
Linalool	0	1000	18	1000	0	500	0	500
Linalyl acetate	0	>1000	0	>1000	0	>1000	0	500
Linalyl cinnamate	0	>1000	0	>1000	0	>1000	0	>1000
Lovage oil	14	NT	0	NT	14	NT	NT	NT
LRG No. 182 (ethoxycyclohexanone)	11	NT	0	NT	0	NT	NT	NT
LRG No. 1181 (neo-, isomenthones)	13	NT	12	NT	0	NT	NT	NT
Lyral	16	1000	13	>1000	12	1000	14	1000

(continued)

TABLE 8.6 Continued

Chemicals	*S. aureus*		*E. coli*		*Candida albicans*		*Diphtheroid*	
	Zone diameter (mm)	MIC (ppm)	Zone diameter (mm)	MIC (ppm)	Zone diameter (mm)	MIC (ppm)	Zone diameter (mm)	MIC (ppm)
Mace, whole extract	12	>1000	0	>1000	0	>1000	0	500
Maltol	0	>1000	16	>1000	16	1000	14	>1000
Mandarin oil	0	1000	0	>1000	0	500	0	>1000
Menthol, USP	11	500	11	500	10	500	0	500
Methallyl pentenoate	14	NT	0	NT	0	NT	NT	NT
p-Methoxy hydrotropic aldehyde	18	1000	26	1000	15	500	16	500
Methyl anthranilate	12	>1000	14	>1000	16	1000	0	1000
p—Methoxy phenoxy acetaldehyde	18	1000	20	>1000	16	1000	19	500
Methyl benzoate	0	>1000	0	>1000	0	1000	0	1000
p-Methoxy phenoxy acetaldehyde dimethyl acetal	0	>1000	0	>1000	11	1000	0	1000
Methyl cinnamate	11	1000	12	>1000	13	500	0	500
α-Methyl cinnamic aldehyde	20	1000	17	500	18	500	0	500
Methyl cyclocitrone	10	NT	0	NT	0	NT	NT	NT
Methyl eugenol	12	1000	11	1000	16	500	12	500
Methyl heptenol	12	NT	13	NT	0	NT	NT	NT
Methyl hexyl acetaldehyde	34	>1000	22	>1000	15	1000	16	500
Methyl isoeugenol	0	>1000	0	>1000	0	>1000	0	100
Methyl lavender ketone	13	100	13	1000	18	500	15	500
Methyl β-naphthyl ketone	0	NT	0	NT	15	NT	NT	NT
Methyl octin carbonate	0	>1000	0	>1000	0	500	0	500
Methyl octyl acetaldehyde	18	NT	0	NT	0	NT	NT	NT
2-Methyl-2-pentenoic acid	19P	1000	35	1000	18	1000	25	1000
Methyl-*p*-cresol	0	>1000	0	1000	0	500	0	>1000

Methyl phenyl ethyl alcohol	13	>1000	19	>1000	16	1000	0	>1000
Methyl *p*-toluate	0	1000	0	1000	0	500	0	1000
Miel Blac, Delaire	11	NT	17P	NT	0	NT	NT	NT
Mousse, absolute verte Maroc	21	50	15	1000	18	500	18	500
Muguet aldehyde	40	NT	0	NT	14	NT	NT	NT
Muscagene	12	NT	0	NT	0	NT	NT	NT
Musk ambrette	0	>1000	0	>1000	0	>1000	13P	500
Musk ketone	0	>1000	0	>1000	0	>1000	12P	100
Musk xylol	0	>1000	0	>1000	0	>1000	11P	50
Myrac aldehyde	17	NT	0	NT	33	NT	NT	NT
Myrrh coeur	13	1000	0	>1000	0	>1000	12	1000
Myrtenal	13	NT	10	NT	0	NT	NT	NT
Myrtle oil, Charabot	12	NT	11	NT	0	NT	NT	NT
Naame (Schiff base)	12	NT	0	NT	0	NT	NT	NT
Narcisse ketone	0	NT	0	NT	13	NT	NT	NT
Narcitol	14	NT	18	NT	13	NT	NT	NT
Nerol	14	500	13	1000	41	500	12	500
Nerolidol	15	NT	0	NT	0	NT	NT	NT
Neroly blanc	13	NT	10	NT	0	NT	NT	NT
Nortonkalactone	29	>1000	13	>1000	17	>1000	0	>1000
Nutmeg oil	0	500	0	>1000	0	500	0	1000
Oakmoss essence	22	50	0	>1000	12	1000	14	500
Ocimene	22	500	14	1000	13	500	16	500
Ocmea (Schiff base)	13	50	10	500	13	50	13	100
Opoponax oil	10	NT	0	NT	0	NT	NT	NT
Orange oil, Fla.	0	500	20	>1000	0	500	22	500
Orange terpeneless, absolute	11	NT	0	NT	0	NT	NT	NT
Orange terpenes	12	NT	16	NT	0	NT	NT	NT
Orenyle	14	NT	0	NT	0	NT	NT	NT
Origanum oil, Spanish	33	500	24	500	13	500	16	500

(continued)

TABLE 8.6 Continued

Chemicals	*S. aureus*		*E. coli*		*Candida albicans*		*Diphtheroid*	
	Zone diameter (mm)	MIC (ppm)	Zone diameter (mm)	MIC (ppm)	Zone diameter (mm)	MIC (ppm)	Zone diameter (mm)	MIC (ppm)
Orivone	11	NT	0	NT	39	NT	NT	NT
Oxyphenylon	0	>1000	12	>1000	13	>1000	12	>1000
para-Cresol	20	>1000	29	1000	16	1000	22	1000
para-Cresyl acetate coeur	0	>1000	0	>1000	12P	1000	0	>1000
para-Isopropyl hydratropic aldehyde	13	NT	0	NT	12	NT	NT	NT
para-Methyl benzyl acetate	11	NT	0	NT	0	NT	NT	NT
para-Methyl dimethyl benzyl carbinol	14	>1000	14	1000	13	1000	14	1000
para-tert-Butyl cycohexanone	11	1000	0	>1000	27P	500	0	500
para-tert-Butyl-*meta*-cresol	85	50	26	500	85P	50	22	100
para-Tolyl alcohol	12	>1000	16	>1000	17	>1000	12	>1000
Patchouli oil, dark	12	100	0	>1000	0	>1000	13	500
Patchouli oil, light	0	500	0	>1000	0	>1000	13	500
Peach aldehyde coeur	12	NT	0	NT	14	NT	NT	NT
Pepper oil, black	0	1000	0	>1000	0	>1000	0	>1000
Peppermint	0	NT	0	NT	10	NT	NT	NT
Persicol (γ-undecalactone)	0	NT	0	NT	11	NT	NT	NT
Petinerol	0	1000	12	1000	14	500	0	1000
Petitgrain S.A.	0	>1000	0	>1000	0	500	0	1000
Petitgrain terpenes	14	500	11	>1000	10	500	12	>1000
Phellandrene	18	NT	18	NT	0	NT	NT	NT
Phenoxyl ethyl propionate	0	>1000	0	>1000	12	500	0	1000
Phenyl acetaldehyde	21	100	40	1000	33	500	18	100
Phenyl ethyl acetate	0	NT	25P	NT	0	NT	NT	NT
Phenyl ethyl alcohol	0	>1000	16	>1000	0	>1000	0	>1000

Phenyl ethyl cinnamate	0	>1000	0	>1000	0	>1000	0	>1000
Phenyl ethyl phenyl acetate	13	>1000	0	>1000	0	>1000	0	500
Phenyl propyl alcohol	12	>1000	18	>1000	16	1000	14	1000
Phenyl propyl aldehyde	12	500	27	500	14	1000	0	500
Phixia	16	>1000	15	>1000	12	1000	14	1000
Piconia	13	NT	0	NT	0	NT	NT	NT
Pimento berry oil	16	500	17	1000	18	500	14	500
Pine needle oil, Siberian	0	500	0	>1000	0	1000	0	1000
Pine oil	12	1000	14	>1000	11	500	12P	1000
Propylene glycol, USP	0	>1000	0	>1000	0	>1000	0	>1000
Rosacene	14	1000	12	1000	29	500	0	1000
Rosalva	16	1000	0	>1000	16	100	14	500
Rosemary oil, Spanish Tunisia	0	1000	0	>1000	0	1000	0	>1000
Rosetone	NT	>1000	0	>1000	0	>1000	0	>1000
Rosin gum	12	NT	0	NT	0	NT	NT	NT
Sandalwood	11	50	0	>1000	*	>1000	11	500
Santalol	13	500	0	>1000	0	>1000	13	500
Sauge sclaree, absolute	12	500	0	>1000	0	>1000	11	500
Sesquiterpenes PC	13	500	12P	>1000	0	>1000	12	500
Spearmint oil	0	1000	0	>1000	0	500	0	1000
Spruce oil	0	500	0	>1000	0	1000	0	1000
St. Guaiol	13	500	0	>1000	0	>1000	12	500
St. John's bread, conc. 10%	0	NT	15P	NT	0	NT	NT	NT
Styrax alva essence	21	NT	0	NT	0	NT	NT	NT
Styrax clarified, extra	11	>1000	0	>1000	0	500	11	>1000
Surfleurs hay	11	NT	0	NT	0	NT	NT	NT
Tabac absolute	10	NT	0	NT	0	NT	NT	NT
Tangerine oil, Fla.	0	500	0	>1000	0	500	0	>1000
Terpineol	12	1000	19	1000	20P	1000	0	1000
Thuja oil	0	500	0	>1000	0	500	0	1000

(continued)

TABLE 8.6 Continued

Chemicals	*S. aureus*		*E. coli*		*Candida albicans*		*Diphtheroid*	
	Zone diameter (mm)	MIC (ppm)	Zone diameter (mm)	MIC (ppm)	Zone diameter (mm)	MIC (ppm)	Zone diameter (mm)	MIC (ppm)
Thyme, white	25	500	27	500	14	500	16	500
Tiglyl piperidide	0	NT	13	NT	11	NT	NT	NT
Tolu resin, absolute, 50%	17	>1000	17	>1000	11	1000	13	1000
Tonalid	0	>1000	0	>1000	0	>1000	0	500
Tonka, absolute	11P	>1000	14	>1000	14P	>1000	0	>1000
trans-Decahydro beta naphthol	15	1000	15	1000	22P	500	16	500
trans-3-Pentenyl acetone	0	NT	18	NT	0	NT	NT	NT
Treemoss, absolute, French, 50%	18	100	0	>1000	0	1000	14	500
Trimethyl cyclohexanol	0	NT	11	NT	0	NT	NT	NT
Trimethyl cyclohexenone	0	>1000	12	1000	10	1000	0	>1000
Trimethyl cyclohexenol	15	NT	13	NT	0	NT	NT	NT
Undecylenic acid	21	NT	0	NT	13	NT	NT	NT
Vanillin	16	>1000	22	>1000	19	1000	12P	>1000
Veltol plus	0	>1000	20	>1000	12	>1000	0	>1000
Veramoss	0	500	0	>1000	12	500	12	100
Verdural extra	13	NT	13	NT	0	NT	NT	NT
Violettone A, colorless	16	NT	12	NT	0	NT	NT	NT
Wintergreen oil	0	>1000	*	>1000	0	500	0	1000
Yaracetal	0	NT	13	NT	0	NT	NT	NT
Ylang concrete	0	>1000	0	>1000	0	1000	0	1000
Zingerone	11	>1000	12	>1000	11	>1000	0	>1000
Controls Hexachlorophene	13	0.10	12	50	0	50	0	100

Source: Ref. 32.

are toxic. However, it must be conceded that products that have been used for eons must, at worst, have only subtle rather than acute toxic effects.

Many of the spice extracts and essential oils have been used with greater diligence in food than in cosmetics. Research into the toxicity of a number of essential oils has been begun by the FAO/WHO Expert Committee on Food Additives. Regulations restricting the use of aromatic substances were made in 1955 (37).

One of the few papers dealing directly with the toxicity of essential oils was written by Chopra et al. (38). Two examples were presented. The LD_{50} of the oil of *Alpinia galanga* for guinea pigs was 0.68 ml/kg, while that of Acorus calamus was 0.25 ml/kg. The oils, when given to guinea pigs in sublethal doses (<1/3 of the LD_{50}), produced no pathological changes that could be attributed to the oils.

In 1968, Hamburger and Boger reviewed in 1968 the carcinogenicity of essential oils, flavors, and spices (39). Fortunately, relatively few essential oils have been shown to cause cancer in animals. Their paper summarizes information on the carcinogenic activity of safrole, citrus oil (d-limone), turpentine oil (1-pinene), eucalyptus oil (phellandrene), bergamot oil, and red pepper.

The results of testing are not always clear or simple, and the reader interested in this problem should read the full review. In addition to governmental agencies interested in this problem, the fragrance industry stepped forward to take measures to assure the safety of chemicals. In 1966, the prinicipal suppliers of aroma and fragrances established the Research Institute for Fragrance Materials (RIFM). The sole purpose of this council is to assure the safety of perfumery raw materials. At present the RIFM is supported by more than 70 companies representing nearly 97% of the industry in the United States and Europe. In addition to the RIFM, another national group was established in 1973. The International Fragrance Association (IFRA) has the task of drawing up a set of rules of procedure on the production, handling, and safety of aroma chemicals and related materials. These are two nongovernment agencies that can provide the interested formulator or microbiologist with information on the safety of various aroma chemicals. The cosmetic industry deserves much credit for establishing such self-policing efforts.

Tea is one of the most widely consumed beverages in the world. It is classified into four types: green, black, oolong, and purer teas. All of these teas are prepared from the leaves of *Camellia sinensis* and its varieties, with the differences in the classes resulting from the processing.

A recent review by Hamilton-Miller (40) indicates that tea inhibits a number of organisms, including *S. aureus*, *S. typhi*, *Pseudomonas* spp., and *Trichophyton mentagrophytes*. The antimicrobial activity is attributed to polyphenolic compounds, including the catechins, which are isoflavanoids, and the theoflavins and theorubigins, which are oxidation and polymerization products of simple isoflavanoids. The MICs for specific isoflavanoids ranged from 73 to 573 μg/ml for *S. aureus* and from 183 to >1140 μg/ml for *E. coli.* Quercitin (a flavonol) had an MIC of 37 μg/ml for *S. aureus*, but was inactive against *E. coli.*

Nakahara et al. (41) reported that oolong tea extract inhibited the water-insoluble glucan-synthesizing enzyme glucosyl transferase I of *Streptococcus mutans*. This inhibited glucan synthesis by this organism and decreased adherence of the growing cells. Kubo et al. (42) also studied the antimicrobial activity of green tea flavor components. They reported that sesquiterpene hydrocarbons δ-cadinene and caryophyllene were active against *S. mutans* and that this antimicrobial activity was enhanced in combination with indole. They noted that jasmine is sometimes added to tea for flavor and that jasmine contains a large quantity of indole. This suggests a role for green tea flavor compounds as antimicrobial agents for foods and cosmetic products.

8.4 SUMMARY

Spices, essential oils, and plant extracts have been used in product preservation for centuries. The active principles include alcohols, phenols, esters, acids, aldehydes, terpenes, and polyphenols. Aroma chemicals are multifunctional in that many have antimicrobial action besides contributing to the fragrance and/or flavor of products in which they are used.

In view of the antimicrobial properties of fragrances, it is reasonable to suggest that the selection of fragrances for cosmetic and drug products be based not only on the desired scent but also upon its inherent germicidal properties. Aroma chemicals with antimicrobial properties may be used to reduce or replace chemical preservatives in products and may be considered to be a part of the preservative system. It is suggested that formulators consider the use of fragrance materials in a systems approach to create self-preserving products.

REFERENCES

1. F. K. Oldham, F. E. Kelsey, and E. M. K. Geiling. *Essentials in Pharmacology.* Lippincott, Philadelphia, 1951.
2. S. Garland. *The Complete Book of Herbs and Spices.* Viking Press, New York, 1979.
3. S. L. Leadbetter. Natural antimicrobials agents. *Food Focus* (13):1–65 (1991).
4. G. Urdang. In: E. Guenther, ed. *The Essential Oils*, Vol. 1. Van Nostrand, New York, 1948, p. 3.
5. S. A. Malcolm and E. A. Sofowora. Antimicrobial activity of selected Nigerian folk remedies and their constituents plants. *Lloydia 32*:512–517 (1969).
6. M. Chamberland. Les essece. Au point due vue leurs proprietes antisepti. *Ann. Inst. Pasteur Paris 1*:153–164 (1887).
7. W. H. Martindale. Antiseptic powers of essential oils. *Perfum. Essent. Oil Rec. 1*:266 (1910).

8. H. C. Hoffman and A. C. Evans. The use of spices as preservatives. *J. Ind. Eng. Chem. 3*:835–838 (1991).
9. D. I. Macht and W. M. Kunkel. Concerning the antiseptic action of some aromatic fumes. *Proc. Soc. Exp. Biol. Med. 18*:68 (1920).
10. F. C. Dyche-Teague. Bactericidal value of some commercial perfumes, *Perfum. Essent. Oil Rec.* 15:6–8, 40–43, 81–83 (1924).
11. J. J. Bryant. Detergent and antiseptic value of perfume material, i.e. essential oils and synthetics, in toilet soap. *Perfum. Essent. Oil Rec. 15*:426–431 (1924).
12. S. Rideal, E. Rideal, and A. Sciver. Investigation into the germicidal powers and capillary activities of certain essential oils. *Perfum. Essent. Oil Rec. 19*:285–305 (1928).
13. G. M. Dyson. Physiological aspects of the essential oils. *Perfum. Essent. Oil Rec. 21*:287–314 (1930).
14. R. M. Gattefosse. *Aromatherapie*. Girardot, Paris, 1937.
15. R. M. Gattefosse. Essential oils as vegetable hormones. *Parfum Mod. 31*:57, 59 (1937).
16. Y. Naves. Evaluation of the bactericidal power of essential oils by determination of their phenol coefficients. *Parfums France 13*:273–284 (1935).
17. R. S. Potter. Antiseptic value of flavors. *Flavours 2*:7–12 (1939).
18. T. McLachlen. Microorganisms in cosmetics. *Soap Perfum. Cosmet. 19*:666–672, 674 (1946).
19. M. Rolle and A. Mayr. Conifer oils as disinfectants against tuberculosis. *Monatsh. Tierheilkd. 4*:421–432 (1952).
20. J. C. Maruzzella and M. B. Lichtenstein. The in vitro antibacterial activity of oils. *J. Am. Pharm. Assoc. Sci. Ed. 45*:378–381 (1956).
21. J. C. Maruzzella and P. A. Henry. The in vitro antibacterial activity of essential oils and oil combinations. *J. Am. Pharm. Assoc. 47*:294–296 (1958).
22. J. C. Maruzzella, J. Balter, and A. Katz. The action of perfume-oil vapors on fungi. *Am. Perfum. Aromat. 74*:21–22 (1959).
23. J. C. Maruzzella. The antifungal properties of essential oil vapours. *Soap Perfum. Cosmet. 33*:835–837 (1960).
24. J. C. Maruzzella and E. Bramnick. The antibacterial properties of perfumery chemicals. *Soap Perfum. Cosmet. 35*:743–745 (1961).
25. J. C. Maruzzella. The germicidal properties of perfume oils and perfumery chemicals. *Am. Perfum. 77*:67–70 (1962).
26. J. C. Maruzzella. An investigation of the antimicrobial properties of absolutes. *Am. Perfum. Cosmet. 78*:19–20 (1963).
27. H. Munzing and H. Schels. Potential replacement of preservatives in cosmetics by essential oils. *J. Soc. Cosmet. Chem. 23*:841 (1972).
28. R. Richards and R. McBride. Enhancement of benzalkonium chloride by aromatic alcohols. *J. Pharm. Sci. 62*:2035–2037 (1973).
29. J. J. Kabara, D. M. Swieczkowski, A. J. Conley, and J. P. Truant. Fatty acid derivatives as antimicrobial agents. *Antimicrob. Agents Chemother. 2*:23–28 (1972).
30. R. D. Fletcher, J. R. Gilbertson, A. C. Albers, and S. D. White. Inactivation of mycoplasma by long-chain alcohols. *Antimicrob. Agents Chemother. 19*:917–921 (1981).
31. R. J. Crout, J. R. Gilbertson, J. D. Gilbertson, D. Platt, H. H. Langkamp, and R. H. Connamacher. Effect of linolenyl alcohol on the in vitro growth of the oral bacterium *Streptococcus mutans. Arch. Oral Biol. 27*:1033–1037 (1982).

32. J. A. Morris, A. Khettry, and E. W. Seitz. Antimicrobial activity of aroma chemicals and essential oils. *J. Am. Oil Chem Soc. 56*:595–605 (1979).
33. J. C. Maruzzella and N. A. Sicurella. Antibacterial activity of essential oil vapors. *J. Am. Pharm. Assoc. Sci. Ed. 49*:692 (1960).
34. W. Sturm. Deosafe fragrances: fragrances with deodorizing properties. *Cosmet. Toiletr. 94*:35–48 (1979).
35. U. F. Houstein, J. Herrmann, U. Hoppe, W. Engel, and G. Sauermann. Growth inhibition of coryneform bacteria by mixture of three natural products—farnesol, glyceryl monolaurate and phenoxyethanol. *J. Soc. Cosmet. Chem. 44*:211–220 (1993).
36. H. Isacoff. Preservation with essential oils and aroma chemicals. In: M. G. DeNavarre, ed. *The Chemistry and Manufacture of Cosmetics*. Continental Press, 1975.
37. Flavoring substances. *Br. Food J.* (Nov./Dec.):175–183 (1973).
38. L. C. Chopra, N. Khajuria, and C. L. Chopra. Antibacterial properties of volatile principles from *Alpinia galanga* and *Acorus calamus*. *Antibiot. Chemother.* 7:378–383 (1957).
39. F. Hamburger and E. Boger. The carcinogenicity of essential oils, flavors and spices: a review. *Cancer Res. 28*:2372–2374 (1968).
40. J. M. T. Hamilton-Miller. Antimicrobial properties of tea (*Camellia sinensis* L.). *Antimicrob. Agents Chemother. 39*:2375–2377 (1995).
41. K. Nakahara, S. Kawabata, H. Ono, K. Ogura, T. Tanaka, T. Ooshima, and S. Hamada. Inhibitory effect of oolong tea polyphenols on glucosyltransferases of mutans streptococci. *Appl. Environ. Microbiol. 59*:968–973 (1993).
42. I. Kubo, H. Muroi, and M. Himejima. Antimicrobial activity of green tea flavor components and their combination effects. *J. Agric. Food Chem. 40*:245–248 (1992).

9

Chelating Agents as Preservative Potentiators

Jon J. Kabara
Technology Exchange, Inc.
Galena, Illinois, and
Michigan State University
East Lansing, Michigan

Chelating agents, or sequestrants, are ingredients that are capable of complexing with and inactivating metallic ions. Chelating agents are used in foods, drugs, and cosmetics to prevent the adverse effects of metal ions on the stability and appearance of cosmetic products. While any chelating agent will have an effect, chemists have considered ethylenediaminetetraacetic acid (EDTA) to be the classic metal-chelating agent. EDTA will be used as a primary example of a preservative potentiator in the material presented.

The mechanism by which EDTA and similar chelating agents work is by capturing metal ions—by forming noncovalent bonds to the cations with unshared electrons on nitrogen atoms and carboxylate oxygens. Due to the molecular shapes of chelating agents, only the antilone pairs of carboxylate oxygens (i.e., the unshared electrons on the side opposite the carbonyl group) interact with the metal ions in the complex.

Chelating agents of the EDTA type have been used since the late 1930s. Originally they were developed to counteract the deleterious effects of water hardness and heavy metal ions on dyestuffs used in textile manufacture; however, they found use in wide-ranging applications from cosmetics and pharmaceuticals to foods. Because EDTA and its related derivatives are safe, inexpensive, effective, and used in small amounts compared to the overall formulation, they have been added to cosmetic and toiletry formulations as a form of "insurance" against spurious metal ion contamination, as well as for rather ill-defined reasons.

There are several desirable benefits obtained from the addition of chelating agents to cosmetic and toiletry formulations. Typical uses of chelating agents such as EDTA have been reviewed extensively (1) and include the following applications: to maintain clarity, to protect fragrance components, to stabilize polymeric thickeners, to prevent rancidity and off-odors, to stabilize color additives, to prevent discoloration in products containing thioglycolate or *p*-aminobenzoic acid, and to retard sulfite decomposition. In addition, chelating agents increase preservative efficacy against gram-negative bacteria, especially fluorescent pseudomonads such as *Pseudomonas aeruginosa*. The potentiating effects of chelating agents on biocides and other properties of chelating agents are examined in this chapter to provide the basis for their use in self-preserving products. Again, although most of the discussion centers around EDTA, all chelating agents give similar effects.

9.1 CHELATORS AS MULTIFUNCTIONAL AGENTS

Compounds that have phosphate or phosphonate groups and compounds that have two carboxyl groups or carboxyl and hydroxyl groups in close proximity may function as classical chelating agents, in the sense described by Marshall et al. (2). Several chelating agents that have been used in cosmetic and products and listed in the CTFA *Cosmetic Ingredients Directory* appear in Table 9.1 (3).

TABLE 9.1 Chelating Agents Used in Cosmetic and Drug Products

Aminotrimethylene phosphonic acid	Pentapotassium triphosphate	TEA-EDTA
Calcium disodium EDTA	Pentasodium aminotrimethylene phosphonate	Tetrahydroxypropyl ethylenediamine
Citric acid	Pentasodium pentenate	Tetrapotassium pyrophosphate
Cyclohexanediamine tetraacetic acid	Pentasodium triphosphate	Tetrasodium EDTA
Diammonium citrate	Pentetic acid	Tetrasodium etidronate
Dipotassium EDTA	Sodium citrate	Tetrasodium pyrophosphate
Disodium EDTA	Sodium dihydroxyethylglycinate	Tripotassium EDTA
Disodium pyrophosphate	Sodium gluconate	Trisodium EDTA
EDTA	Sodium hexametaphosphate	Trisodium HEDTA
Etidronic acid	Sodium metaphosphate	Trisodium NTA
Glucuronic acid	Sodium metasilicate	Trisodium phosphate
HEDTA	Sodium trimetaphosphate	

Source: Ref. 3.

Amino acid chelating agents, such as EDTA, are anionic and react with cationic alkaline earth and heavy metal ions (i.e., copper, manganese, and iron ions) to form anionic or sometimes neutral metal-chelate complexes (4). Sequestration of trace amounts of metal ions in this manner reduces the effective concentration of the metal ions and prevents them from entering into undesirable reactions.

Many dyestuffs used for coloring products are susceptible to shade changes and/or precipitation by heavy metal ions. Chelation of metal ions helps reduce hazing and color changes due to the reaction of metal ions with formula components. Metal ions act as pro-oxidants and initiate free-radical-mediated reactions involved in the auto-oxidation of aldehydes/ketones in fragrances and fatty acids. Auto-oxidation of compounds with carbonyl groups changes the character of fragrances and/or causes the development of rancid odors and darkening in products that contain fatty acids. Again, sequestering metal ions greatly reduces the tendency for metal ions to participate in auto-oxidation reactions and improves the stability of the formula (5).

Chelating agents also contribute to the preservative system of a formula. Although EDTA is capable of functioning as an antimicrobial, more importantly it is a potentiator of preservative action when combined with other commonly used cosmetic preservative materials. This has been discussed by several workers (6–11) and is reviewed in the following paragraphs.

9.2 ANTIMICROBIAL ACTIVITY OF EDTA

The antibacterial activity of EDTA was reported in the late 1950s by Repaske (12) and MacGregor and Elliker (13). MacGregor and Elliker found that *P. aeruginosa* readily adapts to forms that are resistant to quaternary ammonium compounds and that EDTA eliminated this acquired resistance. These workers demonstrated that EDTA exerts a lytic action on the outer layers of the walls of gram-negative bacteria. The EDTA-induced lysis releases intracellular solutes and partially solubilizes the cell envelope of sensitive bacteria.

The minimum inhibitory concentration (MIC) value for EDTA against *P. aeruginosa* depends on the strain and growth conditions of the culture. The bactericidal action of EDTA against *P. aeruginosa* is quite significant but incomplete when used alone. For instance, 250 ppm of EDTA alone gave 99–99.9% kill of *P. aeruginosa*, while 100 ppm of EDTA killed between 57 and 99.5% at pH 7.4 (13). MIC values up to about 30 mM for EDTA at pH 7 have been reported for this bacterium (14,15). Although sensitivity to EDTA is a general characteristic of fluorescent pseudomonads, other gram-negative bacteria reported to be susceptible include *Achromobacter* spp., *Alcaligenes faecalis*, *Azotobacter vinlandii*, *Escherichia coli*, *Salmonella* spp., *Spirillum itersonii*, *Vibrio cholerae*, and *Vibrio succinogenes*. Yeasts and fungi also are sensitive to EDTA (16,17).

Although gram-positive bacteria do not exhibit significant sensitivity to

EDTA, all organisms require iron. The growth of microorganisms is sensitive to iron concentration, and microbial growth increases with increasing availability of iron. Fass and Barnishan (18) reported on the antibiotic susceptibilities of bacteria. Liao (19) observed that some enzyme production and transport in *P. fluorescens*, *P. viridiflava*, and *P. putida* required divalent calcium ions. In addition, the virulence of many pathogenic bactericidal is correlated with their ability to synthesize siderophores, which are naturally occurring iron chelators (see Sec. 2.6).

The use of chelating agents reduces the availability of iron for microbial growth; however, sterility should not be expected from EDTA alone. The inherent lack of broad-spectrum bactericidal activity and the incompleteness of total bacterial kill limits the use of EDTA as the sole agent in self-preserving systems. Nevertheless, the use of EDTA in combination with other multifunctional agents with antimicrobial action affords the formulator an opportunity to reduce and/or eliminate classical preservatives.

9.3 ACTION OF EDTA ON *PSEUDOMONAS*

The seriousness of gram-negative bacterial contamination of cosmetics is well documented (20–22). Yablonski and Goldman (20) reported that these organisms account for approximately 98% of all contaminants found in shampoo products and raw materials used in compounding them. Organisms in the genus *Pseudomonas* lead the list of gram-negative bacteria that cause problems in manufacturing plants and finished cosmetic products.

P. aeruginosa is a ubiquitous organism that has very simple growth requirements. Certain strains have been found to be capable of growth in distilled water (23,24). Of all gram-negative bacteria encountered in cosmetic products, *P. aeruginosa* generally causes the most concern. This organism is able to survive contact with concentrations of antibacterials many times greater than that required to kill other species (13). Also, it has numerous survival strategies, which enable it to readily adapt to more resistant forms (see Chapter 2). Gould (25) described it as follows: "Biologically active to an extreme degree, it can live on simple substrates, survive for long periods under [a] wide range of environmental conditions apart from severe dehydration, and adapt to the prescence of high concentrations of most antibacterial substances."

The cosmetic industry is constantly vigilant in its efforts to eliminate *Pseudomonas* contamination. *P. aeruginosa* has caused severe eye infections and blindness (26,27). Upon entrance to the corneal epithelium, this organism can be totally destructive to the eye and will often cause rapid, severe, and permanent damage. Even minute abrasions on the cornea surface can lead to corneal destruction when infected with this organism. Indeed, it has been shown that the corneal damage to a rabbit's eye caused by a commercial shampoo increased the likelihood of infection by this organism (28). Shampoos and skin cleansers have been found to be especially vulnerable to contamination by pseudomonads. [*Note*: Concentrations in most formulas (0.05%) frequently are too low to be effective for potentiation of preservative action.] While the potential for serious eye injury is of great concern to the cosmetic formulator, the emulsion breakdown and spoilage due to lipases and other enzymes released by *Pseudomonas* cells may be a serious problem in itself. Use of chelating agents would be expected to reduce microbial growth because metal ions are of crucial importance in many cellular reactions (29).

Russell (23) observed that EDTA has considerable microbiological importance because it causes an increase in permeability, which makes bacterial cells more susceptible to antimicrobial compounds. The ability of EDTA to increase cell wall permeability increases the sensitivity of the bacteria to a wide variety of antimicrobial compounds, including compounds with widely differing chemical structures and modes of action. Voss (30) suggested that the greater resistance of gram-negative bacteria may due to the greater complexity of their cell wall structure, which acts to exclude the antibacterial agent from the interior of the cell. Wilkinson (31) reviewed the sensitivity of *P. aeruginosa* to EDTA, particularly in regard to its effect on the structure of the cell envelope. In addition, EDTA has been used in studies on cell wall chemistry of gram-negative bacteria and in classification of the pseudomonads.

EDTA is believed to exert a lytic action on the cell wall structure of gram-negative bacteria, and especially on *P. aeruginosa*. The lysis of cells may be measured by the decrease in turbidity of cell suspensions along with a corresponding release of intracellular solutes. Eagon and Carson (32) found that a 99% loss of viability corresponded to about a 50% decrease in turbidity. The extraction of metal cations—especially calcium and magnesium—by EDTA is followed rapidly by partial solubilization of the cell membrane (31).

The lipopolysaccharide (LPS) core of *P. aeruginosa* is highly substituted with polyphosphate residues, which possess metal cation-binding capacity. It is believed that these polyphosphate-rich residues, located either on or within the inside surface of the outer layer of the outer membrane, bind to other membrane components (i.e., proteins and phospholipids) through cation bridges. EDTA removes these cations, which, in turn, destabilizes the outer membrane/cell wall (33) to release a complex of protein and LPS, thereby "permeabilizing" the outer membrane and cell wall to make the cell membrane more accessible to chemicals in the microenvironment of the cell.

Additional evidence for the mechanism of action of EDTA comes from studies on the activity of other amino polycarboxylic acid–chelating agents on *P. aeruginosa* (34,35). Table 9.2 lists several chelating agents that have been found to be active against *P. aeruginosa* (36). They are shown in decreasing order of antimicrobial activity, which is correlated with their ability to chelate magnesium.

TABLE 9.2 Chelating Agents Active Against P. aeruginosa

Cyclohexanediaminetetraacetic acid (CDTA)
Ethylenediaminetetraacetic acid (EDTA)
Diethylenetriaminepentaacetic acid (DTPA)
Hydroxyethylethylenediaminetriacetic acid (HEEDTA)

9.4 POTENTIATION OF PRESERVATIVE ACTION WITH EDTA

Lawrence (37) observed that calcium and magnesium in hard water reduced the antimicrobial activity of quaternary germicides. It quickly became accepted practice to add EDTA to quaternary ammonium germicides to improve performance under actual use conditions. Klimek (38) found that EDTA increased the antimicrobial activity of cationic germicides, even in distilled water, so it became apparent that EDTA was causing some effect on bacterial cells that rendered them more sensitive to the germicidal action. Several workers have reported that this highly adaptive organism was able to grow and remain viable for extended periods in solutions of benzalkonium chloride and cetrimonium chloride (14,15). Since MacGregor and Elliker found that the use of EDTA prevented *P. aeruginosa* from developing resistance to quaternary ammonium compounds, EDTA is now strongly recommended as a potential additive to quaternary ammonium salts and other surfactants.

Hart (9) noted that chelating agents potentiate many different types of preservatives. The preservatives reported to be enhanced by EDTA are listed in Table 9.3 and include cationics (quaternary ammonium salts and other cationic compounds), anionics (parabens, sorbic acid, potassium sorbate), phenolic compounds (*tert*-butyl hydroxyanisole, triclosan, chloroxylenol, salicylic acid), and nonionics (formaldehyde donors, phenethyl alcohol, monolaurin).

Chlorhexidine salts such as chlorhexidine gluconate (CHG) are used as preservatives in cosmetic formulations. CHG is a cationic compound and is susceptible to the development of resistant pseudomonads. Burdon and Whitby (39) reported contamination of chlorhexidine–cetrimonium bromide stock solutions with *P. aeruginosa*. The addition of EDTA eliminates resistance of these bacteria and lowers the MICs for chlorhexidine (34,40). A 1993 survey listing ingredients in products voluntarily reported to the FDA listed 27 products that contained this preservative (41).

EDTA increases the antimicrobial action of formaldehyde-releasing agents. Quaternium-15 is widely used in cosmetics because of its broad-spectrum antimicrobial activity. Izzat and Bennett (36) reported that EDTA potentiated the antimicrobial activity against gram-negative bacteria. Use of EDTA with quaternium-15 provides enhanced preservative efficacy, which allows the formu-

TABLE 9.3 Preservatives Known to be Enhanced by EDTA

Cationic preservatives		
	Quaternary ammonium salts	
		Benzalkonium chloride
		Benzethonium chloride
		Myristalkonium chloride
		Cetylpyridinium chloride
		Cetrimonium bromide
		Lauryl pyridinium chloride
		Methylbenzethonium chloride
		Lauryl isoquinolinium bromide
	Other cationic compounds	
		Quaternium 15
		Chlorhexidine digluconate
		Chlorhexidine dihydrochloride
Anionic preservatives		
	Parabens	
		Methylparaben
		Propylparaben
		Butylparaben
		Ethylparaben
		Benzylparaben
		Sodium methyl paraben
	Other phenolic compounds	
		tert-Butyl hydroxyanisole
		Di-*tert*-butyl methylparaben
		Triclosan
		Chloroxylenol
		Sodium o-phenyl phenate
		Salicylic acid
		Resorcinol
		Phenol
	Miscellaneous anionic preservatives	
		Sorbic acid
		Potassium sorbate
Nonionic preservatives		
	Imidazolidinyl urea	
	2-Bromo-2-nitropropane-1,3-diol	
	DMDM hydantoin	
	Phenethyl alcohol	
	Monolaurin	

Source: Ref. 9.

lator to use a lower concentration of quaternium-15 and still have satisfactory product preservation. Rosen (W. E. Rosen, unpublished work) noted that EDTA increased the antimicrobial efficacy of imidazolidinyl urea. In a challenge test with 10^6 pseudomonads/ml, neither 0.3% imidazolidinyl urea nor 0.3% EDTA alone killed the pseudomonads within 3 days. The combination of 0.15% imidazolidinyl urea and 0.15% EDTA killed the test organisms. DMDM hydantoin has gained in popularity in recent years. Shull and Bennett (42) found that EDTA enhances the antimicrobial activity of DMDM hydantoin because one part of DMDM hydantoin and one part of EDTA was more active than two parts of DMDM hydantoin alone in their studies. Lentsch (43) observed that 2-bromo-2-nitropropane-1,3-diol can be made more effective against *P. aeruginosa* through potentiation with EDTA.

Esters of *p*-aminobenzoic acid (i.e., parabens) are known to be active against a wide range of gram-positive bacteria and fungi, but they are less active against gram-negative bacteria, especially the pseudomonads (44). EDTA frequently is used in combination with parabens to provide broad-spectrum antimicrobial activity. Orth et al. (45) reported that antimicrobial synergy could be determined in multicomponent systems when the slope of the survival curve (i.e., rate of death of the population of test organisms) is a larger negative number than the sum of the slopes of each of the components. These workers used the survival curve slope method to study anti-*Pseudomonas* synergy of methylparaben (MP) and acrylic acid homopolymer/copolymers (i.e., carbomer thickening agents). Preservative efficacy testing of nonionic lotions containing 0.2% MP and 0.2% acrylic acid homopolymer/copolymers revealed anti-*Pseudomonas* synergy against fluorescent pseudomonads (*P. aeruginosa*, *P. putida*, *P. fluorescens*, and *P. stutzeri*). Their work indicated that chelation of divalent metal ions by the acrylic acid homopolymer/copolymers was responsible for the observed synergy and that addition of 0.1% $CaCl_2$ to the test lotion caused significant increases in decimal reduction times (D-values) and eliminated the anti-*Pseudomonas* synergy.

Orth and coworkers (45) observed similar patterns of synergy in tap water with 0.2% MP + 0.01% Na_2EDTA. Data in Table 9.4 present the sterilization times obtained with the above fluorescent pseudomonads, *Pseudomonas* spp., *Staphylococcus aureus*, and *E. coli*. These data were used to determine antimicrobial synergy, as shown in Table 9.5. Antimicrobial synergy was observed against all fluorescent pseudomonads and one strain of *P. cepacia*, but not against the second strain of *P. cepacia*, *S. aureus*, or *E. coli*. Although no attempt was made to optimize the concentrations of EDTA or MP used in these studies, it is apparent that the concentrations used were highly effective against *P. aeruginosa*, *P. putida*, *P. fluorescens*, and *P. stutzeri*. It is expected that higher concentrations would be required for permeabilization synergy of MP and EDTA with the other *P. cepacia* and *E. coli*.

TABLE 9.4 Determination of Sterilization Times of Several *Pseudomonas* Species in MP, Na_2EDTA, and MP + Na_2EDTA Solutions

Test organism	0 hr	2 hr	4 hr	24 hr	48 hr	ST	MPST
0.2% MP solutions:							
P. aeruginosa 9027	+	+	+	+	+	—	>48
P. aeruginosa 27853	+	+	+	+	+	—	>48
P. aeruginosa 9721	+	+	+	+	+	—	>48
P. aeruginosa 10145	+	+	+	+	+	—	>48
P. cepacia 13945	+	+	+	+	–	48	—
P. cepacia 25416	+	+	+	+	+	—	>48
P. fluorescens 13525	+	+	+	+	+ [2]	—	>48
P. putida 12633	+	+	+	+	–	48	—
P. stutzeri 17588	+	+	+	+	+	—	>48
Pseudomonas sp. 9230	+	+	+	+	+	—	>48
S. aureus 6538	+	+	+	+	+	—	>48
E. coli 8739	+	+	+	+	+	—	>48
0.01% Na2EDTA solutions:							
P. aeruginosa 9027	+	+	+	+	+	—	>48
P. aeruginosa 27853	+	+	+	+	+	—	>48
P. aeruginosa 9721	+	+	+	+	+	—	>48
P. aeruginosa 10145	+	+	+	+	+	—	>48
P. cepacia 13945	+	+	+	+	+	—	>48
P. cepacia 25416	+	+	+	+	+	—	>48
P. fluorescens 13525	+	+	+ [s]	+	+	—	>48
P. putida 12633	+	+	+ [s]	+ [2]	—	48	—
P. stutzeri 17588	+	–	+ [1]	—	—	24	—
Pseudomonas sp. 9230	+	+	+	+	+	—	>48
S. aureus 6538	+	+	+	+	+	—	>48
E. coli 8739	+	+	+	+	+	—	>48
0.2% MP + 0.01% Na_2EDTA solution:							
P. aeruginosa 9027	+	+ [s]	–	–	–	4	—
P. aeruginosa 27853	+	+ [s]	+ [2]	–	–	24	—
P. aeruginosa 9721	+	+ [s]	–	–	–	4	—
P. aeruginosa 10145	+	+ [1]	–	–	–	4	—
P. cepacia 13945	+	+	+	+	–	48	—
P. cepacia 25416	+	–	–	–	–	2	—
P. fluorescens 13525	+	–	–	–	–	2	—
P. putida 12633	+	–	–	–	–	2	—
P. stutzeri 17588	–	–	–	–	–	0.1[a]	—
Pseudomonas sp. 9230	+		+ [1]	–	–	24	—
S. aureus 6538	+	+	+	+	+	—	>48
E. coli 8739	+	+	+	+	+	—	>48

(continued)

Table 9.4 Continued

Deionized water (control): Test organism	0 hr	2 hr	4 hr	24 hr	48 hr	ST	MPST
P. aeruginosa 9027	+	+	+	+	+	—	>48
P. aeruginosa 27853	+	+	+	+	+	—	>48
P. aeruginosa 9721	+	+	+	+	+	—	>48
P. aeruginosa 10145	+	+	+	+	+	—	>48
P. cepacia 13945	+	+	+	+	+	—	>48
P. cepacia 25416	+	+	+	+	+	—	>48
P. fluorescens 13525	+	+	+	+	+	—	>48
P. putida 12633	+	+	+	+	+	—	>48
P. stutzeri 17588	+	+	+	+	+	—	>48
Pseudomonas sp. 9230	+	+	+	+	+	—	>48
S. aureus 6538	+	+	+	+	+	—	>48
E. coli 8739	+	+	+	+	+	—	>48

+, growth on TSALT; –, no growth on TSALT; numbers in brackets (i.e., [1] and [2]) indicate the number of colonies growing on TSALT; [s] is used to designate scant growth (i.e., 8–15 colonies growing on TSALT). ST, sterilization time in hr. MPST, minimum possible sterilization time in hr.

[a] *P. stutzeri* was not recovered, so ST was set at 0.1 hr because this is the approximate time required for setting up the series of samples after inoculation.

APC of the saline suspensions inocula:

P. aeruginosa 9027= 1.1×10^7/ml.
P. aeruginosa 9721= 1.8×10^7/ml.
P. aeruginosa 27853= 1.5×10^7/ml.
P. aeruginosa 10145= 2.0×10^7/ml.
P. cepacia 13945= 1.7×10^7/ml.
P. cepacia 25416= 9.9×10^6/ml.
P. fluorescens 13525= 1.6×10^7/ml.
P. putida 12633= 1.6×10^7/ml.
P. stutzeri 17588= 1.7×10^7/ml.
Pseudomonas sp. 9230= 6.3×10^6/ml.
S. aureus 6538= 1.4×10^7/ml.
E. coli 8739= 6.3×10^6/ml.

Source: Ref. 45.

Stothart and Beecroft (46) observed a potentiation of the action of EDTA and a phenolic germicide (benzyl-*p*-chlorophenol). A 250-fold increase in activity against *P. aeruginosa* was observed in their studies. This report takes on added importance, because *P. aeruginosa* has been reported to be growing in phenolic and chloroxylenol disinfectant solutions (47,48). The activity of chloroxylenol, mixed phenols, and O-benzyl-*p*-chlorophenol against *P. aeruginosa* is strongly potentiated by EDTA (23,49–52). It is reported that these compounds have increased activity against *E. coli* and *S. aureus* when EDTA is present. It would be expected that this effect depends on the concentration of the particular antimicrobial and EDTA (as noted for MP + EDTA in Table 9.5).

EDTA has been reported to potentiate the antimicrobial action of other preservatives. The antimicrobial action of triclosan against gram-negative bacteria including *P. aeruginosa*, *E. coli*, and *Klebsiella aerogenes* is strongly potentiated by EDTA (21,53). Sodium myristyl sulfate, a conventional anionic surfactant, had little activity against *Salmonella typhi*; however, use of EDTA produced a 40-fold increase in the bactericidal action against this organism (54).

TABLE 9.5 Use of Sts and MPSTs to Determine Antimicrobial Synergy for Several Species of *Pseudomonas* in 0.2% MP, 0.01% Na_2EDTA, and 0.2% MP + 0.01% Na_2EDTA

0.2% MP solution: Test organism	ST	MPST	D-value	MPD-value	Slope	MPSlope
P. aeruginosa 9027	—	>48	—	>9.5	—	<–0.105
P. aeruginosa 27853	—	>48	—	>9.3	—	<–0.108
P. aeruginosa 9721	—	>48	—	>9.1	—	<–0.109
P. aeruginosa 10145	—	>48	—	>9.1	—	<–0.110
P. cepacia 13945	48	—	9.2	—	–0.109	—
P. cepacia 25416	—	>48	—	>9.6	—	<–0.104
P. fluorescens 13525	—	>48	—	>9.2	—	<–0.108
P. putida 12633	48	—	9.2	—	–0.108	—
P. stutzeri 17588	—	>48	—	>9.2	—	<–0.109
Pseudomonas sp. 9230	—	>48	—	>10	—	<–0.100
S. aureus 6538	—	>48	—	>9.3	—	—
E. coli 8739	—	>48	—	>10	—	—
0.01% Na_2EDTA solution: Test organism	**ST**	**MPST**	**D-value**	**MPD-value**	**Slope**	**MPSlope**
P. aeruginosa 9027	—	>48	—	> 9.5	—	<–0.105
P. aeruginosa 27853	—	>48	—	> 9.3	—	<–0.108
P. aeruginosa 9721	—	>48	—	> 9.1	—	<–0.109
P. aeruginosa 10145	—	>48	—	> 9.1	—	<–0.110
P. cepacia 13945	48	—	9.2	—	–0.109	—
P. cepacia 25416	—	>48	—	> 9.6	—	<–0.104
P. fluorescens 13525	—	>48	—	> 9.2	—	<–0.108
P. putida 12633	48	—	9.2	—	–0.108	—
P. stutzeri 17588	24	—	4.6	—	–0.218	—
Pseudomonas sp. 9230	—	>48	—	>10	—	<–0.100
S. aureus 6538	—	>48	—	> 9.3	—	<–0.107
E. coli 8730	—	>48	—	>10	—	<–0.100
0.2% MP + 0.01% Na_2EDTA solution: Test organism	**ST**	**MPST**	**D-value**	**MPD-value**	**Slope**	**MPSlope**
P. aeruginosa 9027	4	—	0.8	—	–1.260[a]	—
P. aeruginosa 27853	4	—	0.8	—	–1.294[a]	—
P. aeruginosa 9721	24	—	4.6	—	–0.219[a]	—
P. aeruginosa 10145	4	—	0.8	—	–1.325[a]	—
P. cepacia 13945	48	—	9.2	—	–0.109	—
P. cepacia 25416	2	—	0.4	—	–2.498[a]	—
P. fluorescens 13525	2	—	0.5	—	–2.102[a]	—
P. putida 12633	2	—	0.4	—	–2.602[a]	—
P. stutzeri 17588	0.1	—	>0.02	—	<–52.304[a]	—
Pseudomonas sp. 9230	24	—	5.0	—	–0.200	—
S. aureus 6538	—	>48	—	> 9.3	—	<–0.107
E. coli 8739	—	>48	—	>10	—	<–0.100

ST, sterilization time in hr; MPST, minimum possible sterilization time in hr; D-value, D-value in hr; MPD-value, minimum possible D-value in hr; Slope, slope of the survivor curve, in hr^{-1}; MPSlope, maximum possible slope of the virtual survivor curve, in hr^{-1}.

[a]Synergy observed, because the slope of the survivor curve was a larger negative number than the sum of the slopes (or MPSlopes) for the same concentrations of MP and Na_2EDTA taken separately.

Source: Ref. 46.

A wide variety of lipid materials have been examined for antimicrobial activity (55). Of these numerous compounds, the monoglyceride of lauric acid (monolaurin) has been found to be most active. This compound is principally active against gram-positive bacteria, yeasts, and molds. The principal limitation to this nontoxic preservative is its lack of activity against gram-negative bacteria. It has been found that EDTA potentiates the activity of monolaurin against gram-negative bacteria (56–58). The data in Table 9.6 illustrate potentiation of antimicrobial action against *P. aeruginosa* by a hand lotion with monolaurin.

By way of contrast, the addition of MP to the lotion with monolaurin did not kill this organism in 7 days (59). It has been demonstrated that EDTA lowers the MIC of monolaurin against yeasts. Shibasaki and Kato (58) found that EDTA tripled the percentage uptake of monolaurin into *E. coli* cells, which it normally does not affect. Broad-spectrum preservation is claimed for mixtures of monolaurin, a phenolic such as a paraben ester or tert-butyl hydroxyanisole (BHT) and EDTA (56). (See Chapter 5 for additional information on potentiation of antimicrobial action of monolaurin with EDTA.)

Richards and McBride (10) concluded that the antimicrobial activity of phenethyl alcohol against *P. aeruginosa* is potentiated by EDTA. Reports such as these illustrate that chelating agents increase the antimicrobial action of many preservatives and/or multifunctional formula components. The potentiation of antimicrobial action is believed to be due to permeabilization synergy, in which

TABLE 9.6 Effect of Preservative on Oil-Water Hand Lotion Emulsified with Monolaurin Against *Pseudomonas aeruginosa* Inoculation

	Time[a]		
Preservative	1 day	2 days	7 days
Control with monolaurin only	+	+	+
EDTA, 0.1%	+	+	–
EDTA, 0.2%	–	–	–
EDTA, 0.3%	–	–	–
Methylparaben, 0.1%	+	+	+
Methylparaben, 0.2%	+	+	+
Methylparaben, 0.3%	+	+	+
Methylparaben, 0.1%, + EDTA, 0.1%	+	±	–
Methylparaben, 0.2%, + EDTA, 0.2%	–	–	–
Methylparaben, 0.3%, + EDTA, 0.3%	–	–	–

[a]+, growth; ±, slight growth; –, no growth.
Source: Ref. 44.

the chelating agent assists the passage of another compound—an antimicrobial agent—through the cell wall or outer membrane (7,45). This provides the rationale for use of appropriate concentrations of chelating agents in self-preserving formulas, because chelating agents render the cell wall/outer membrane more permeable to the possible antimicrobial action of various components in the formula.

9.5 PRESERVATIVES UNAFFECTED BY EDTA

While the above discussion suggests that chelating agents such as EDTA will potentiate the action of virtually any antimicrobial agent, there are reports to the contrary. Reybrouek and van de Voorde (50) reported that EDTA had no effect on the antimicrobial activity of an iodophor, hydrogen peroxide, formaldehyde, or glutaraldehyde. Also, Richards and McBride (10) found that combination of EDTA and chlorobutanol appeared to have no advantage over chlorobutanol alone. It has been the author's experience that reported failures are due to use of too low a concentration of the chelator in the formula. Use of 0.05% EDTA may be too low to produce the desired permeabilization synergy, and the EDTA concentration should be increased to 0.1–0.3% in some systems.

There are other possible explanations for failure to observe potentiation of antimicrobial action. Experimental conditions, of course, could be the basis of the negative results. Free, uncomplexed EDTA is essential for potentiation of preservative action, and calcium and magnesium chelates of EDTA are inactive (60). If sufficient calcium or magnesium ions are present in the formula, due to their incidental addition via formula components, they may form a complex with the chelator and leave no free EDTA. Another possible explanation is that the formula may contain other chelating agents, such as citric acid or acrylic acid homopolymer/copolymers (i.e., carbomer resins), so that addition of EDTA would produce no additional antimicrobial benefit. These considerations lead us to conclude that the reported failure of EDTA to potentiate preservative action is due to insufficient concentration of EDTA or the presence of other agents in the formula that bind/interfere with thc chelator.

9.6 TOXICOLOGY OF EDTA

EDTA and its salts exhibit a low order of animal toxicity and have been used for many years without incident. The LD_{50} of disodium EDTA when administered to rabbits is 47 mg/kg (intravenous) and the oral LD_{50} is 2.3 g/kg (61). Oral dosing of disodium EDTA at levels up to 5% of body weight per day over a 2-year period produced no observable effects on gross or histopathological examination of organs or tissues. At the highest dose level, toxic symptoms of diarrhea and reduced weight gain were noted in some of the treated groups (62).

Published medical records show that there were no adverse side effects when

81 human patients were given 2000 infusions of disodium EDTA over a 2-year period (63). EDTA has no reported carcinogenicity effects (64). Repeated-insult skin patch tests on quinea pigs have demonstrated that EDTA is not a skin sensitizer (65) and that the small quantities of EDTA in cosmetic products do not represent any appreciable risk of human skin sensitization.

EDTA has been shown to reduce the toxicity of acidic acetate/citrate buffered solutions on corneal epithelial tissue, but it enhanced the toxicity of alkaline solutions. Epithelial tissue culture studies of this type are useful for developing safety guidelines (66).

9.7 SUMMARY

Chelating agents are multifunctional ingredients that are used to sequester metal ions to improve color and odor stability and potentiate preservative action. Although the discussion in this chapter focused on the use of EDTA as a potentiator of the antimicrobial action of preservatives in cosmetic and drug formulations, EDTA is not the only chelating agent that can be used for increasing the efficacy of preservative systems. Acidulants such as citric, lactic, and polyphosphoric acids have chelating ability and also show potentiation of preservative action. Similarly, acrylic acid homopolymer/copolymers have chelating properties and have anti-*Pseudomonas* synergy when used with MP.

Chelating agents play an important role in product preservation because they increase the antimicrobial action and antimicrobial spectrum of a product's preservative system. The true utility of chelating agents is found in formulas with a fairly neutral pH that have a high water content (i.e., where the water activity is >0.97) because these formulas frequently are susceptible to growth of pseudomonads. Chelating agents are believed to increase antimicrobial action by limiting the availability of essential minerals, such as iron, and via permeabilization synergy, in which the chelating agent sequesters divalent metal ions to decrease the integrity of the gram-negative bacterial cell envelope and assist the passage of another compound—with an antimicrobial action through the cell wall or outer membrane.

Permeabilization synergy is not restricted to chelating agents and compounds normally considered to be cosmetic preservatives. The use of EDTA as part of a product preservative system may enable the formulator to reduce and/or eliminate classical preservatives. The use of EDTA in combination with other multifunctional agents with antimicrobial action generally will increase preservative efficacy and may allow development of self-preserving products.

REFERENCES

1. J. R. Hart. EDTA-type chelating agents in personal care products. *Cosmet. Toiletr.* *98*:54–58 (1983).

2. L. Marshall, K. Parris, J. Rebek, Jr., S. V. Luis, and M. I. Burguete. A new class of chelating agents. *J. Am. Chem. Soc. 110*:5192–5193 (1988).
3. J. M. Nikitakis, ed. *CTFA Cosmetic Ingredient Handbook*, 1st ed. The Cosmetic, Toiletry and Fragrance Association, Inc., Washington, DC, 1988, pp. 61, 62.
4. J. R. Hart. Chelating Agents: new possibilities in detergents and specialties. *Soap/Cosmet/Chem. Spec. 62*(5):38,41,48 (1986).
5. J. R. Hart. Chelating agents in cosmetic and toiletry products. *Cosmet. Toilet. 93*(12):28–30 (1978).
6. D. S. Orth, C. M. Lutes, S. R. Milstein, and J. J. Allinger. Determination of shampoo preservative stability and apparent activation energies by the linear regression method of preservative efficacy testing. *J. Soc. Cosmet. Chem. 38*:307–319 (1987).
7. S. P. Denyer, W. B. Hugo, and V. D. Harding. Synergy in preservative combinations. *Int. J. Pharm. 25*:245–253 (1985).
8. J. J. Kabara. GRAS antimicrobial agents for cosmetic products. *J. Soc. Cosmet. Chem. 31*:1–10 (1980).
9. J. R. Hart. Chelating agents as preservative potentiators. In: J. J. Kabara, ed. *Cosmetic and Drug Preservation: Principles and Practice.* Marcel Dekker, Inc., New York, 1984, pp. 323–337.
10. R. M. E. Richards and R. J. McBride. The preservation of ophthalmic solutions with antibacterial combinations. *J. Pharm. Pharmacol. 24*:145–148 (1972).
11. P. A. Lambert. Resistance to non-antibiotic antimicrobial agents. In: W. B. Hugo and A. D. Russell, eds. *Pharmaceutical Microbiology*, 4th ed. Blackwell Scientific Publications, Oxford, England, 1988, pp. 288–297.
12. R. Repaske. Lysis of gram-negative organisms and the role of Versene. *Biochem. Biophys. Acta 30*:225–232 (1958).
13. D. R. MacGregor and P. R. Elliker. A comparison of some properties of strains of *Pseudomonas aeruginosa* sensitive and resistant to quaternary ammonium compounds. *Can. J. Microbiol. 4*:499–503 (1958).
14. R. Weiser, A. W. Asscher, and J. Wimpenny. In vitro reversal of antibiotic resistance by ethylenediamine tetraacetic acid. *Nature 219*:1365–1366 (1968).
15. F. W. Adair, S. G. Geftic, and J. Gelzer. Resistance of *Pseudomonas* to quaternary ammonium compounds. II. Cross-resistance characteristics of a mutant *Pseudomonas aeruginosa. Appl. Microbiol. 21*:1058–1063 (1969).
16. K. G. Indge. The effects of various anions and cations on the lysis of yeast protoplasts by osmotic shock. *J. Gen. Microbiol. 41*:425–432 (1968).
17. C. Y. Li and J. J. Kabara. Effects of lauricidin on *Fomes annosus* and *Phellinus weirii.* In: J. J. Kabara, ed. *The Pharmacological Effect of Lipids.* American Oil Chemists Society, Champaign, IL, 1979, pp. 44–50.
18. R. J. Fass and J. Barnishan. Effect of divalent cation concentrations on the antibiotic susceptibilities of nonfermenters other than *Pseudomonas aeruginosa. Antimicrob. Agents Chemother. 16*:434–438 (1979).
19. C. Liao. Cloning of pectate lyase gene pel from *Pseudomonas fluorescens* and detection of sequences homologous to *pel* in *Pseudomonas viridiflava* and *Pseudomonas putida. J. Bacteriol. 173*:4386–4393 (1991).
20. J. I. Yablonski and C. L. Goldman. Microbiology of shampoos. *Cosmet. Toiletr. 91*:45–52 (1975).

21. K. H. Wallhauser. The problem of preserving cosmetics. *Cosmet. Toiletr.* *91*:45–51 (1976).
22. R. A. Cowen and B. Steiger. Why a preservative must be tailored to a specific product. *Cosmet. Toiletr.* *92*:15–20 (1977).
23. A. D. Russell. Ethylenediaminetetraacetic acid. In: W. B. Hugo, ed. *Inhibition and Destruction of the Microbial Cell.* Academic Press, London, 1971.
24. L. A. Carson, M. S. Favero, W. W. Bond, and N. J. Petersen. Morphological, biochemical and growth characteristics of *Pseudomonas cepacia* from distilled water. *Appl. Microbiol.* *25*:476–483 (1973).
25. J. C. Gould. *Pseudomonas pyocyanae* infections in hospitals. In: *Infection in Hospitals: Epidemiology and Control.* Blackwell Scientific Publications, Oxford, United Kingdom, 1963.
26. L. A. Wilson, W. Kuehne, S. W. Hall, and D. G. Ahern. Microbial contamination in ocular cosmetics. *Am. J. Ophthalmol.* *71*:1298–1302 (1971).
27. L. A. Wilson and D. G. Ahern. *Pesudomonas*-induced corneal ulcers associated with contaminated eye mascaras. *Am. J. Ophthalmol.* *84*:112–119 (1977).
28. F. N. Marzulli, J. R. Evans, and P. D. Yoder. Induced *Pseudomonas* keratitis as related to cosmetics. *J. Soc. Cosmet. Chem.* *23*:89–97 (1972).
29. J. H. Dawson. Probing structure-function relations in heme-containing oxygenases and peroxidases. *Science* *240*:433–439 (1988).
30. J. G. Voss. Effect of inorganic cations on bactericidal activity of anionic surfactants. *J. Bacteriol.* *86*:207–211 (1963).
31. S. G. Wilkinson. Sensitivity to ethylenediaminetetraacetic acid. In: M. R. W. Brown, ed. Resistance of *Pseudomonas aeruginosa.* Wiley, London, 1975.
32. R. G. Eagon and K. J. Carson. Lysis of cell walls and intact cells of *Pseudomonas aeruginosa* by ethylenediaminetetraacetic acid and by lysozyme. *Can. J. Microbiol.* *11*:193–201 (1965).
33. K. Matsushita, O. Adachi, E. Shinagawa, and M. Ameyama. Isolation and characterization of outer and inner membranes from *Pseudomonas aeruginosa* and effect of EDTA on the membranes. *J. Biochem.* *83*:171–181 (1978).
34. H. Hague and A. D. Russell. Effect of chelating agents on the susceptibility of some strains of gram-negative bacteria to some antibacterial agents. *Antimicrob. Agents Chemother.* *6*:200–206 (1974).
35. N. A. Roberts, G. W. Gray, and S. G. Wilkinson. The bactericidal action of ethylenediaminetetraacetic acid on *Pseudomonas aeruginosa. Microbios* *2*:189–208 (1970).
36. I. N. Izzat and E. O. Bennett. The potentiation of the antimicrobial activities of cutting fluid preservatives by EDTA. *Lubr. Eng.* *35*:153–159 (1979).
37. C. A. Lawrence. *Surface-Active Quaternary Ammonium Germicides.* Academic Press, New York, 1950.
38. J. W. Klimek. The quaternary interests. Chemical Speciality Manufacturer's Association, Proceedings, 1950, pp. 109–115.
39. D. W. Burdon and A. L. Whitby. Contamination of hospital disinfectants with *Pseudomonas* species. *Br. Med. J.* *2*:153–155 (1967).
40. M. R. W. Brown and R. M. E. Richards. Effect of ethylenediaminetetraacetate on the resistance of *Pseudomonas aeruginosa* to antibacterial agents. *Nature* *207*:1391–1393 (1965).
41. FDA database. Frequency of preservative use. *Cosmet. Toiletr.* *108*:47–48 (1993).

42. S. E. Shull and O. Bennett. Antimicrobial hydantoin derivative compositions and method of use. U.S. Patent 4,172,140, Glyco Chemicals, Inc. (October 23, 1979).
43. S. E. Lentsch. Control of mastitis and compositions therefore. U.S. Patent 4,199,602, Economics Laboratory, Inc. (April 22, 1980).
44. E. M. Scott and S. P. Gorman. Chemical disinfectants, antiseptics and preservatives. In: W. B. Hugo and A. D. Russell, eds. *Pharmaceutical Microbiology*, 4th ed. Blackwell Scientific Publications, Oxford, England, 1988, pp. 226–252.
45. D. S. Orth, C. M. Lutes Anderson, D. K. Smith, and S. R. Milstein. Synergism of preservative system components: use of the survival curve slope method to demonstrate anti-*Pseudomonas* synergy of methyl paraben and acrylic acid homopolymer/ copolymers in vitro. *J. Soc. Cosmet. Chem. 40*:347–365 (1989).
46. S. N. H. Stothart and G. C. Beecroft. Antiseptics and disinfectants. Br. Patent No. 858,030 (1961).
47. C. H. Jellard and G. M. Churcher. An outbreak of *Pseudomonas aeruginosa* in the newborn. *J. Hyg.* (Cambridge) *65*:219–228 (1967).
48. G. A. J. Ayliffe, E. J. L. Lowbury, J. G. Hamilton, J. M. Small, E. A. Ashshov, and M. T. Parker. Hospital infection with *Pseudomonas aeruginosa* in neurosurgery. *Lancet 12*:365–368 (1956).
49. G. W. Gray and S. G. Wilkinson. The action of ethylenediaminetetraacetic acid on *Pseudomonas aeruginosa. J. Appl. Bacteriol. 28*:153–164 (1965).
50. G. Reybrouek and H. van de Voorde. Effect of ethylenediaminetetraacetic acetate germicidal action of disinfectants against *Pseudomonas aeruginosa. Acta Clin Belg. 24*:32–41 (1969).
51. G. Smith. Ethylenediaminetetraacetic acid and the bactericidal efficiency of some phenolic disinfectants against *Pseudomonas aeruginosa. J. Med. Lab. Technol. 27*:203–206 (1970).
52. A. D. Russell and J. R. Furr. The antibacterial activity of a new chloroxylenol preparation containing ethylenediaminetetraacetic acid. *J. Appl. Bacteriol. 43*:253–260 (1977).
53. A. K. Panezai. Ger. Offen. No. 2,351,386 to Beecham Group, Ltd. (April 25, 1974).
54. L. H. Muschel and L. Gustafson. Antibiotic detergent, and complement sensitivity of *Salmonella typhi* after ethylenediaminetetraacetic acid treatment. *J. Bacteriol. 95*:2010–2013 (1968).
55. J. J. Kabara. Fatty acids and derivatives as antimicrobial agents—a review. In: J. J. Kabara, ed. *The Pharmacological Effect of Lipids*. American Oil Chemists Society, Champaign, IL, 1979, pp. 1–14.
56. J. J. Kabara. Synergistic microbicidal composition and method. U.S. Patent No. 4,067,997 to Med-Chem Laboratories (January 10, 1978).
57. J. J. Kabara. GRAS antimicrobial agents for cosmetic products. *J. Soc. Cosmet. Chem. 31*:1–10 (1980).
58. I. Shibasaki and N. Kato. Combined effects on antibacterial activity of fatty acids and their esters against gram-negative bacteria. In: J. J. Kabara, ed. *The Pharmacological Effect of Lipids*. American Oil Chemists Society, Champaign, IL, 1979, pp. 15–22.
59. J. J. Kabara and C. M. Wernette. Cosmetic formulas preserved with food-grade chemicals. *Cosmet. Toiletr. 97*:77–84 (1982).
60. L. Leive. Studies on the permeability change produced in coliform bacteria by ethylenediaminetetraacetate. *J. Biol. Chem. 243*:2373–2380 (1968).

61. S. Shibata. Toxicological studies of EDTA (disodium ethylenediaminetetraacetate). *Nippon Yakurigaku Zasshi. 52*:113–119 (1956).
62. L. E. Meltzer, J. R. Kitchell, and F. Palmon. The long-term use, side effects, and toxicity of disodium ethylenediaminetetraacetic acid (EDTA). *Am. J. Med. 242*:11–17 (1961).
63. S. S. Yang and M. S. Chan. Summaries of toxicological data: toxicology of EDTA. *Food Cosmet. Toxicol. 2*:763–767 (1964).
64. Bioassay of trisodium ethylenediaminetetraacetate trihydrate (EDTA) for possible carcinogenicity. National Cancer Institute. Technical Report Series No. 11, U.S. Department of Health, Education and Welfare. National Institutes of Health, Washington, DC, 1977.
65. J. W. Henck, D. D. Lockwood, and K. J. Olsen. Skin sensitization potential of trisodium ethylenediaminetetraacetate. *Drug Chem. Toxicol. 3*:99–103 (1980).
66. D. R. Meyer and J. P. McCulley. Differential effect of chelation on the pH tolerance of corneal epithelium in tissue culture. *Ophthalmic. Res. 23*:204–212 (1991).

10

The Role of Packaging in Product Preservation

Daniel K. Brannan
Abilene Christian University
Abilene, Texas

10.1 INTRODUCTION AND BACKGROUND

Use of packaging as part of the preservative system is often overlooked when attempting to formulate preservative-free and self-preserving consumer products. The concept of "the product is the preservative" usually refers to the formula being inherently hostile without added biocides. One typically thinks of the package as providing convenience of use, appeal to the consumer, and protection from contamination and deterioration until it gets to the customer (1); however, the package and the closure or cap also contribute to keeping a product free of microorganisms during consumer use. This chapter explores the historical and regulatory basis for considering the package as an adjunct for protecting products. It will be shown that packaging plays a highly significant role in preventing contamination during consumer use. After reading this chapter, the reader should understand that the package and the formula both constitute the preservative system.

10.1.1 History of Cosmetics and Cosmetic Microbiology

Archaeologists believe that humans have used cosmetics for nearly 50,000 years. In the caves where Grimaldi man (Upper Paleolithic Aurignacian tradition) was unearthed outside of the village of Grimaldi, Italy, the skeletons of *Homo sapiens* were sprinkled with red ocher. The Grimaldi burials with the bodies decorated in red ocher date to the latter part of the Pleistocene Epoch, about 20,000 years ago (2). Shamanism and rituals pervaded society then. A key characteristic of being human is our apparent altruism toward others—an altruism that causes us to anticipate how our actions today affect our species and other species future. Rituals help reduce the anxiety of anticipating a sometimes bleak future. During these rituals and burials, primitive humans often painted their bodies with cosmetic paints to transform themselves into entities no longer associated with self and so, perhaps, allowed at least a temporary escape from contemplating a frightening future. Even today, we masquerade behind masks of powder, paint, and perfume in hopes the world will never find us and discover our true selves. Cosmetics are our cures for the disease of being whom we really are.

More recently, we have found that the Babylonians, Hebrews, and Egyptians nearly 6000 years ago used a green eyeliner made from malachite to emphasize the eyes. This eyeliner may have been the first preserved cosmetic, since malachite green is an excellent antibacterial dye (3). Even then the package was considered little more than a container for holding the product. Typically, a clay jar with or without a lid was good enough. Microbial contamination of cosmetics by mold spoilage, however, did not become recognized until about 50 years ago (4). During the 1960s, the contamination of cosmetics by *Escherichia, Klebsiella, Enterobacter, Serratia,* and *Pseudomonas* spp. (5,6) coupled with consumer-desired changes in formulations caused a demand for more effective and easier to use preservatives. By the mid-1970s, eye cosmetics came under special scrutiny since there were several cases of blindness due to *Pseudomonas*-contaminated mascaras (7). The products reached the consumer without microbial contamination, but could not withstand consumer use during which microbes were inadvertently added into the product (8,9).

The main goal over the next few years was to develop preservative efficacy tests (PETs) to predict consumer contamination risk. In 1975 and in 1985 the Food and Drug Administration (FDA) gave contracts to develop such PETs. In 1990, the Cosmetics, Toiletries, and Fragrance Association (CTFA) published the results of a survey to determine if companies tried to validate their PET data by correlating it with consumer use data (10). Nearly all companies claimed they had correlation programs in place (11). Recently, several of the compendial tests were compared for their ability to predict "in-use" contamination in the FDA-sponsored study (12). The major criticism of this work was that the in-use test was merely *simulated.* Nevertheless, all of the PETs were ranked against a single standard, permitting one to glean useful information from the study. The purpose of this work was not to prove a particular test's ability to predict potential for consumer contamination. As a result, no test was shown to be capable of doing so.

Finally, in 1990 the FDA established a joint program with CTFA and the Association of Official Analytical Chemists (AOAC) to develop a standard PET.

Rather than rely on a compendial method, the collaborative study by the CTFA, AOAC, and FDA intends to develop a statistically based protocol that is reproducible and reliable. One major limitation of the method is that no correlative consumer use studies will have been conducted to prospectively validate the method. In addition, the role of the package in protecting the product will also not have been determined.

Two PET methods have published validation data claiming to predict the in-use potential for consumer contamination (13,14). However, neither of these studies accounts for the fact that the container and its closure play a significant role in preventing contamination. No PET can ever approximate consumer use. Only in-use consumer studies can provide that perspective.

10.1.2 Importance of Designing Protective Packaging

The goal of preventing microbial contamination and subsequent growth in cosmetic or drug products is based on the vision of delivering to the consumer a product that is of high quality, is of appropriate identity, strength, and purity, and is safe. Consumers rely predominantly on the sensory properties of a product—its appearance, tactile, and fragrance characteristics—to decide if they will purchase it a second time (15). Microbial spoilage can decompose products so that they are no longer aesthetically pleasing, incapable of performing as claimed, and possibly even a health hazard to those that use them. Most of the time, microbial contamination is controlled by manufacturing cleanliness and making the product hostile, either through the use of preservatives or by formulating with inherently hostile ingredients as advocated in this book. With today's consumer focused on the concept of "harmful" chemical additives, the use of "natural" products is growing. These are very difficult to preserve with traditional preservatives (16). The ability to produce inherently hostile formulas without preservatives or to manufacture susceptible products under sterile environments but in packages that prevent or even eliminate contamination during use is desirable to meet this perceived consumer need.

Clearly, the first function of the package is simply to contain the product. The package is expected to not leak and to be inert with respect to the product it contains. It should also be robust enough to withstand transport and use without breaking or leaking. In addition, we expect the package to protect the product from the environment. It should prevent the entry of water, air (where damaging), dust, light, and microorganisms. During use, the package needs to provide a mechanism for dispensing the product without allowing contamination to occur. The dispensing mechanism should also allow for closure of the package so that the product inside will not become contaminated. Other considerations for packages include their ability to prevent child use, to prevent tampering, and to have clear labeling for sale and use of the product.

As a result, three factors play a role in keeping microorganisms from contaminating a product: 1) sanitary or sterile manufacture [e.g., voluntary compliance with current Good Manufacturing Practices (cGMP)] in order to deliver a microbial-free product to the consumer, 2) hostility of the product (either due to inherent factors or preservatives) in order to inhibit both growth from transient microorganisms introduced during manufacturing and contamination introduced during consumer use, and 3) packaging (e.g., container and closure) in order to prevent introducing microorganisms into the product after packing and during use.

Perhaps two analogies will be helpful to emphasize the critical role of the package. The role of the package in preventing microbial contamination can be thought of as being similar to the immune system in preventing the human from becoming ill. The package is the first line of defense—the integument of the product. The hostility of the formula is the second line of defense—similar to the macrophages of the cellular portion of the immune system. Finally, the preservative used is the third line of defense, acting similarly to the humoral factors such as antibodies in the immune response. Another analogy might be the steps taken in a hospital to prevent nosocomial infections. The package is like the physical controls used in a hospital to prevent infections from starting. The hostility of the product is like the disinfection practices, hand-washing, and the sterilizing of instruments. Finally, the preservative is like the last resort of using antibiotics to cure a nosocomial infection. If the first two steps are applied rigorously, then the use of antibiotics in prophylactic treatment is unnecessary. In a similar way, if the package is designed to prevent contamination and the product is made inherently hostile, then the use of preservatives can be made unnecessary.

10.2 REGULATORY BASIS FOR PROVIDING PROTECTIVE PACKAGING

More and more cosmetic manufacturers desire ways to make products more resistant to microbial contamination. The growth of consumer demands for "all natural" and "preservative-free" products also contributes to this desire. However, the scope of available preservatives is rather limited. In addition to designing packages with the ability to ward off contamination during use, the cosmetic industry may need to look into the use of aseptic packaging, especially where preservatives are not wanted. The cosmetic industry lags behind the pharmaceutical industry in the area of aseptic packaging, much like the pharmaceutical industry lags behind the cosmetic industry with respect to use of preservatives. The cosmetic industry relies on preservative systems and sanitary packaging for aqueous products in multiple-use containers where some contamination may occur. Rarely are single-dosage forms of product used.

From a regulatory point of view, the packaging issues began with the Fair

Packaging and Labeling Act of 1973–1975. Prior to this rule, which controls label claims, ingredient listing, and product safety warnings, the U.S. Congress passed the Food, Drug and Cosmetic Act of 1938, as the 1907 Drug Act proved too weak to ensure effective drug and food safety. Even though a stronger bill had been proposed in 1933, it did not pass both House and Senate until 1937, when a company marketed an oral tonic that contained a poisonous ingredient. Congress passed the act in 1938. This gave the FDA a means of regulating and defining cosmetics. It also defined adulteration and allowed the FDA to request recalls. If the manufacturer fails to conduct the recall, the FDA can seize the product. After several cases of blindness resulted in 1974 from use of mascaras contaminated with *Pseudomonas*, the FDA turned from being concerned that products reached the consumer in clean conditions to wondering if the products could withstand microbial insults added during consumer use. In an unpublished 1989 House Appropriations Committee progress report (obtainable through Freedom of Information), the FDA stated the need for the "development and use of microbial testing procedures which ensure that cosmetic preservation systems effectively protect products from hazardous microbial contamination during . . . consumer use." Absent from this need was the request for testing procedures to ensure that cosmetic *packages* effectively protected products from hazardous microbial contamination during consumer use. This omission is particularly puzzling, since packages arguably play as important a role as does the preservative in protecting a product from consumer contamination during use.

Other regulations impacting drug packaging concerns include: 1) Section 501(a)(3) of the Act, which allows one to pronounce a product adulterated if the container is composed of substances that may injure the health of the consumer, 2) Section 502(i)(1), indicating misbranding if the container is made, formed, or filled such that the amount of the contents is misleading, and 3) Section 602, where similar rules regarding misbranding of cosmetics are promulgated. In CFR 21 §211.94, we are told that containers and closures shall not be reactive, additive, or absorptive so as to alter the safety, identity, strength, quality, or purity of the drug beyond the official or established requirements. CFR 21 §211.94(b) specifically requires that the container *closures provide adequate protection against* foreseeable external *factors in* storage and *use that can cause* deterioration or *contamination* of the product (emphasis mine). Clearly, packaging as a protective mechanism from microbial contamination should be a major concern from a regulatory point of view.

As far as packaging concerns that impact the submission of investigational new drugs and new drug applications, there is plenty to do. Most of the concerns outlined in the guidelines (17) do not address microbiological issues, but rather focus on physical characteristics of composition, chemical resistance, light transmission, compatibility, and leaching/migration tests. However, in Chapter 3 of these guidelines, the FDA does request a challenge of the integrity of the stopper where multiple-entry penetration into the vial is expected. The proof of this integrity must be done on the actual container/closure and the actual drug in question under real-world conditions. Perhaps this requirement should be interpreted as preventing the entry of microorganisms as part of the definition of integrity.

The role of the package in preventing microbial contamination is currently not high on the regulatory agenda. However, there are rules that could easily be interpreted as requiring that packages and closures be proven to prevent contamination during use. If this happens, what test methods do we have? The only logical test methods are consumer-use studies to demonstrate how the closure protects a product. Simulated in-use studies would not be very effective in truly understanding how the closure protects the product from contamination.

10.3 EMPIRICAL BASIS FOR PROVIDING PROTECTIVE PACKAGING

In order to demonstrate the empirical basis for providing protective packaging, one must understand that this concern originates in the FDA's desire "to remove from the market any cosmetic that poses an unreasonable risk of injury because of inadequate preservation to withstand contamination under customary conditions of use" (18). This notice was given in 1977. None of the studies conducted before or since this notice were adequate for the establishment of guidelines or regulations by the FDA for preservative testing protocols. After summarizing the studies done to date on correlating PETs with consumer use results, the only known study on how packaging prevents microbial contamination of consumer products will be detailed.

10.3.1 Correlation of Consumer Use with In Vitro PET Results

Preservative adequacy of consumer products is typically determined using preservative efficacy tests (19,20). Several studies have been conducted to show that a PET predicts consumer contamination. One showed that one-third of the mascara formulas that failed a PET would actually be contaminated due to consumer use (21). The reason why the other products were not contaminated was not addressed. It can be hypothesized that either the challenge test was too rigorous, the applicator brush did not permit transfer of organisms after use to the tube of mascara, or the consumer used the product in such a way as to not permit contamination during use. In a study using eyeshadows, the microbial content after use was determined. Although the authors did conduct challenge testing on two of the eyeshadows, they did not attempt to correlate the incidence of in-use contamination with the challenge test results (22). Another study using shampoos attempted to correlate minimum inhibitory concentration results with a simulated "in-use" test (23). Not surprisingly, very poor correlation was found to exist since the authors used a poor method to assess preservative efficacy and compared it against a poor test to assess the incidence of in-use contamination.

Another study compared several different PETs for their ability to predict "in-use" contamination (24). The major criticism of this work is that the in-use test was merely simulated. The subjects dabbed their fingers in the product after rubbing their armpits. The significance of ranking PETs against their ability to predict how well a product can withstand finger dabbing after armpit rubbing is certainly not a valid representation of actual consumer use conditions. Nevertheless, since all of the PETs were ranked against a single, albeit flawed, standard, one can still derive useful information. For example, nearly all of the compendial tests separated poorly preserved from well-preserved ones. With some of the more conservative tests, marginally preserved products were classified the same as poorly preserved ones, while the more liberal tests allowed marginally preserved products to rank with well-preserved ones. The CTFA test exhibited the tendency to rank all three (poor, marginal, and well) correctly against the flawed but useful standard of the simulated in-use test. This study does not, however, support the use of the CTFA test to enforce recalls. Again, the criticism can be offered that it used an invalid simulated in-use test and did not account for packaging effects.

Two studies showed that a modified version of the CTFA's PET could predict risk of consumer contamination (13,14). In a study by Brannan et al. (13), both a shampoo and a skin lotion were preserved at three different levels (well preserved, adequately preserved, and poorly preserved) as determined using a PET test, which was a modified CTFA test but included dilutions of the product (30, 50, 70, and 100%) in order to rank them into the three classifications. The products, once ranked by the PET into the three categories of preservative efficacy, were placed into in-use testing. After use of the products, the subjects returned the products and microbial content testing was conducted on each unit. Product units were considered contaminated if >100 CFU/g was observed or if gram-negative bacteria at any level were detected at initial receipt and 4–7 days postreceipt. The product contamination rates due to consumer use showed good correlation with the PET results except that the skin lotion, which was classified as marginally preserved, did not become contaminated during use. Thus the authors concluded that the test may have been overly conservative (too rigorous) with respect to these types of products under their particular type of consumer use. The researchers also recognized that they had purposely designed the test with closures that would permit easy exposure of the product to consumer use: screw-tops and wide-mouth jars. As a result, they appropriately concluded that "if container design provides adequate protection, even poorly preserved products could withstand consumer use" (13).

A study by Lindstrom and Hawthorne described a program showing how a PET can be refined by bench-marking its in vitro results against actual consumer use results (14). After receiving a few product units back from use showing contamination, the PET was modified to include a higher inoculum concentration, more stringent microbial reduction requirements, and the development of an accelerated preservation test. The results of these modifications to the PET were then compared against consumer use results again, and complete agreement between the in vitro PET and consumer use testing was reached. Although Lindstrom and Hawthorne did not point out the critical role that the package plays in protection from contamination, they did indicate that consumer-use testing is a routine practice to assess the true ability of a package to remain uncontaminated. This was intimated by the final sentence of the discussion: "The consumer-use panels . . . now include testing of . . . any applicator or packaging changes." Most of the more knowledgeable people working with cosmetic and drug preservation will readily admit that in-use or in-container testing is important to assess the ability of a product to resist contamination (25).

As stated earlier, the FDA's view is that the purpose of a PET is to predict consumer contamination potential (26,27). The FDA has never been successful in doing this. One of the main reasons why such a test would be desirable in the FDA's eyes is that they could request a recall of products failing such a test. A collaborative study is now being concluded between CTFA, AOAC, and FDA to develop a standardized PET complete with multilab comparisons and statistical analysis. No published data are as yet available. However, the method will be flawed with respect to the FDA's desire to have a method for recall purposes. It is flawed because it does not predict a product's ability to withstand microbial insult during consumer use. It will not have this ability because of two issues. First, the study will not have any correlative, "real-world," in-use studies using the same products for which the PET was conducted. Such an omission is fortunate for the industry, since such a PET could be mistakenly used in an attempt to enforce a recall on those products that fail it. Second, and most important, the study does not account for the role that packaging plays in preventing contamination. Packaging can be shown to totally protect some products—even products that completely fail a PET and contain absolutely no preservative. One might also make the observation that PET data only apply to freshly made products, not shelf-aged products. However, this would be a very weak argument, tantamount to admitting that you market a product that becomes a risk over time.

10.3.2 The Role of Packaging in Product Preservation

Some PETs use proprietary "in-house" organisms unavailable to others. This fact alone would preclude the development of a "standard" PET that could be used to enforce a recall. More importantly, since no PET can account for the protective effects of a package or dispensing device during use, even the concept and logic of thinking that a PET can be used to predict consumer contamination potential is flawed. It is illogical to think that there can be a "standard" PET that can be used to enforce any kind of regulatory action.

In order to demonstrate that this statement is valid, we will summarize the only study published with the express purpose of testing the hypothesis that "the dispensing closure used for containers plays an important role in preventing contamination of cosmetics during use." This study showed that the type of closure will prevent or minimize microbial contamination of a completely unpreserved product (28).

In the study, two unpreserved product types were evaluated (shampoo and skin lotion) in packages with three perceived levels of protection from in-use contamination due to the closure (Fig. 10.1). The shampoo was put into bottles with a closure that was considered to provide no protection (standard screw-cap with a 24-mm-diameter opening), a flip-cap that was perceived as capable of providing a high level of protection (a hinged cap that covered a 6-mm-diameter opening when not in use), and a slit-top, the protective abilities of which were unknown. It provided no direct opening for the product but instead allowed the product to escape through a slit in the top of a neoprene gasket when the bottle was squeezed. The slit-top had a flip-cap that could be flipped down to cover the gasket when not in use. The skin lotion was placed into bottles with the same two caps (open screw-top and flip-cap) and into a pump-top.

After confirming that the two unpreserved products had no inherent hostility to microorganisms using a modified CTFA preservative efficacy test, the authors placed the products into consumer-use testing. All of the products were aseptically packed, and the absence of microorganisms was confirmed by standard plate count methods. Unexposed product controls were also maintained and rechecked at the end of the study to ensure that no microbial growth had occurred as a result of contamination during aseptic packing. After consumer use for 2–3 weeks, the products were collected and microbial content testing conducted immediately and 4–7 days later. A product was deemed contaminated if >100 CFU/g of the product was observed or if gram-negative bacteria at any level were seen at both plating times.

Of the three caps used for the shampoo, the flip-cap provided complete protection, with 0% of the products becoming contaminated. This was in contrast to the screw-cap and slit-cap design, which allowed 21—29% of the units to become contaminated. In addition to these data, the authors also factored the level of contamination (CFU/g) of the product and the diversity of the contaminants into an overall rating of contamination extent (Table 10.1). Based on these two additional types of data, the protective abilities of the shampoo caps could be ranked as providing the least to the most protection: screw-cap < slit-cap < flip-cap. Using similar considerations for the skin lotion, the abilities of the dispensing devices to protect the product could be ranked as follows: screw-cap < flip-cap < pump-top. Thus, the authors showed that contamination from consumer use could be controlled by container designs that provide for minimal exposure of the product to consumer and environmental microbial insult. Additional studies are needed to

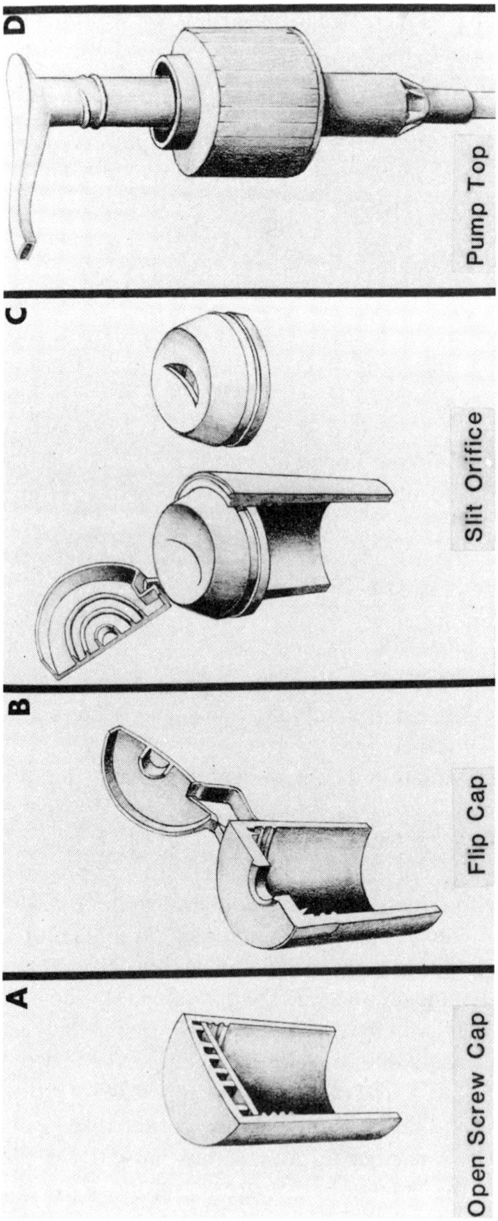

Figure 10.1 Closure types used to determine how protection from contamination during use could be afforded an unpreserved product due to closure alone. (A) Open screw-cap used for shampoo and skin lotion; (B) flip-cap used for shampoo and skin lotion; (C) slit-cap used for shampoo; (D) pump-top used for skin lotion. (From Ref. 28.)

TABLE 10.1 Comparison of Contamination Incidence, Average Contamination Level, and Diversity of Microbial Contaminants in Unpreserved Cosmetic Products with Various Closures After Consumer Use

Product/Closure	In-use contamination incidence (%)	Log-normalized average contamination (CFU/g)	Diversity[a] of contaminants
Shampoo			
Screw cap	29	5.375	High
Slit cap	21	3.294	Low
Flip cap	0	0	NA
Skin lotion			
Screw cap	71	4.417	High
Flip cap	39	3.75	Low
Pump	10	5.977	Low

[a]Measured by Shannon diversity index.
Source: Ref. 28.

further determine the role that packaging plays in protection from microbial contamination during consumer use. Especially needed are packaging studies with products that require applicators.

10.4 THE NEED FOR ASEPTIC MANUFACTURING AND USE OF COMPATIBLE MATERIALS

In addition to realizing that physical design of containers and closures is important for protecting products from consumer use contamination, it must also be recognized that the composition of the closure and the container may affect the integrity of the product, and thus the materials used in the package and closure should be compatible with the product (29,30). Preservative manufacturers, in comparison with other raw material suppliers, rarely include a statement of container compatibility. During long-term storage, a preservative may be adsorbed into the container material or inactivated by the dyes used in the plastic or lost because of the volatility of the preservative. In one instance, 50% of the parabens used to preserve mascara were absorbed into the polyethylene container after storage and could only be extracted from the container by 50% methanol (S. Richards, personal communication). The microbiologist should always test long-term stability of the product in the container expected for marketing. Aging tests can be done under accelerated conditions (high temperatures). One should include chemical analyses for the preservative or preserving factors in the formula as well as conducting PETs. If this is done, a failure in the PET can be traced to a breakdown or loss of the preservative or preserving factors in the formula. If the preservative or preserving factors are still present in the formula but the product still fails a PET, then other factors are acting to keep the organisms from contacting the biocidal factors in the formula. One factor may be the preservative partitioning out of the water phase into an oil phase of the product.

When considering containers, one should also realize that some closures can even inactivate the preservative (31). Alternatively, even the package itself appears to exhibit biocidal activities in some cases. An ointment contaminated with several pathogens and packed into metal tubes killed off the bacteria inoculated into it. When this same contaminated product was inoculated into glass jars, the inoculum survived (32). The authors attributed this result to the exclusion of oxygen in the metal tube, but the fact that many metals themselves are biocidal should not be ruled out. Studies like this illustrate that preservation is a function of both the formula and the container—not just the biocide.

Preservation during use even extends to the label on the package and the instructions for use on the label. The obvious area in which preservation applies to the label is in the use of adhesives and sealants, which also must be preserved against microbial growth (33). A less obvious consideration is the instructions for use. A good way to create self-preserving products would be to instruct consumers to discard unused products after a certain time period, as do some contact lens care products that are manufactured under aseptic conditions. "If the product is not completely used up after 3 months of continual use, the remaining product must be discarded" would be a logical instruction for informing the consumer that the trade-off of a preservative-free product is that she should discard old mascaras and other cosmetic products. The wise microbiologist inserts himself or herself into the product development team early on to help decide upon the formula, the package, and even specific instructions for use on the label.

Finally, we must also admit that our abilities to aseptically package products can be continually improved if we choose to rely primarily on the package to provide the protection from contamination during consumer use. People are one of the primary sources of microbial contamination in cosmetic and pharmaceutical manufacturing, because they shed anywhere from 100,000 to 5 million particles 0.5 μm or larger per minute (34). Thus, it is critical to remove airborne particles from the atmosphere around the filling and closing operation. The processes of packing must take place under stringently clean conditions (35,36). Clean air is produced by filtering the air through high-efficiency particulate air (HEPA) filters. Building the room that receives the air is also critical. The materials used in constructing it must not shed particles. The room should have no ledges, and windows (nonopening) should be flush to the inside of the room. Any needed ledges or transoms should be sloping with at least a 45° angle. The doors should be limited access and provide air-tight seals. All joints in walls, floors, and ceilings should also be sealed to prevent leakage of air and bacteria. Aseptic packing requires that all the items involved be sterilized: package, product, equipment, and the environment under which packing takes place. Rarely does this approach occur in a cosmetic plant. This approach is more often found in drug plants that make sterile pharmaceuticals. Packaging can be supplied by the manufacturer to the drug or cosmetic manufacturer already cleaned and sterilized, or the manufacturer may clean and sterilize them during aseptic filling. Alternatively, the manufacturer may decide to become their own package supplier by using blow-form-fill-seal machines for aseptic packaging. The other aspects of aseptic packing including terminal sterilization are far beyond the scope of a single chapter. Rather, volumes have been produced on the topic (37,38).

10.5 FUTURE FOR PACKAGING IN PRESERVATIVE-FREE AND SELF-PRESERVING PRODUCTS

If the use of preservative-free and self-preserving ("inherently hostile") formulas are to be realized, then we must also factor into their development the use of better protective packaging and, in particular, delivery devices that do not allow for the product to become contaminated during use. Companies are becoming more aware of this approach. Flip-caps for shampoos and pump-tops for lotions are becoming almost standard. Perhaps we could even develop a flip cap that automatically opens when the bottle is tipped and closes when it is set upright. For pharmaceuticals, prefilled syringes for some vaccines are being introduced to avoid the risk of contamination that conventional methods offer (39). However, we should be looking at other unique ways of delivering products in a manner that minimizes consumer contact with the product. For mascaras and other eye makeup, we should search out applicator wands and brushes that are themselves impregnated with biocides or at least resistant to contamination (40). Sterile, single-use containers should not be our only option except as testers at makeup counters and for some pharmaceuticals (41). Some companies are already marketing products without preservatives in sterile air-tight containers that dispense products in a way that avoids contamination (42).

Manufacturers should also consider delivering shampoos and body lotions in aerosol mousses with the claim that they prevent exposure to the body's own microorganisms. With the advent of "bag-in-can" technology, a wider range of creams, gels, lotions, and pastes can now be delivered in an aerosol form that allows in-use application from a virtually contamination-proof container (43). At least one company is already doing this to deliver petroleum jelly, a product that is virtually anhydrous and thus cannot (and does not need to) be preserved but still needs protection from dust that carries bacterial spores (44) and moisture.

Packaging suppliers should develop pump-tops that completely prevent not only siphoning back of the product upon upstroke, but even the ability of microorganisms to "grow back" into the product. Some companies are already doing this with nasal decongestants where the dispenser provides a controlled dose in a one-way spray operation to prevent contamination of the solution in the bottle (45).

All of this technology could be promoted as sanitary packaging. The general public needs to be educated to look for these kinds of products since they not only allow us to reduce or eliminate preservatives but also provide protection from microorganisms that can grow in or on the product during use (e.g., bars of in-use soap vs. liquid soap).

The production of protective packaging involves an interesting irony. While consumers demand a preservative-free product that does not become contaminated during use, they also want the packaging to be environmentally friendly. A new plastic resin, called Biopol, has been developed by ICI Bio Products (46). The resin is used to make bottles that are completely biodegradable. While there is a need to protect the product from microbial contamination until used, we want the package providing that protection to then be degraded by the microorganisms that it once protected the product against! When discarded, the package that protects the product from microorganisms is then degraded by microorganisms. But the irony does not stop there. Biopol is made by *Alcaligenes eutrophus*, a bacterium! A package made by a bacterium that protects against bacteria and that is degraded by bacteria is, indeed, a singular discontinuity.

Once again, microbes are both the bane and boon of our quest for survival. And so, the little creatures that provide earth with its homeostatic mechanism once again provide us with the unique opportunity to anticipate the future and respond with an altruism that is so characteristic of truly being human: safely providing others with the things they need.

REFERENCES

1. Cosmetic packaging quality control. *Household Pers. Prod. Ind.* (April): 36 (1995).
2. I. Tattersall. *The Human Odyssey: Four Million Years of Human Evolution*. Prentice Hall, New York, 1993.
3. F. Gunn. *The artificial face: a history of cosmetics. In: Multimedia Encyclopedia*, ver. 5.01. Grolier, Inc. and The Software Toolworks, Novato, CA, 1974.
4. J. C. Curry. Cosmetic microbiology—50 years of change. *Household Pers. Prod. Ind.* 8:44–52 (1983).
5. L. J. Morse, H. L. Williams, F. P. Grenn, Jr., E. E. Eldridge, and J. R. Rotta. Septicemia due to *Klebsiella pneumoniae* originating from a hand cream dispenser. *N. Eng. J. Med. 277*: 472–473 (1967).
6. W. C. Noble and J. A. Savin. Steroid cream contaminated with *Pseudomonas aeruginosa. Lancet 1*: 347–349 (1966).
7. L. A. Wilson and D. G. Ahearn. *Pseudomonas*-induced corneal ulcers associated with contaminated eye mascaras. *Am. J. Ophthalmol. 84*: 112–119 (1977).
8. L. A. Wilson, A. J. Julian, and D. G. Ahearn. The survival and growth of microorganisms in mascara during use. *Am. J. Ophthalmol. 79*: 596–601 (1975).

9. N. L. Dawson and D. J. Reinhardt. Microbial flora of in-use, display eye shadow testers and bacterial challenges of unused eye shadows. *Appl. Environ. Microbiol. 42*: 297 (1981).
10. CTFA Microbiology Committee. CTFA survey: preservative test methods companies use. *Cosmet. Toiletr. 105*: 79–82 (1990).
11. R. J. Spielmaker. Report of a cosmetic industry survey concerning correlation of preservative challenge and consumer use test results. *J. Soc. Cosmet. Chem. 36*: 376 (1985).
12. J. K. Farrington, E. L. Martz, S. J. Wells, C. C. Ennis, J. Holder, J. W. Levchuk, K. E. Avis, P. S. Hoffman, A. D. Hitchens, and J. M. Madden. Ability of laboratory methods to predict in-use efficacy of antimicrobial preservatives in an experimental cosmetic. *Appl. Environ. Microbiol. 60*: 4553–4558 (1994).
13. D. K. Brannan, J. C. Dille, and D. J. Kaufman. Correlation of in vitro challenge testing with consumer use testing for cosmetic products. *Appl. Environ. Microbiol. 53*: 1827–1832 (1987).
14. S. M. Lindstrom and J. D. Hawthorne. Validating the microbiological integrity of cosmetic products through consumer-use testing. *J. Soc. Cosmet. Chem. 37*: 481–488 (1986).
15. Investigating cases of product contamination. *Cosmet. Toilet. Manuf. Supp.* (June): 37 (1994).
16. A. H. Shaw. Preserving the future. *Soap Cosmet. Chem. Sp. 68*: 26–28 (1992).
17. Guideline for Submitting Documentation for Packaging for Human Drugs and Biologicals. U.S. Department of Health and Human Services, Food and Drug Administration, Washington, D.C., February 1987.
18. Notice of intent to propose regulations and request for information on preservation of cosmetics coming in contact with the eye. *Fed. Reg. 45*: 54837–54838 (1977).
19. Cosmetics, Toiletries, and Fragrance Association. Evaluation of methods for determining preservative efficacy. *CTFA Cosmet. J. 5*: 2–7 (1973).
20. Cosmetics, Toiletries, and Fragrance Association. Determination of adequacy of preservation of cosmetic and toiletry formulations. In: *CTFA Technical Guidelines*. CTFA, Inc., Washington, DC, 1983.
21. D. G. Ahearn, J. Sanghvi, G. J. Haller, and L. A. Wilson. Mascara contamination: in use laboratory studies. *J. Soc. Cosmet. Chem. 29*: 127–131 (1978).
22. N. L. Dawson and D. J. Reinhardt. Microbial flora of in-use, display eye shadow testers and bacterial challenges of unused eye shadows. *Appl. Environ. Microbiol. 42*: 297–302 (1981).
23. R. A. Cowen. Relative merits of ‘in use’ and laboratory methods for the evaluation of antimicrobial products. *J. Soc. Cosmet. Chem. 25*: 307–323 (1974).
24. J. K. Farrington, E. L. Martz, S. J. Wells, C. C. Ennis, J. Holder, J. W. Levchuk, K. E. Avis, P. S. Hoffman, A. D. Hitchins, and J. M. Madden. Ability of laboratory methods to predict in-use efficacy of antimicrobial preservatives in an experimental cosmetic. *Appl. Environ. Microbiol. 60*: 4553–4558 (1994).
25. J. R. Sabourin. Evaluation of preservation for cosmetic products. *Drug Cosmet. Ind. 147*: 24–27 (1990).
26. T. T. Tran and A. D. Hitchins. Microbial survey of shared-use cosmetic test kits available to the public. *J. Ind. Microbiol. 13*: 389–391 (1994).
27. T. T. Tran, F. J. Hurley, M. Shurbaji, and L. B. Koopman. Adequacy of cosmetic

preservation: chemical analysis, microbial challenge and in-use testing. *Int. J. Cosmet. Sci. 16*: 61–76 (1994).
28. D. K. Brannan and J. C. Dille. Type of closure prevents microbial contamination of cosmetics during consumer use. *Appl. Environ. Microbiol. 56*: 1476–1479 (1990).
29. M. S. Parker. Preservation of pharmaceutical and cosmetic products. In: A.D. Russell, W. B. Hugo, and G. A. J. Ayliffe, eds. *Principles and Practice of Disinfection, Preservation and Sterilization*, 2d ed. Blackwell Scientific Publications, London, 1992, pp. 335–350.
30. T. J. McCarthy. Interaction between aqueous preservative solutions and their plastic containers. *Pharm. Weekbl. 105*: 557–563, 1139–1146 (1970).
31. L. Lachman, T. Urbanyl, and S. Weinstein. Stability of antibacterial preservatives in parenteral solutions. IV. Contribution of rubber closure composition on preservative loss. *J. Pharm. Sci. 52*: 244–249 (1963).
32. Q. M. Abu Shaqra and N. T. Hasari. Fate of some types of bacteria in hydrocortizone acetate ointment. *Int. Biodeterioration 23*: 115–120 (1987).
33. M. Friedman, F. Casati, and J. Stamm. Selecting the proper biocide for adhesives and sealants. *Adhesives Age* (Feb.): 29–32 (1988).
34. T. McCarthy. Sources of microbial contamination. *Pharm. Cosmet. Rev. 18*: 31–32 (1991).
35. Aseptic advances ensure clean conditions for pharmaceuticals. *Manufact. Chem. 66*(6): 13 (1995).
36. D. K. Brannan. Cosmetic microbiology. In: J. Lederberg, ed. *Encyclopedia of Microbiology*, Vol 1. Academic Press, Inc., San Diego, 1993, pp. 593–603.
37. R. F. Morrissey and Y. I. Prokopenko, eds. *Sterilization of Medical Products*, Vol V. Polyscience Publications, Inc., Canada (1991).
38. G. B. Phillips and W. S. Miller, eds. *Industrial Sterilization, International Symposium, Amsterdam 1972*. Duke University Press, Durham, NC (1973).
39. Merieux DT vaccine: Merieux UK introduces prefilled syringe diphtheria/tetanus vaccine, first in UK in 27 years. *Chem. Drug.* (October): 608 (1991).
40. R. J. Edmundson and C. G. Baker. Technical aspects of formulating lash and brow makeup products. *Cosmet. Toiletr.* (April): 39–42 (1979).
41. Anchor Advanced Products offering Testix cosmetic sampling system. *FDC Reports Rose Sheet* (July 16): (1990).
42. Monteil Paris launches ICE Monteil skin cleansing and moisturizing products in US market. *Cosmet. Toiletr.* (June):16 (1993).
43. Aerosols International—Developing the products of the future. *Cosmet. Toiletr. Manuf. Suppliers* (September): 24 (1992).
44. Aerosols petroleum jelly reduces sanitary woes. *Food Drug Pac.* (March): 6, 69 (1985).
45. Boehringer Ingelheim's Nostril and Nostrilla marketed in a unique dispenser system which provides a controlled dose. *Drug Cosmet. Industr.* (January): 61 (1983).
46. R. Gilbert. Bottles for natural shampoos fade away: containers biodegrade quickly. *Food Drug Pac. 56*: 4 (1992).

11

Preservative-Free and Self-Preserving Cosmetic and Drug Products: The Future

Jon J. Kabara
Technology Exchange, Inc.
Galena, Illinois, and
Michigan State University
East Lansing, Michigan

Donald S. Orth
Neutrogena Corporation
Los Angeles, California

Preservative-free and self-preserving cosmetic and drug products are not new. Many products have been marketed for years, often without conscious thought as to why they did not require use of much, if any, chemical preservatives. Products such as antidandruff shampoos with salicylic acid at a pH of 3, sterile eye drop solutions, and antibiotic ointments may have no preservatives. These products are satisfactorily preserved as a result of their physicochemical composition and/or packaging, and experience has shown that this approach is a successful means of delivering safe, uncontaminated products to consumers.

This book has been written to help cosmetic scientists and microbiologists think about formulation in a new way. It may be possible to continue using standard preservative systems for domestic use; however, globalization requires that we think beyond current ways of doing things and find new, better ways. In many respects, preservative-free and self-preserving products offer advantages for formulation because they reduce and/or eliminate chemical preservatives that are sources of skin irritation and contact sensitization. Self-preserving products reduce the costs of using chemical preservatives, meet the demands of a growing segment of consumers who demand natural (i.e., preservative-free) products, encourage the use of contamination-resistant packaging, and allow the use of global formulas without the usual regulatory issues surrounding product preservatives.

The goal of this book is not to provide formulas that do not contain classical preservatives. Rather, it is our aim to provide the scientific basis for developing preservative-free and self-preserving products. The principles of preservation are introduced and discussed in several chapters to show how physicochemical conditions [i.e., low pH, low water activity (a_w)], classes of chemicals (i.e., acids, salts, chelating agents, surfactants, aroma chemicals, alcohols, botanicals, etc.), and prevention of contamination (i.e., by adherence to GMPs during manufacturing or use of protective packaging) may be used to create adequately preserved products that prevent microbial contamination/growth and/or kill contaminating organisms. It is important to understand that the ability of a formula to resist contamination often depends on the *preservative system*—not on one or two specific chemicals. Considering the multifunctional aspects of ingredients is helpful in formulating self-preserving products. This chapter summarizes the principles of preservation and illustrates their application for the creation of preservative-free and self-preserving cosmetic and drug products.

11.1 COMPONENTS THAT MAKE UP PRESERVATIVE SYSTEMS

Formulation chemists generally give little thought to the fact that some formulas are easier to preserve than others. The ease with which antiperspirant/deodorants and hair conditioners are preserved often goes unnoticed because more time is usually required to determine the best preservative systems to use in conventional shampoos and emulsions. Perhaps it is human nature, the pace of the work in the formulation lab, or the crisis *modus operandi* that prevents one from thinking about why some things are easy—because the limited time available must be devoted to solve problems with the difficult formulas. One should reflect and ask why these formulas are easier to preserve.

Not surprisingly, the physicochemical composition of these products prevents microbial growth/survival. The *preservative system* of a product includes both compounds with known antimicrobial activity and formula components that may contribute directly or indirectly to the antimicrobial activity (1,2). If a formula does not contain chemical preservatives, then the physicochemical make-up of the formula, which includes factors such as product pH, a_w, nutrient availability, surfactant concentration, sequestering agents, alcohols, and interfering materials, determines whether the formula is self-preserving (see Chapter 1).

In 1990, a major manufacturer of preservatives in the United Kingdom commissioned a survey to learn about consumer awareness of preservatives. Marie

R. Frost (3), a consultant to the cosmetic industry, presented the following summary of this survey:

1. Seventy percent of consumers considered natural ingredients to be important.
2. Seventy percent of the consumers associated preservatives with food.
3. Forty percent of consumers were aware of preservatives in cosmetics and toiletries.
4. Twenty five percent of consumers actively expressed concern at the inclusion of preservatives in cosmetics and toiletries.

The responses to this survey suggested that there was little understanding of the function of preservatives in ensuring the safety of products. Frost felt that the response in the United States probably would be different because American consumers are generally more aware of preservative-free foods.

In formulating preservative-free products, the manufacturer's primary responsibility must be to the safety of the consumer. One cannot simply remove chemical preservatives from a formula and expect to have no problems. Frost reported that a toothpaste formula without preservatives showed fungal growth after 14 days in inoculated samples. Adding aroma chemicals (i.e., peppermint oil, fennel oil, cinnamon oil) and substituting propylene glycol for sorbitol created a product that reduced levels of bacteria, yeasts, and molds to <10 CFU/g in ≤ 7 days. (See Chapter 8 for additional information on aroma chemicals.)

11.1.1 Acid pH

All living microorganisms have characteristic ranges of pH that permit growth (see Chapter 2). Many cosmetic and drug formulas including anionic shampoos, antidandruff shampoos, facial cleansers, nonionic emulsions (i.e., sunscreens, hand and body lotions, face creams, night creams, etc.), and acne treatment gels have pH values around neutrality (pH 5–8). Many microorganisms of interest to the cosmetic and pharmaceutical industries generally are aerobic and have pH optimal for growth around neutrality. High-acid products (pH 3–4.5) become more unfavorable for microbial growth and survival as pH decreases. Antiperspirants with 25% aluminum chlorohydrate or 20% aluminum zirconium tetrachlorohydrate, hair conditioners with quaternary ammonium compounds, alpha-hydroxy acid (AHA) products with glycolic, lactic, malic and/or tartaric acids, acne treatment gels or antidandruff shampoos with salicylic acid, creams/lotions using quaternary emulsifiers, and shampoos/conditioners containing biomimetic phospholipids (i.e., linoleamido phosphadidyl PG-dimonium chloride) generally are high-acid products. These products generally are much easier to preserve than are similar products with pH values in the neutral range. In some cases, the preservative system created by the acid pH and other formula components is sufficient for these products to be adequately preserved without the need for additional chemical preservatives.

11.1.2 Alkaline pH

The ability of microorganisms to grow/survive also becomes increasingly difficult as the pH increases above the neutrality range. Products with pH values greater than pH 9 often require little or no chemical preservatives, whereas similar formulas in the neutral pH range required added preservatives to be adequately preserved. The reason for this is that extreme pH conditions—either acid or alkaline—make it necessary for microorganisms to expend energy for maintenance of intracellular pH. Alkaline products include bar soaps and liquid soaps containing neutralized fatty acids which have a pH in the range of 8–10, and highly alkaline products include hair relaxers/perm neutralizers with thioglycollate or reducing agents and depilatory creams may have a pH in the region of 10–14. As with the high-acid products, the highly alkaline products are much easier to preserve than are similar products with pH values in the neutral range.

11.1.3 Water Activity

A number of water-soluble compounds including salts, sugars, amino acids, peptides, and short-chain fatty acids/alcohols require water for hydration and reduced a_w. The use of any of these materials will help lower the a_w of the formulation and make water unavailable for microorganisms (see Chapter 3). Antiperspirants containing 25% aluminum chlorohydrate have reduced a_w and lower pH. The addition of up to 40% glycerin in a hand cream lowers the a_w to 0.83, which is below that which would allow growth of bacteria that are known to cause problems in cosmetic products. The low a_w and the lack of nutrients also makes hand cream difficult for yeast and molds to spoil. Relatively anhydrous products, including mineral oil, baby oil, cornstarch or talc foot powders, press powders, anhydrous sunblocks, bar soaps, and perfumes, require no preservatives because the low a_w prevents microbial growth. Nevertheless, manufacturers may add preservatives to some of these products to help prevent microbial growth if experience has shown that the consumer adds water during normal use/abuse.

11.1.4 Surfactants

A number of surfactants have antimicrobial action because they reduce surface tension and destabilize cell membranes. Different classes of microorganisms appear to be affected differently by surfactants. Anionic surfactants including sodium lauryl sulfate, ammonium lauryl sulfate, and sodium laureth sulfate are the most common detergents used in shampoos, cleansers, and body washes. Gram-positive cocci, such as *Staphylococcus aureus*, are less tolerant to these anionic surfactants than are coliforms (i.e., *Escherichia coli*) and some *Bacillus* species. Nonionic surfactants such as polysorbate 80 (Tween 80) often are thought to interfere with preservative action; however, nonionic esters of fatty acids, such as glyceryl monolaurate (Lauricidin®) and sucrose laurate, are reported to be active primarily against gram-positive bacteria (see Chapter 5).

Many fatty acids have antimicrobial activity. Lauric acid, palmitoleic acid, and polyunsaturated fatty acids are reported to be active against gram-positive bacteria, whereas short-chain fatty acids (i.e., alkyl chain < 8 carbon atoms) are reported to have activity against both gram-positive and gram-negative bacteria (see chapter 5). The antimicrobial action of true liquid soaps is due, in part, to the fatty acid salts, as well as the free alkalinity, the low a_w, and other inhibiting materials in the formula (i.e., aroma chemicals, chelators, antioxidants, etc.).

Cationic surfactants such as benzalkonium chloride (BAC), cetyl pyridinium chloride, and cetyl trimethylammonium bromide are quats that have activity against gram-positive bacteria and some gram-negative bacteria; however, gram-negative bacteria—especially the pseudomonads—are more resistant to quats than are gram-positive bacteria. Quats contribute to the preservative system in hair conditioners, mouthwashes, antiperspirant/deodorants, and cationic emulsions (see Chapter 4). Biomimetic phospholipids such as linoleamidopropyl phosphatidyl PG-dimonium chloride have multiple cationic groups that provide a high degree of substantivity to proteinaceous substrates as well as antimicrobial activity. These compounds are reported to be effective against a variety of bacteria, yeasts, and molds and may aid the preservative system, especially in acidic products (see Chapter 6).

Ahlström et al. (4) reported on *soft quats*—long-chain alkanoylcholines that are antimicrobial agents and enzymatically degradable into products that are significantly less toxic than the original substance. Amphiphilic quats are surface-active and are known to have antimicrobial activity due to their membrane-disruptive effects. The high degree of affinity of these agents for biological membranes results in low selectivity, which results in broad spectrum of activity against gram positive and gram negative bacteria, fungi and certrain viruses (see Chapter 4).

11.1.5 Antioxidants and Chelating Agents

Although the primary justification for the use of antioxidants in cosmetic and drug formulations is that they delay the onset and/or slow the rate of autoxidation by inhibiting the formation of free radicals (initiation) or by interrupting propagation of free radical formation, they are also known to exert preservative action (5). Butylated hydroxyanisole (BHA), butylated hydroxytoluene (BHT), tertiary butylhydroquinone (TBHQ), and propyl gallate (PG) are the most common phenolic antioxidants used in foods and cosmetics. TBHQ has been found to be the most effective phenolic antioxidant and is effective against bacteria, yeasts, and molds. BHA, BHT, PG, and TBHQ frequently have additive or synergistic antimicrobial activity when used with preservatives or chelating agents including tetrasodium EDTA, citric acid, and disodium etidronate (see Chapters 7 and 9). Antioxidants and chelating agents are multifunctional chemicals that may be used to enhance the activity of the preservative systems of cosmetic and drug products.

11.1.6 Aroma Chemicals and Alcohols

Spices and essential oils have been used for centuries to help prevent deterioration of products. The active extracts typically contain alcohols, phenols, esters, acids, aldehydes, and terpenes that have antimicrobial activity. The usefulness of ethyl and isopropyl alcohol is recognized in the tentative final monograph for topical antimicrobial antiseptic drug products for over-the-counter human use, which allows the use of 60–95% alcohol by volume in antiseptic handwashes, patient preoperative skin preparations, and surgical scrubs and 70–91.3% isopropyl alcohol by volume in patient preoperative skin preparations (6). Microbial growth is prevented in aqueous systems containing 25% alcohol; however, lower levels of alcohol (5–10%) may have additive or synergistic antimicrobial action with other formula components. For example, wines with 9–12% alcohol frequently become contaminated and spoil within a few weeks after opening when stored at room temperature. Fortified wines, such as port and sherry, which have alcohol contents of 18–22%, may be stored months to years at room temperature without microbial spoilage. The alcohol content, pH, organic acids, antioxidants, and polyphenolic compounds (tannins) contribute to the preservative system of wines. El-Kady et al. (7) reported that a number of essential oils including clove, thyme, sandalwood, cinnamon, lemon, and eucalyptus have antimicrobial activity (see Chapter 8). It is possible to learn from this and use similar physicochemical conditions for creating preservative-free and/or self-preserving cosmetic and drug products.

Richards and McBride (8) noted the enhancement of antimicrobial action of BAC by aroma chemicals including 2-phenylethanol, 3-phenylpropan-1-ol, and benzyl alcohol. Fragrance manufacturers are using "deo" fragrances in an effort to counter microbial growth and malodor production. Haustein et al. (9) reported on a mixture of three natural compounds (farnesol, glyceryl monolaurate, and phenoxyethanol) that killed corynebacteria associated with axillary maloder. Aroma chemicals are multifunctional in that they may have antimicrobial action besides contributing to the fragrance and/or flavor of products (see Chapter 8).

11.1.7 Botanicals and Herbals

A number of natural products have become available to the cosmetic and drug industries in the past few years. In 1989, Toda et al. (10) reported that extracts of tea inhibited and killed *S. aureus*, *S. epidermidis*, *Salmonella typhimurium*, *S. enteritidis*, *Shigella flexneri*, *S. dysenteriae*, and *Vibrio cholerae* and *Vibrio* species. The antimicrobial properties of tea polyphenols have been reported by several authors (11–13). Mukhtar et al. (14) noted that the polyphenol fraction of green tea prevents oxidant-induced damage to cells and provides protection against cancer induction. It appears that tea polyphenols are multifunctional ingredients that may provide antimicrobial and antioxidant effects in products and possibly to tissues to which they are applied.

Tea tree oil is a volatile oil obtained by steam distillation of freshly harvested foliage and terminal branchlets of *Melaceuca alternifolia*, a plant indigenous to New South Wales, Australia.Tea tree oil has been used as a natural antiseptic and fungicide and possesses antimicrobial activity against a wide range of gram-positive and gram-negative bacteria and fungi. Minimum inhibitory concentrations (MICs) for these organisms typically were in the range of 0.5–1%, although the MIC for *P. aeruginosa* was 2% (15). Bassett et al. (16) performed a clinical study with 124 patients and reported that both 5% tea tree oil gel and 5% benzoyl peroxide lotion had a significant effect in ameliorating patients' acne by reducing the number of inflamed and noninflamed lesions (open and closed comedones).

As early as the eighteenth century B.C. in Egypt, lichen extracts were used to treat topical infections. Lauterwein et al. (17) tested secondary metabolites of different species of lichen and found that usnic acid and vulpinic acid were active against *S. aureus*, *Enterococcus faecalis*, *Enterococcus faecium*, and some species of *Bacteroides* and *Clostridium* and had MICs of 2–32 μg/ml. Usnic acid generally was more active than vulpinic acid, and neither was effective against gram-negative bacteria or fungi.

The flowers of *Prunus spinosa* L. (rose petals) have been used as folk remedies for purgatives and diuretics, in herbal teas, and in anti-inflammatory medications. Sakar et al. (18) reported the isolation and antimicrobial activity of flavonoids from *P. spinosa* L. flowers and identified the active materials as kaempferol-3-O-α-L-arabinofuranoside and kaempferol-3-O-α-L-rhamnopyranoside. These flavonoids had moderate antimicrobial activity against *E. coli*, *Salmonella typhimurium*, *S. aureus*, *Sarcina lutea*, and *Candida utilis*. Reports such as this suggest that flavonoids may have application for augmenting the preservative systems in products.

11.1.8 Microbial Products with Antimicrobial Properties

Curiously, several natural products produced by microorganisms have antibiotic/antimicrobial activity. Familiar antibiotics produced by molds and actinomycetes are penicillin and actinomycin. Mupirocin (pseudomonic acid A) is an antibiotic produced by *Pseudomonas fluorescens*. Mupirocin is active against all species of staphylococci, most species of streptococci, and some gram-negative bacteria (19, 20). In vitro tests show that this antibiotic has MICs of 0.03–16 μg/ml against *S. aureus* (20). Mupirocin (2%) has been used topically in cream formulations for the treatment of skin infections. In addition, microorganisms produce a number of biologically active materials including peptide antibiotics (i.e., nisin, a peptide antibiotic produced by *Streptococcus lactis*; gramicidin A, an ionophore that affects cation permeability of gram-positive bacteria membranes), bacteriocins, short-chain fatty acids (lactic acid, acetic acid, and propionic acid), and alcohols (21).

11.1.9 Antimicrobial Amphiphilic Peptides

Antimicrobial peptides such as magainin from frogs and defensins from neutrophils are known to have broad antimicrobial activity due to their membrane-disrupting ability (22–24). Haynie et al. (24) reported on the antimicrobial activities of amphiphilic peptides covalently bonded to a water-insoluble resin. They found that immobilized peptides typically reduced the number of viable cells by ≥5 log units, which suggests that the surface action by magainins and structurally related antimicrobial peptides were sufficient for consideration in preparing antimicrobial articles such as catheters and food preservation systems (24). This suggests the possibility of using these natural chemicals in cosmetic and drug products.

11.1.10 Enzymatic Antimicrobial Systems

Enzymatic systems have been used to help preserve products. Glucose oxidase has been used with glucose to consume oxygen in closed systems, thereby reducing the oxidation/reduction potential, which decreases the rate of growth or stops growth of aerobic organisms, such as molds and *Bacillus* species, and oxidative organisms, such as *Pseudomonas* species (21). Ashworth and Turton (25) reported on Myavert C, a dual-enzyme system whose CTFA nomenclature is lactoperoxidase (and) glucose oxidase (and) glucose. Since the oxidation of glucose results in the formation of gluconic acid and concomitant reduction in pH, it has been found helpful to use a buffer system, such as a citric acid/sodium citrate buffer. Gas-impermeable packaging is desirable to prevent reentry of the oxygen that is consumed as glucose is converted to gluconic acid. Enzymes are natural, and the lactoperoxidase system has similarities to that used by oral flora to generate H_2O_2, which has antimicrobial action (21). Use of this type of enzyme system may enable claims of "natural" to be made for the product.

11.2 PACKAGING OF SELF-PRESERVING PRODUCTS

All cosmetic and drug products require appropriate packaging to deliver an uncontaminated, aesthetically pleasing product to the consumer and, for aqueous products in multiple-use containers, to protect the formula from contamination, dilution, and (unintentional) abuse during normal conditions of use. The role of packaging is especially important for preservative-free/self-preserving formulas because one cannot simply add more preservative to help prevent contamination.

The most straightforward system for preventing contamination is the use of "unit-dose" packaging, such as a small packet that is opened for a single use and then discarded. If the product in this package can be sterilized, it may be possible to have a truly preservative-free product, much the same as dextrose and saline intravenous solutions are provided for use by patients in hospitals. If an aqueous product is intended to be used repeatedly, then it must have protective packaging that prevents entry of microorganisms into the product and/or a preservative system that prevents growth of microorganisms that find their way into the product. Brannan and Dille (26) reported that different types of closures provide different levels of protection for both shampoos and skin lotions. The flip-cap provided the greatest degree of protection for shampoos, and the flip-cap was better than the slit-cap. Shampoo bottles with the standard screw-cap showed the highest incidence of contamination in their studies. Similarly, they found that contamination of skin lotion was highest when containers had the standard screw-cap, and that it was statistically different from contamination with the flip-cap. Here, the pump provided their test lotion with the best protection from microbial contamination.

Packaging developments, including use of sterilizing membranes fitted into dispensing caps, viritually prevent the possibility of microbial contamination (27–29). New cost-effective innovations in packaging are needed to minimize consumer contact with the product and the likelihood of contamination during normal use and abuse, such as use of shampoos and conditioners during showering (see Chapter 10).

11.3 GLYCERYL MONOLAURATE—A MULTIFUNCTIONAL INGREDIENT

Glyceryl monolaurate is a nonionic emulsifier that enhances percutaneous absorption and has antimicrobial activity (30–36). It was formulated into a topical hydrocortisone ointment containing 36% water, 25% white petrolatum, 12% glycerol, 10% glyceryl monolaurate, 1% sodium deoxycholate, and 0.5% hydrocortisone. The hydrocortisone in this preparation permeated through human cadaver epidermis with a flux of 70 mg/cm^2/46 hr at 35°C (30). Andrews and Kure (33) patented shampoos for disinfecting, cleansing, conditioning, and moisturizing human hair and the skin and coat of an animal. Their invention used an antimicrobial shampoo that contains glyceryl monolaurate as the antimicrobial agent, a chelating agent, a cleansing agent, a conditioner, and a moisturizer. They noted that the preferable concentration of monoester was in the range of 1–5%. Gross et al. (34) used glyceryl monolaurate, fatty alcohol, and quaternary ammonium compounds in a hair conditioner.

Interest in the use of glyceryl monolaurate has included a study of its effect on the production of toxins by *S. aureus*. Projan et al. (37) studied the effect of glycerol monolaurate on inhibition of β-lactamase, toxic shock syndrome toxin-1, and other staphylococcal exoproteins and reported that use of initial test concentrations of 2%

glyceryl monolaurate inhibits the synthesis of most staphylococcal toxins and other exoproteins, apparently without affecting growth. Their work should be verified because additional amounts of glycerol monolaurate were added during the experiments to make up for its hydrolysis by staphylococcal esterase, and it seems unlikely that a mild, nontoxic ester like glyceryl monolaurate could interfere with exoprotein synthesis and not with the synthesis of proteins required for cell growth.

Thornfeldt (31) proposed a therapeutically effective dermatological formulation for the treatment of skin diseases and tumors. This ointment formulation contained no preservative and had the following composition:

Monolaurin (lauricidin): 17.5%
Isopropanol: 10%
Propylene glycol: 7%
Aquaphor: 100%

Other self-preserving formulations including shampoos, hair conditions, an enhancer of transdermal delivery, and wound cleansers (34–36). Norgel [i.e., a polymeric gel of acrylic acid and methacrylic acid (Sederma)] and Carbopol (polyacrylic acid or acrylic acid copolymers) are multifunctional agents that can eliminate or reduce the amount of preservative needed to protect a product against microbial contamination. Norgel diluted in water to levels of 40–50% is still bactericidal; however, yeasts and molds are able to survive in these dilutions. At least 60% Norgel is necessary to be effective against all microorganisms. It is believed that the primary effect of Norgel is to lower the a_w so that there is insufficient free water for survival and growth of microorganisms. The use of 50% polyglyceryl methacrylate in a self-preserving O/W skin cream is shown in Table 11.1.

TABLE 11.1 Formulation of O/W Skin Care Cream Using 50% Norgel

Ingredient	w/w
Polyoxyethylene glycol 1500	1.50
Stearoxy dimethicone	3.00
Dimethicone	3.00
Octyl palmitate	6.00
Cyclomethicone	5.00
Beeswax	0.75
Octyl dodecanol	1.50
Isostearyl neopentanoate	6.00
Water	27.10
Polyglyceryl methacrylate	50.00
Fragrance	0.15

Acrylic acid homopolymer/copolymers have been shown to have chelating properties and exhibit synergistic anti-*Pseudomonas* activity with methylparaben (38). A prototype nonionic lotion formula that contained acrylic acid homopolymer/copolymer and methylparaben is shown in Table 11.2. Although this formula had rapid anti-*Pseudomonas* activity for fluorescent pseudomonads (i.e., *P. aeruginosa, P. putida, P. fluorescens*, and *P. stutzeri*), it was not adequately preserved for *S. aureus* or *E. coli*. This work was instructive because it demonstrated that formula components contribute to the preservative system. Figure 11.1 compares D-values for *P. aeruginosa, P. cepacia, P. fluorescens*, and *P. putida* in the nonionic lotion containing 0.2% methylparaben, 0.2% acrylic homopolymer (Carbapol 934), or both 0.2% methylparaben and 0.2% 934. The synergistic effect of the methylparaben and 934 is evident.

Leng and Parrott (39) developed an antiperspirant composition that contained amine oxides to reduce or eliminate the need for classical preservatives. The TEA stearate lotion shown in Table 11.3 contained 2.5% active C_{18} DMAO (stearic dimethyl amine oxide). The preservative system of this formula was improved by the substitution of glyceryl laurate (Lauricidin) for glyceryl monostearate. Like et al. (40) observed that the addition of a chelating agent (e.g., disodium EDTA) may potentiate the preservative system in a formula. They found that addition of 500–2000 ppm of certain quaternaries was effective as a preservative and that high concentrations (2000 ppm) of certain quaternaries produced only mild/transient irritation in the Draize skin irritation test. The antimicrobial action of certain betaines and amine oxides was useful in controlling microbial growth (41).

Use of multifunctional ingredients allows development of formulas with reduced levels of preservatives, and self-preserving formulas can be made without the need for classical preservatives. Again, it must be emphasized that safe products under normal conditions of use and/or reasonably expected abuse must be provided to the consumer—no compromise on safety is acceptable.

TABLE 11.2 Nonionic Lotion Formula[a]

Part	Formula components	Weight (%)
A	Mineral oil	7.50
	Glyceryl stearate and PEG 100 stearate	3.50
B	Water	30.00
	Acrylic acid homopolymer/copolymer	0.20
	TEA 99%	0.30
C	Methyl paraben	0.20
	Water	58.30
		100.00

[a]pH of lotion adjusted to pH 7.0 (± 0.1) by addition of TEA (38).

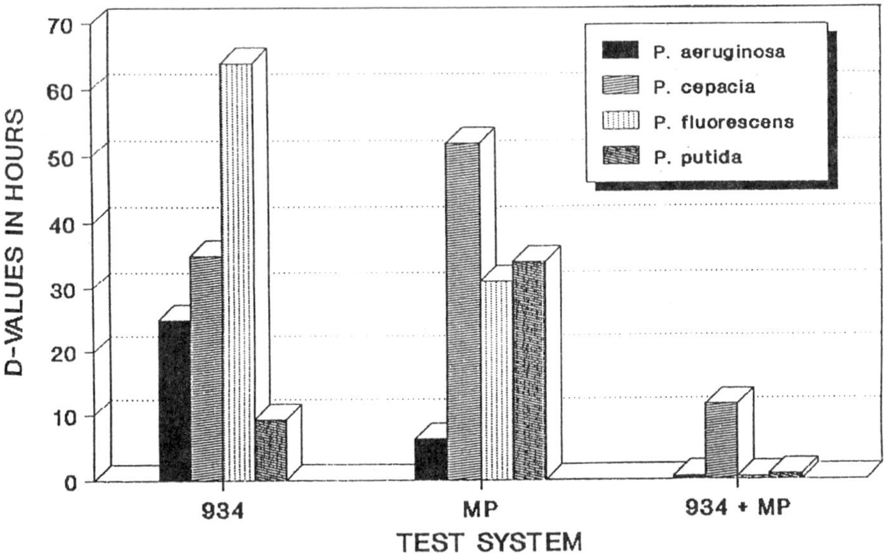

Figure 11.1 Comparison of D-values for *P. aeruginosa* 9027, *P. cepacia* 13945, *P. fluorescens* 13525, and *P. putida* 12633, determined by preservative efficacy testing of the nonionic lotion containing 0.2% MP, 0.2% 934, or both 0.2% MP and 0.2% 934. (From Ref. 38.)

11.4 PRESERVATIVE EFFICACY TESTING OF PRESERVATIVE-FREE AND SELF-PRESERVING PRODUCTS

Although self-preserving formulas may be capable of killing challenge organisms fast enough to meet acceptance criteria and/or satisfactorily prevent their growth during preservative efficacy testing, these formulas must be packaged to maintain

TABLE 11.3 Formulation for TEA Stearate Lotion

Lanolin alcohol	5.0
Acetylated lanolin alcohol	2.0
Glyceryl monostearate	1.0
Proplylene glycol	1.0
Stearic acid	—
Triethanolamine	—
C_{18} DMAO	2.5
Water	88.5
	100.0%

Source: Ref. 39.

their physicochemical composition. Furthermore, preservative-free products that do not kill microorganisms rapidly due to low or high pH, high surfactant concentration, high alcohol concentration, etc. may be able only to exert microbiostatic action through low a_w, lack of nutrients, an inert atmosphere, and other factors and must be packaged in containers/closure systems that maintain the integrity of the formula and prevent the introduction of contaminating microorganisms.

While formulation technology can be used to create a hostile environment for microorganisms, special care must be given to manufacturing conditions. Raw materials used in products should have the minimum microbial loads attainable (<10/g) and manufacturing conditions should adhere to GMPs to prevent the introduction of contaminating microorganisms during production or filling of the product. Aseptic conditions and/or terminal sterilization is necessary for preservative-free products. Strict quality controls and the use of hazard analysis critical control point (HACCP) to identify and control critical points in processing may be necessary for successful manufacturing of these products.

11.5 MICROBIAL LIMITS FOR PRESERVATIVE-FREE AND SELF-PRESERVING PRODUCTS

In general, aqueous personal care products should contain no viable microorganisms, which means that they should have an aerobic plate count (APC) of <10/g. Conventionally preserved aqueous products that contain viable microorganisms may not remain safe because we do not understand all of the factors involved in microbial dormancy and outgrowth. Products in distribution channels and in warehouses are subject to a number of conditions—vibration, temperature fluctuations, and possibly other physical changes—that may cause the effective concentration of a specific preservative to change due to partitioning into and out of the aqueous phase with changes in the environment. As noted earlier (2), a laboratory can give no assurance that conditions governing dormancy will not change to permit growth; therefore, aqueous products that can support microbial growth must contain no viable vegetative organisms. The best course of action is to design a preservative system to render the product self-sterilizing to meet appropriate acceptance criteria and to institute microbial controls of the manufacturing operations to ensure that finished aqueous products meet microbial limits of APC < 10/g.

An exception to this is preservative-free and self-preserving products. It may be acceptable for aqueous products to contain low numbers of nonobjectionable microorganisms, provided that it has been determined that the physicochemical composition of the product will not support growth of these organisms and that packaging ensures the integrity of the formula during use by consumers. For example, an oil-in-water emulsion with 40% glycerin has an a_w of about 0.83, which is below the minimum a_w required for growth of bacteria known to cause problems in cosmetic and drug products (see Chapter 3). Even if this product were to become contaminated, it would not support growth of bacteria such as *S. aureus*, *E. coli*, *P. aeruginosa*, and *B. cereus*. The packaging must protect the product from dilution with water, which would raise the a_w, and/or introduce foreign materials or organisms into the product, because these changes may destabilize emulsions and cause the a_w to change. Other examples include AHA products with very low pH values (pH 3) or hair curl relaxers at pH 12–14 (42), because these pH extremes prevent growth of virtually all organisms that may cause problems in cosmetic and drug products.

Adequacy of preservation of conventional products and preservative-free and self-preserving products includes preservation during manufacturing and while the product is in the hands of the consumer. Orth et al. (43) reported on the use of the required D-value (RDV) for evaluating the effect of formula, packaging, and consumer use/abuse on product preservation. The three variables used to determine the RDV of a product are 1) the preservative system of the formula, which may be expressed as the maximum D-value for specific test organisms (i.e., target organisms) in the formula, 2) the packaging factor, which is based on a risk assessment of whether the packaging will allow contamination of the product, and 3) a consumer use/abuse factor, which is based on a risk assessment of the potential for microbial contamination of the product while in the hands of the consumer. The RDV concept relates these factors by the following simple expression:

$$\text{RDV} = \frac{\text{Maximum acceptable D–value for target organization}}{\text{(Packaging factor) (Consumer use/abuse factor)}}$$

Orth et al. (43) noted that the target organisms(s) should be selected from those that present the highest risk of product contamination.

It is generally accepted that conventionally preserved cosmetics in multiple-use containers are not necessarily sterile; however, preservative systems are used to make them self-sterilizing. The goal of preservative efficacy testing is to determime the minimum concentration of preservative(s) required to meet acceptance criteria for a product. It is necessary that the preservative system be capable of self-sterilizing the product quickly enough to prevent adaptation of contaminating microorganisms. The linear regression method (44) uses the following rationale for setting acceptance criteria of ≤4 hours for pathogens: if the product were to be contaminated accidently with up to 10^6 organisms/g during use by a consumer, it would self-sterilize (i.e., go through a 6-log reduction in APC in 24 hours) and not pose a potential for infection of the user 24 hours later. Similarly, the use of acceptance criteria of ≤28 hours for nonpathogenic bacteria, yeasts, and molds was based on the ability of the product to self-sterilize within one week if it were to be contaminated with up to 10^6 nonpathogenic organisms/g. Finally, the product must be bacteriostatic/bactericidal for spores (2,44). Preservative-free products (i.e., nonpreserved sterile formulas that are aseptically packaged and in dispensing systems that prevent contamination) may not meet these acceptance criteria. Self-preserving products (i.e., formulas that have a preservative system, but do not have chemical preservatives) may not meet these criteria. It is evident that preservative-free formulas must be manufactured and packaged to prevent contamination. Although self-preserving products may not meet the above criteria, testing must have demonstrated that they do not permit growth of microorganisms—they must be bacteriostatic/ bactericidal and fungistatic/fungicidal (2).

Consumer use/abuse testing should be considered a part of a complete evaluation of the suitability of a preservative system. Orth and Milstein (45) recommend that the risk of unintentional contamination by consumers be evaluated for each product. The likelihood of consumer contamination of a product as a result of unintentional abuse or misuse is relatively unlikely for aerosol sprays. The risk is a little greater for emulsion products in containers with flip-top caps or pump dispensers. The chance for product abuse is high for products that are used in the shower, such as those with removable caps or lids, or with caps that may catch water and allow it to run down the pump into the product. Removing the cap from a bottle of shampoo during use in a shower may expose the product to dilution and contamination with tap water. Similarly, removal of the lid from a jar of face cream and dipping into the product with a moist finger may expose the cream to dilution and/or contamination (46). Neither of these situations is desirable for any cosmetic or drug product, and protective packaging should be used wherever consumer use/abuse may compromise the integrity of the formula.

In-use studies and abuse tests are instructive, and microbiological examination of samples after consumer use and/or abuse may provide insights concerning the need to reformulate and/or to modify product packaging. The evaluation of returned samples should include the APC to determine the extent of microbial contamination that occurred during use.

11.6 ARE PRESERVATIVE-FREE AND SELF-PRESERVING PRODUCTS NATURAL?

A product is the sum-total of its components. We have seen expanding interest in marketing products as "natural" over the past 5–10 years, and this trend appears to be increasing. "Natural" claims may be made for the formula, but what about the packaging, which contains plastic, glass, or metal components? The enzymes used in Myavert® [lactoperoxidase (and) glucose oxidase (and) glucose] currently are obtained from living systems. Natural materials currently in use include glycerin and salts, which lower the a_w, short-chain fatty acids (i.e., lactic, acetic, propionic, etc.) and citric acid, which lower the pH, aroma chemicals and alcohols, which interfere with metabolism and/or destabilize membranes, glyceryl monolaurate [Lauricidin®], which is an emulsifier and an antimicrobial agent, chelating agents such as phytic acid and citric acid, which sequester metal ions, and other plant, animal, and biotechnology-derived materials that may add to the preservative system of a formula. The silver chloride/titanium dioxide composite contained in JMAC™ are obtained from mineral sources. All ingredients used in a formula must of coure be natural for the product to be positioned as natural. Furthermore, it should be recognized that claims for natural ingredients may not be justified just because a material occurs in nature, because most ingredients used in cosmetic and drug products are processed (i.e., purified, concentrated, and/or derivitized). It is apparent that claims of naturalness often depend on a number of factors.

11.7 SUMMARY

This book has provided information on the principles of preservation and insights into their application in creating preservative-free and/or self-sterilizing cosmetic and drug products. These products indicate the direction in which our industry is moving in that they are less irritating, less sensitizing, less costly, less of a regulatory problem for introduction in countries around the world, and they fulfill the desire of consumers for natural, preservative-free products.

It is not expected that new regulations and/or test methods will be required for preservative-free products. Self-preserving aqueous cosmetic and drug formulas should meet the acceptance criteria required for adequately preserved products. Although preservative-free formulas may not meet these acceptance criteria, adherence to GMPs during manufacturing, use of aseptic processing and terminal sterilization, and use of protective packaging make the product safe for its intended use (and foreseeable abuse) by consumers.

Preservative-free formulas require container/closure systems that protect the formula from the time it is manufactured until it is used up by the consumer. It is apparent that there will be opportunities for better attention to GMPs (to prevent contamination during manufacturing) and for innovative, new packaging (to prevent contamination during use by consumers). Although this may be accomplished by use of unit dose (i.e., single-use) products, it is believed that the greatest opportunities may be found in paradigm shifts that allow preservative-free aqueous formulas to remain uncontaminated in multiple-use delivery systems. The RDV will be instructive in evaluation of the ability of a formula to withstand contamination by specific microorganisms, the degree of protection provided by the packaging, and the risk of product contamination when the product is used by the consumer. Consumer use testing must be integrated into product development programs to demonstrate that products remain uncontaminated and present no unacceptable safety risks to consumers.

Understanding how the principles of preservation may be applied to creation of preservative-free and/or self-preserving products is only the first step on a new and exciting road to better and safer products that deal with consumer and environmental concerns. Additional steps include product development work, product testing, implementation of validated cleaning and sanitization programs that truly comply with GMPs, development of data that support product claims and advertising, and determining marketing/sales strategies. The development of new, preservative-free and/or self-preserving cosmetic and drug products presents a challenge to formulators and microbiologists. Not all current formulas may be changed into preservative-free or self-preserving products in their current form, and opportunities exist for innovators who accept the challenge to look beyond current technologies and apply the principles of preservation to the development of preservative-free and/or self-preserving products.

REFERENCES

1. J. J. Kabara. Food-grade chemicals in a systems approach to cosmetic preservation. In: J. J. Kabara (Ed). *Cosmetic and Drug Preservation: Principles and Practice.* Marcel Dekker, New York, 1984, pp. 339–356.
2. D. S. Orth. *Handbook of Cosmetic Microbiology.* Marcel Dekker, New York, 1993, pp. 1–118.
3. M. Frost. Preservative-free products. *Society of Cosmetic Scientists*, Midland Hotel, Derby, March 25, 1993.
4. B. Ahlström, M. Chelminska-Bertilsson, R. A. Thompson, and L. Edebo. Long-chain alkanoylcholines, a new category of soft antimicrobial agents that are enzymatically degradable. *Antimicrob. Agents Chemother. 39*:50–55 (1995).
5. J. J. Zeelie and T. J. McCarthy. Antioxidants—multifunctional preservatives for cosmetics and toiletry formulations. *Cosmet. Toiletr. 98*(12):51–55 (1983).
6. Food and Drug Administration. Topical antimicrobial drug products for over-the-counter human use; tentative final monograph for health-care antiseptic drug products. *Fed. Reg. 59*:31409–31452 (1994).
7. A. El-Kady, S. S. Mohamed El-Maraghy, and E. Mostafa M. Antibacterial and antidermatophyte activities of some essential oils from spices. *Qatar Univ. Sci. J. 13*(1):63–69 (1993).
8. R. Richards and R. McBride. Enhancement of benzalkonium chloride by aromatic alcohols. *J. Pharm. Sci. 62*:2035–2037 (1973).
9. U. F. Haustein, J. Herrmann, U. Hoppe, W. Engel, and G. Sauermann. Growth inhibition of coryneform bacteria by a mixture of three natural products—farnesol, glyceryl monolaurate and phenoxyethanol: HGQ. *J. Soc. Cosmet. Chem. 44*:211–220 (1993).
10. M. Toda, S. Okubo, R. Hiyoshi, and T. Shimamura. The bactericidal activity of tea and coffee. *Lett. Appl. Microbiol. 8*:123–125 (1989).
11. J. M. T. Hamilton-Miller. Antimicrobial properties of tea (*Camellia sinensis* L.). *Antimicrob. Agents Chemother. 39*:2375–2377 (1995).
12. K. Nakahara, S. Kawabata, H. Ono, K. Ogura, T. Tanaka, T. Ooshima, and S. Hamada. Inhibitory effect of oolong tea polyphenols on glucosyltransferases of mutans streptococci. *Appl. Environ. Microbiol. 59*:968–973 (1993).

13. I. Kubo, H. Muroi, and M. Himejima. Antimicrobial activity of green tea flavor components and their combination effects. *J. Agric. Food Chem. 40*:245–248 (1992).
14. H. Mukhtar, S. K. Katiyar, and R. Agrawal. Green tea and skin—anticarcinogenic effects. *J. Invest. Dermatol. 102*:3–7 (1994).
15. International Sourcing Technical Bulletin. *Australian Tea Tree Oil*, 1994.
16. I. B. Bassett, D. L. Pannowitz, and R. S. C. Barnetson. A comparative study of tea-tree oil versus benzoylperoxide in the treatment of acne. *Med. J. Aust. 153*:455–458 (1990).
17. M. Lauterwein, M. Oethinger, K. Belsner, T. Peters, and R. Marre. In vitro activities of the lichen secondary metabolities vulpinic acid, (+)-usnic acid, and (–)-usnic acid against aerobic and anaerobic microorganisms. *Antimicrob. Agents Chemother. 39*:2541–2543 (1995).
18. M. K. Sakar, R. Engelshowe, and A. Ü. Tamer. Isolation and antimicrobial activity of flavonoids from *Prunus spinosa* L flowers. *Hacettepe Univ. J. Fac. Pharm. 12*(2):59–63 (1992).
19. J. V. Hirschmann. Topical antibiotics in dermatology. *Arch. Dermatol. 124*:1691–1700 (1988).
20. A. K. Balin, I. Leong, and D. M. Carter. Effects of mupirocin on the growth and lifespan of human fibroblasts. *J. Invest. Dermatol. 88*:736–749 (1987).
21. D. S. Orth. *Handbook of Cosmetic Microbiology*. Marcel Dekker, New York, 1993, pp. 521–582.
22. R. I. Lehrer, T. Ganz, and M. E. Selsted. Defensins: natural peptide antibiotics from neutrophils. *ASM News 56*:315–318 (1990).
23. M. Zasloff, B. Martin, and H. D. Chen. Antimicrobial activity of synthetic magainin peptides and several analogues. *Proc. Natl. Acad. Sci. USA 85*:910–913 (1988).
24. S. L. Haynie, G. A. Crum, and B. A. Doele. Antimicrobial activities of amphiphilic peptides covalently bonded to a water-insoluble resin. *Antimicrob. Agents Chemother. 39*:301–307 (1995).
25. D. Ashworth and K. Turton. An alternative to removing preservatives from formulations. *Drug Cosmet. Ind. 154*:40,42,44 (1994).
26. D. K. Brannan and J. C. Dille. Type of closure prevents microbial contamination of cosmetics during consumer use. *Appl. Environ. Microbiol. 46*:1476–1479 (1990).
27. J. V. Ranalletta, F. E. Williams, Jr., and R. W. Kanner. Liquid dispenser nozzle assembly. U.S. Patent No. 5,025,957, assigned to Ryder International Corp. (June 25, 1991).
28. F. E. Ryder, R. W. Kanner, and R. Rabenau. Solution delivery nozzle and system with antimicrobial features. U.S. Patent No. 5,154,325, assigned to Ryder International Corp. (Oct. 13, 1992).
29. R. W. Kanner and G. M. Nevelli. Liquid dispenser nozzle assembly. U.S. Patent No. 5,431,310, assigned to Ryder International Corp. (July 11, 1995).
30. E. S. Lee, S. I. Yum, and M. J. N. Michel. Topical pharmaceutical with hydrophobic permeation enhancers. *PTC Int. Appl.*:26 (1993).
31. C. R. Thornfeldt. Topical compositions for treatment of skin diseases and tumors. *PCT Int. Appl.*:26 (1990).
32. K. Tanabe. Controlled-release suppositories. *Jpn. Kubai Tokyo Koho*: 4 (1992).
33. J. F. Andrews and J. T. Kure. Disinfecting shampoo composition for animals. U.S. Patent No. 5,219,887 (June 15, 1993).

34. P. Gross, K. Grundman, and G. Lang. Storage stable pearlescent hair-conditioning composition: glyceryl monolaurate, fatty alcohol, quaternium ammonium compounds. U.S. Patent No. 5,213,792 (May 25, 1993).
35. E. S. Lee, S. I. Yum, and M. J. N. Michel. Topical pharmaceutical with hydrophobic permeation enhancers. *PCT Int. Appl.*: 26 (1993).
36. B. H. McAnalley and D. E. Moore. Wound cleanser. *PCT Int. Appl.*: 52 (1993).
37. S. J. Projan, S. Brown-Skrobot, P. M. Schlievert, F. Vandenesch, and R. P. Novick. Glycerol monolaurate inhibits the production of β-lactamase, toxic shock syndrome toxin-1, and other staphylococcal exoproteins by interfering with signal transduction. *J. Bacteriol. 176*:4204–4209 (1994).
38. D. S. Orth, C. M. Lutes Anderson, D. K. Smith, and S. R. Milstein. Synergism of preservative system components: use of the survival curve slope method to demonstrate anti-*Pseudomonas* synergy of methyl paraben and acrylic acid homopolymer/copolymers in vitro. *J. Soc. Cosmet. Chem. 40*:347–365 (1989).
39. F. J. Leng and D. T. Parrott. Antiperspirant compositions containing amphiphilic substances. *Con. Pat. Appl.*: 41 (1993).
40. B. Like, R. Sorrentino, and A. Petrocci. Utility of amine oxides in o/w cosmetic systems. II. *Prod. Chim. XX*:390–405 (1980).
41. D. Kenny. Antimicrobial compositions for controlling body odor. *Cosmet. Toiletr. 93*:13–14 (1979).
42. P. Obukowho and M. Birman. Hair curl relaxers. *Cosmet. Toiletr. 110*(10):65–69 (1995).
43. D. S. Orth, R. F. Barlow, and L. A. Gregory. The required D-value: evaluating product preservation in relation to packaging and consumer use/abuse. *Cosmet. Toiletr. 107*(12):39–43 (1992).
44. D. S. Orth. Linear regression method for rapid determination of cosmetic preservative efficacy. *J. Soc. Cosmet. Chem. 30*:321–332 (1979).
45. D. S. Orth and S. R. Milstein. Rational development of preservative systems for cosmetic products. *Cosmet. Toiletr. 104*(10):91,92,94–100,102,103 (1989).
46. D. S. Orth. Microbiological considerations in cosmetic formula development and evaluation. I. Microbiological quality of a product. *Cosmet. Toiletr. 104*(5):51–56,58,60,62,63 (1989).

Index